HARCOURT

Math

Intervention Strategies and Activities

Grade 4

Orlando • Boston • Dallas • Chicago • San Diego
www.harcourtschool.com

Printed in the United States of America

ISBN 0-15-320764-7

4 5 6 7 8 9 10 082 2004 2003

CONTENTS

Skills (*continued*)

Number Sense: Money

Number Sense: Whole Number Multiplication

Number Sense: Whole Number Division

Number Sense: Fractions

Number Sense: Decimals

Algebra and Functions

Measurement and Geometry: Measurement

Skills (*continued*)

Measurement and Geometry: Geometry

Statistics and Probability

Using the Intervention Strategies and Activities

The *Intervention Strategies and Activities* will help you accommodate the diverse skill levels of students in your class and will help you better prepare students to work successfully on grade-level content by targeting the prerequisite skills for *each chapter* in the program. The following questions and answers will help you make the best use of this rich resource.

How can I determine which skills or strategies a student or students should work on?

Before beginning each chapter, have students complete the "Check What You Know" page in the Pupil Book. This page targets the prerequisite skills necessary for success in the chapter. A student's performance on this page will allow you to diagnose skill weaknesses and prescribe appropriate interventions. Intervention strategies and activities are tied directly to each of the skills assessed. A chart at the beginning of each chapter correlates the skill assessed to the appropriate intervention materials. The chart appears in the Harcourt Math Teacher's Edition.

In what format are the intervention materials?

Intervention materials are available in different formats, including the following:

A. Cards—which provide skill development on one side and skill practice on the other side. These cards can be used by students in an independent setting such as a learning center. The kit that houses the cards also contains the *Intervention Strategies and Activities Teacher's Guide with Copying Masters*. Please note that students should not write on the Cards.

B. Copying masters—which provide the skill development and skill practice on reproducible pages. These pages in the *Intervention Strategies and Activities Teacher's Guide with Copying Masters* can be used by individual students or small groups. You can also allow students to record their answers on copies of the pages. This guide also provides teaching suggestions for skill development, as well as an alternative teaching strategy for students who continue to have difficulty with the skill.

C. CD-ROM—which provides the skill development and practice in an interactive format. Teaching suggestions and alternative teaching strategies are provided as printable PDF files.

Are manipulative activities included in the intervention strategies?

The teaching strategies in the teacher's materials for the *Intervention Strategies and Activities* do require manipulatives, easily gathered classroom objects, or copying masters from the *Teacher's Resource Book*. Since these activities are designed for only those students who show deficits in their skill development, the quantity of manipulatives will be small. For many activities, you may substitute materials, such as squares of paper for counters, coins for two-color counters, and so on.

How can I organize my classroom so that I have time and space to help students who show a need for these intervention strategies and activities?

You may want to set up a Math Skill Center with a folder for each of your students. Based on a student's performance on the *Check What You Know* page, assign appropriate skills by marking the student's record folder. The student can then work through the intervention materials, record the date of completion, and place the completed work in a folder for your review. You may wish to assign students to a partner or assign a small group to work together, or you may wish to have a specified time during the day to meet with one or more of the individuals or small groups to assess their progress and to provide direct instruction.

How are the activities structured?

Each skill begins with a model or an explanation with a model for each skill. The first section of exercises titled *Try These* provides 2–4 exercises that allow students to move toward doing the work independently. A student who has difficulty with the *Try These* exercises might benefit from the activity for that skill described in this Teacher's Guide before they attempt the *Practice on Your Own* page. The *Practice on Your Own* page provides an additional model for the skill and scaffolded exercises, which gradually remove prompts. Scaffolding provides a framework within which the student can achieve success for the skill. At the end of the *Practice on Your Own*, there is a *Check*. The *Check* provides 3–4 problems that check the student's proficiency in the skill. Guidelines for success are provided in the teacher's materials.

Name ______________________________

Individual Prerequisite Skills Checklist

Chapter	Prerequisite Skill	Prescription	Skill Mastered

Intervention Strategies and Activities Chapter Correlations

Number Sense

Place Value

Skill Number	Skill	Chapter Correlation
1	Place Value to Hundreds	1, 14
2	Place Value to Thousands	1, 14
3	Benchmark Numbers to 100	1
4	Read and Write Numbers to Thousands	1
5	Compare and Order Numbers to Hundreds	2, 29
6	Compare and Order Numbers to Thousands	2, 29
7	Round to Tens and Hundreds	2, 19, 20

Whole Number Addition

Skill Number	Skill	Chapter Correlation
8	2-Digit Addition: Regrouping Ones as Tens	3
9	2-Digit Addition: Regrouping Tens as Hundreds	3
10	3-Digit Addition: Regrouping Ones as Tens	3
11	3-Digit Addition: Regrouping Tens as Hundreds	3
12	3-Digit Addition: Regrouping Hundreds as Thousands	3
13	Order and Zero Properties of Addition	3
14	Grouping Property of Addition	3
15	Fact Families (Addition and Subtraction) 1–10	4
16	Fact Families (Addition and Subtraction) 11–20	4

Whole Number Subtraction

Skill Number	Skill	Chapter Correlation
17	2-Digit Subtraction: Regrouping Tens as Ones	3
18	3-Digit Subtraction: Regrouping Tens as Ones	3
19	3-Digit Subtraction: Regrouping in More than One Place	3
20	3-Digit Subtraction: Regrouping Across Zeros	3

Money

Skill Number	Skill	Chapter Correlation
21	Round Money Amounts	22
22	Add Money	22
23	Subtract Money	22

Whole Number Multiplication

Skill Number	Skill	Chapter Correlation
24	Multiplication Facts to 5	8, 16
25	Multiplication Facts to 10	8, 16
26	Meaning of Multiplication	8
27	Model Multiplication (2-digit by 1-digit numbers)	10, 11
28	Record Multiplication (basic facts)	10
29	Multiply by Multiples of 10	11, 24
30	Multiply 1-, 2-, 3-Digit Numbers by 1-Digit Numbers Regrouping Once	12
31	Multiply 1-, 2-, 3-Digit Numbers by 1-Digit Numbers Regrouping Twice	12
32	Estimate Products	12
33	Multiplication Facts and Patterns	14
34	Multiplication Properties	16
35	Multiplication Patterns (for customary measurement)	23
36	Multiply Three Factors	26

Whole Number Division

Skill Number	Skill	Chapter Correlation
37	Division Facts (divide by 2 and 5)	8, 13
38	Division Facts (divide by 3 and 4)	8, 13
39	Division Facts (divide by 6 and 7)	8, 13
40	Division Facts (divide by 8 and 9)	8, 13
41	Meaning of Division	8
42	Divide with Remainders	15
43	Use Compatible Numbers	15
44	Divide by 1-Digit Numbers	21, 23

Fractions

Skill Number	Skill	Chapter Correlation
45	Model Parts of a Whole	19, 28
46	Model Parts of a Group (or set)	19, 28
47	Count Parts of a Whole	20
48	Count Parts of a Group	20
49	Compare Parts of a Whole	20
50	Fractions With Denominators of 10 and 100	21

Decimals

Skill Number	Skill	Chapter Correlation
51	Model Decimals (Tenths and Hundredths)	21
52	Relate Decimals to Money	21

Algebra and Functions

Skill Number	Skill	Chapter Correlation
53	Missing Addends	4
54	Number Patterns for Addition	4
55	Number Patterns for Subtraction	4
56	Missing Factors	9
57	Number Patterns (for Multiplication)	9
58	Number Patterns (for Division)	9
59	Read a Number Line	29
60	Locate Points on a Coordinate Grid (first quadrant only)	30
61	Use Function Tables	30

Measurement and Geometry

Measurement

Skill Number	Skill	Chapter Correlation
62	Time to the Half Hour and Quarter Hour	7
63	Time to the Minute	7
64	Use a Calendar	7
65	Measure to the Nearest Inch, Half-Inch	23
66	Measure to the Nearest Centimeter	24

Geometry

Skill Number	Skill	Chapter Correlation
67	Identify Angles	17
68	Compare Figures	17
69	Identify Symmetric Figures	17
70	Sort Polygons	25
71	Identify Plane Figures	25, 26
72	Find Perimeter (by counting)	25
73	Find Area (by counting)	25
74	Identify Solid Figures	26
75	Classify Angles (right, greater or less than right)	18

Statistics and Probability

Skill Number	Skill	Chapter Correlation
76	Read Pictographs (use half symbols)	5
77	Tallies to Frequency Tables	5
78	Read Bar Graphs	6
79	Parts of a Bar Graph	6
80	Identify Points on a Grid	6
81	Certain and Impossible	27
82	Identify Possible Outcomes	27

Number Sense

Place Value

Skill 1 Grade 4 — Place Value to Hundreds

OBJECTIVE Read and write place-value to hundreds

MATERIALS base-ten blocks and place-value charts

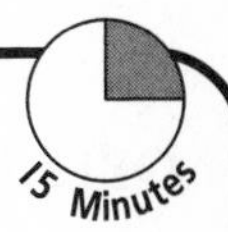

You may wish to review how using a place-value chart can help students recognize the value of each digit in a number.

Direct students' attention to Step 1.

Ask: **How many hundreds blocks are there?** 2 **How many tens rods?** 0 **How many ones cubes?** 3

Point to the place-value chart.

Ask: **What digit is in the hundreds place?** 2 **What digit is in the tens place?** 0 **What digit is in the ones place?** 3

Review the relationships between the numeral 203, the base-ten model of 203, and the number of tens and ones.

Continue with Step 2. Remind students that the value of each digit in a number is determined by its position in the number.

Ask: **What is the value of the digit 2?** 2 hundreds or 200 **What is the value of the digit 0?** 0 tens or 0 **What is the value of the digit 3?** 3 ones or 3

Explain that in Step 3 students use place-value words to write the number.

TRY THESE Exercises 1–2 model the type of exercises students will find on the **Practice on Your Own** page.

- **Exercise 1** Write the value of each digit and place-value words for 432.
- **Exercise 2** Write the value of each digit and place-value words for 809.

PRACTICE ON YOUR OWN Review the example at the top of the page. In Exercises 1–2, students write the value of each digit and place-value words to hundreds, with place-value charts as a visual cue. In Exercises 3–4, students write the value of each digit and place-value words to hundreds, without visual cues.

CHECK Determine if students know the value of digits to hundreds. Success is indicated by 3 out of 3 correct responses.

Students who successfully complete the **Practice on Your Own** and **Check** are ready to move on to the next skill.

COMMON ERRORS

- Students may have difficulty identifying the value of 0.
- Students may not understand that the position of a digit in a number determines its value.

Students who made more than two errors in the **Practice on Your Own**, or who were not successful in the **Check** section, may benefit from the **Alternative Teaching Strategy** on the next page.

Optional

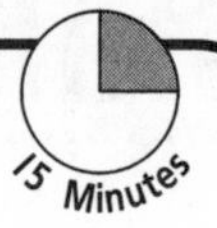

Alternative Teaching Strategy
Model Place Value to Hundreds

OBJECTIVE Read and write place value to hundreds

MATERIALS base-ten blocks, place-value charts, paper

Students work in pairs for this activity. Distribute base-ten blocks and place-value charts to each pair. Point out to students that in this lesson they will find the value of the digits in a number.

Have one student display 2 hundreds blocks, 4 tens rods, and 8 ones cubes.

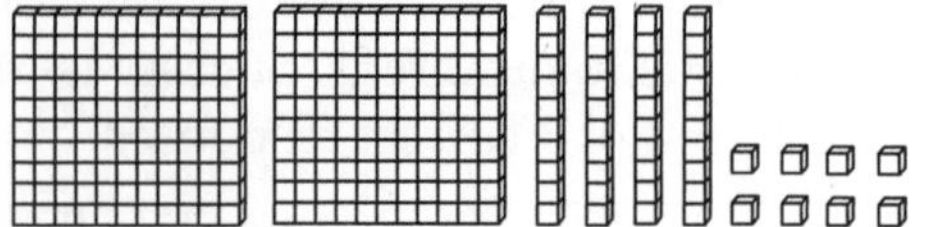

Ask: **How many hundreds blocks are there?** 2 **How many tens rods?** 4 **How many ones cubes?** 8

Guide students to understand that they have 2 hundreds 4 tens 8 ones, so the model shows the number 248.

Then display the place-value chart shown below.

Hundreds	Tens	Ones
2	4	8

Ask: **What digit is in the hundreds place?** 2 **What is its value?** 2 hundreds or 200 **What digit is in the tens place?** 4 **What is its value?** 4 tens or 40 **What digit is in the ones place?** 8 **What is its value?** 8 ones or 8

Say: **You have 2 hundreds 4 tens 8 ones. What number does the place-value chart show?** 248

Repeat the activity with similar examples. When the students show an understanding of place value, suggest that each student write a three-digit number and take turns displaying it, while other students tell the value of each digit and name the number.

Name ______________________ Skill ______________________

Grade 4 **Skill** 1

Place Value to Hundreds

Write the value of each digit in the number 203.

Step 1 Use a place value chart.

Hundreds	Tens	Ones
2	0	3

Step 2 Write the value of each digit.

The value of the **2** is 2 hundreds or 200.
The value of the **0** is 0 tens or 0.
The value of the **3** is 3 ones or 3.

Step 3 Use place value words to write the number.

2 hundreds 0 tens 3 ones = 203

Try These

Write the value of each digit.

1. **432**

Hundreds	Tens	Ones
4	3	2

The value of the **4** is ☐ hundreds or ☐.
The value of the **3** is ☐ tens or ☐.
The value of the **2** is ☐ ones or ☐.
☐ hundreds ☐ tens ☐ ones = 432

2. **809**

Hundreds	Tens	Ones
8	0	9

The value of the **8** is ☐ hundreds or ☐.
The value of the **0** is ☐ tens or ☐.
The value of the **9** is ☐ ones or ☐.
☐ hundreds ☐ tens ☐ ones = 809

Go to the next side.

Name ______________________ Skill ______________________

Practice on Your Own

Skill 1

Write the value of each digit in the number 750.

Hundreds	Tens	Ones
7	5	0

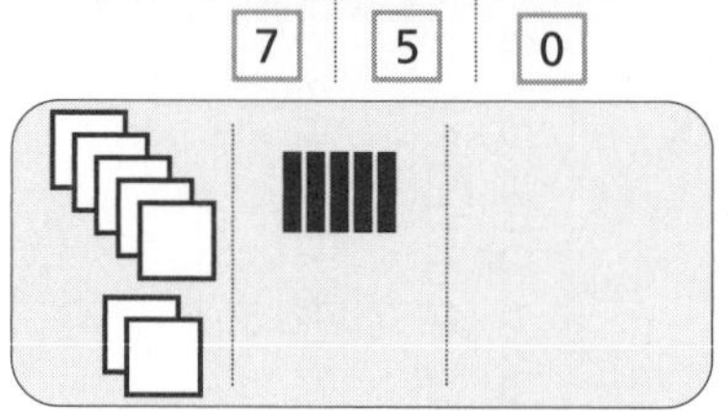

The value of the **7** is 7 hundreds or 700.
The value of the **5** is 5 tens or 50.
The value of the **0** is 0 ones or 0.

7 hundreds 5 tens 0 ones = 750

Write the value of each digit.

1 **956**

Hundreds	Tens	Ones
9	5	6

The value of **9** is ☐ hundreds or ☐.

The value of **5** is ☐ tens or ☐.

The value of **6** is ☐ ones or ☐.

☐ hundreds ☐ tens ☐ ones = 956

2 **600**

Hundreds	Tens	Ones
6	0	0

The value of **6** is ☐ hundreds or ☐.

The value of **0** is ☐ tens or ☐.

The value of **0** is ☐ ones or ☐.

☐ hundreds ☐ tens ☐ ones = 600

3 **720**

The value of **7** is ☐ hundreds or ☐.

The value of **2** is ☐ tens or ☐.

The value of **0** is ☐ ones or ☐.

☐ hundreds ☐ tens ☐ ones = 720

4 **394**

The value of **3** is ☐ hundreds or ☐.

The value of **9** is ☐ tens or ☐.

The value of **4** is ☐ ones or ☐.

☐ hundreds ☐ tens ☐ ones = 394

Check

Write the value of the underlined digit.

5 7<u>1</u>6 ☐ **6** 26<u>0</u> ☐ **7** <u>9</u>87 ☐

Skill 2 Grade 4

Place Value to Thousands

OBJECTIVE Write the value of the digits in numbers to thousands

MATERIALS base-ten blocks

You may wish to use base-ten blocks to model the example.

Begin by having the students look at the place-value chart, and recall that the place-value labels show the value of each digit in a number. Remind them also that they can use the place-value labels to help them write a number in different ways.

Have students look at Step 2. As they read the value of each digit, suggest that students display the appropriate base-ten blocks.

Ask: **What is the value of the digit 1?** **1 thousand or 1,000** **How do you know?** **The 1 is in the thousands place.**

Continue asking similar questions. Guide students as they see that the value of the digit 1 is greater than the value of the digit 2; the value of the digit 2 is greater than the value of the digit 7, and so on.

In Step 3, point out how the place-value words help with writing the number two ways. Have students read the number aloud. Have them note the position of the comma in 1,275.

Ask: **What does the comma separate?** **the hundreds from the thousands**

TRY THESE Exercises 1 and 2 provide practice in using place-value charts to write the value of digits.

- **Exercise 1** Value of the digits in 7,569.
- **Exercise 2** Value of the digits in 6,403.

PRACTICE ON YOUR OWN Review the example at the top of the page. Discuss why the value of 0 hundreds is 0, and why it is necessary to write a zero in that place.

CHECK Determine if students know the names of the places in the place-value chart, and can tell the value of the given digits. Success is determined by 2 out of 2 correct responses.

Students who successfully complete the **Practice On Your Own** and **Check** are ready to move to the next skill.

COMMON ERRORS

- Students may be able to write the number of ones, tens, hundreds, and thousands, but be unable to relate that number to the value of the digit.
- Students may have trouble identifying the value of 0; for example, they may write a number such as two thousand, twenty-seven as 227 instead of 2,027.

Students who made more than two errors in the **Practice On Your Own,** or who were not successful in the **Check** section, may benefit from the **Alternative Teaching Strategy** on the next page.

Optional

Alternative Teaching Strategy

20 Minutes

Use Models to Show Place Value to Thousands

OBJECTIVE Use base-ten blocks to model place value to thousands

MATERIALS base-ten blocks

Write the number 2,438 in a place-value chart. Then guide students as they use base-ten blocks to model the number.

Point to the thousands place. Explain that the 2 in the thousands place means $2 \times 1{,}000$ or 2,000.

Ask: **What is the value of the 2 in the chart?** 2 thousands, or 2,000 **How many thousands blocks will you use?** 2

The 4 in the hundreds place means 4 × 100, or 400. What is the value of the digit 4? 4 hundreds or 400 **How many hundreds blocks will you use?** 4

Continue to ask similar questions as the students model the remaining digits.

As students focus on the model, ask them which digit has the greatest value and which has the least value. Students can see that the thousands place has the greatest value, and the value of the places decrease from left to right.

Relate the model to the number, and write and say the number two ways.

2 thousands, 4 hundreds 3 tens 8 ones
2,438

Repeat the activity several times with other 4-digit numbers. Include examples with zeros in ones, tens, or hundreds places, such as 3,026, 1,407, and 4,580.

When students show an understanding of place value to thousands, have one partner choose a 4-digit number and write it in a place-value chart, while the second partner uses base-ten blocks to model the number. Have the partners record the number two ways. Then have them reverse roles to complete another example.

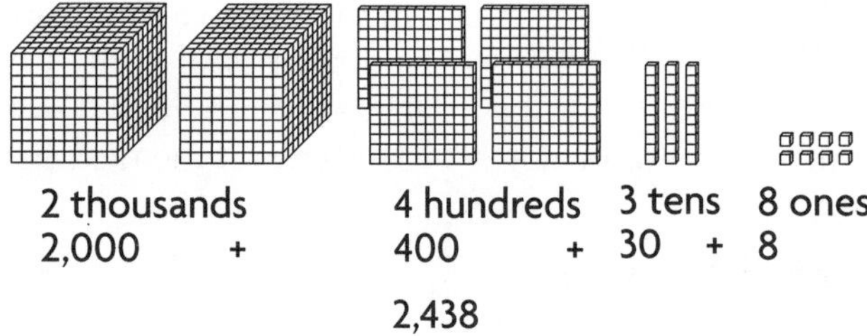

2 thousands 4 hundreds 3 tens 8 ones
2,000 + 400 + 30 + 8
2,438

Name ______________________ Skill ______________________

Grade 4 **Skill** 2

Place Value to Thousands

Write the value of each digit in the number 1,275.

Step 1 Use a place value chart.

TH	H	T	O
1	2	7	5

Step 2 Write the value of each digit.

The value of the **1** is 1 thousand or 1,000.
The value of the **2** is 2 hundreds or 200.
The value of the **7** is 7 tens or 70.
The value of the **5** is 5 ones or 5.

Step 3 Use place value words to write the number.

1 thousand, 2 hundreds 7 tens 5 ones = 1,275

Try These

Write the value of each digit.

1 **7,569**

TH	H	T	O
7	5	6	9

The value of the **7** is ☐ thousands or ☐.

The value of the **5** is ☐ hundreds or ☐.

The value of the **6** is ☐ tens or ☐.

The value of the **9** is ☐ ones or ☐.

☐ thousands ☐ hundreds ☐ tens ☐ ones = **7,569**

2 **6,403**

TH	H	T	O
6	4	0	3

The value of the **6** is ☐ thousands or ☐.

The value of the **4** is ☐ hundreds or ☐.

The value of the **0** is ☐ tens or ☐.

The value of the **3** is ☐ ones or ☐.

☐ thousands ☐ hundreds ☐ tens ☐ ones = **6,403**

Go to the next side.

Name ____________________ Skill ____________________

Practice on Your Own

Skill 2

Example:

Write the value of each digit of the number 2,063.

TH	H	T	O
2	0	6	3

The value of the **2** is 2 thousands or 2,000.
The value of the **0** is 0 hundreds or 0.
The value of the **6** is 6 tens or 60.
The value of the **3** is 3 ones or 3.
2 thousands 0 hundreds 6 tens 3 ones = **2,063**

Write the value of each digit.

1. **3,079**

TH	H	T	O
3	0	7	9

The value of **3** is ☐ thousands or ☐.

The value of **0** is ☐ hundreds or ☐.

The value of **7** is ☐ tens or ☐.

The value of **9** is ☐ ones or ☐.

2. **7,503**

TH	H	T	O
7	5	0	3

The value of **7** is ☐ thousands or ☐.

The value of **5** is ☐ hundreds or ☐.

The value of **0** is ☐ tens or ☐.

The value of **3** is ☐ ones or ☐.

3. **8,290**

TH	H	T	O
8	2	9	0

The value of **8** is ☐ thousands or ☐.

The value of **2** is ☐ hundreds or ☐.

The value of **9** is ☐ tens or ☐.

The value of **0** is ☐ ones or ☐.

4. **6,100**

TH	H	T	O
6	1	0	0

The value of **6** is ☐ thousands or ☐.

The value of **1** is ☐ hundred or ☐.

The value of **0** is ☐ tens or ☐.

The value of **0** is ☐ ones or ☐.

Check

Write the value of the underlined digit.

5. **<u>4</u>,066** ☐

6. **2,0<u>7</u>8** ☐

Benchmark Numbers to 100

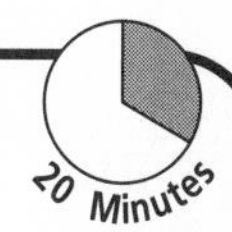

OBJECTIVE Estimate how many squares will cover a figure using benchmark numbers

MATERIALS graph paper

You may wish to have the students verify how many squares are in each benchmark shape by counting. They may also write a multiplication problem to match the number of rows and columns and find the product to verify the number of squares in each.

Direct students' attention to Example A.

Ask: **Which benchmark numbers are about the same length or width as the figure?** 10, 50, and 100 are all about the same length **Which benchmark numbers are too wide?** 25, 50, and 100 are all too wide **How many groups of 10 will cover the figure?** 3 groups **Three groups of 10 are how many squares?** 30 **Will this be less than or greater than the area in the figure?** It will be less than the area in the figure.

Continue to ask similar questions as you work through Example B.

TRY THESE Exercises 1-3 model the type of exercises students will find on the **Practice on Your Own** page.

- **Exercise 1** Benchmark number 10.
- **Exercise 2** Benchmark number 25.
- **Exercise 3** Benchmark number 50.

PRACTICE ON YOUR OWN Review the example at the top of the page. Ask students why benchmark number 100 was used to estimate the number of squares it would take to cover the figure. **Possible response: The figure is large, so it makes sense to use the largest benchmark that will fit on the figure.**

CHECK Make sure that the students select the appropriate benchmark number to estimate.

Success is determined by 3 out of 3 correct responses.

Students who successfully complete the **Practice on Your Own** and **Check** are ready to move on to the next skill.

COMMON ERRORS

- Student may select a benchmark number that is too large.
- Students may not use enough of the benchmark number to cover the figure.

Students who made more than two errors in the **Practice on Your Own,** or who were not successful in the **Check** section, may benefit from the **Alternative Teaching Strategy** on the next page.

Optional

Alternative Teaching Strategy

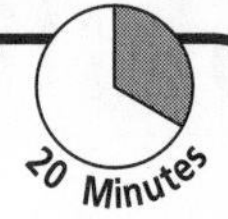

Modeling Benchmark Numbers

OBJECTIVE Use benchmark numbers made from graph paper to estimate the number of squares it takes to cover a figure

MATERIALS benchmark numbers 10, 25, 50, and 100 on grid paper as shown in the lesson, blank figures labeled A through E, Figure A is a rectangle 5 units by 10.8 units, Figures B through E can be any size

Distribute a set of benchmark numbers and blank figures to each student.

Say: **Look at Figure A and the benchmark numbers. Which benchmarks fit in the figure without going over the edges?** Benchmark numbers 10, 25, and 50 **Which is the largest benchmark number you could use?** 50

How many times does the benchmark number 50 fit in the figure? 1 **Does the benchmark match the figure precisely?** No **What is an estimate for the number of squares it will take to cover the figure?** 50

Figure A

Benchmark Numbers

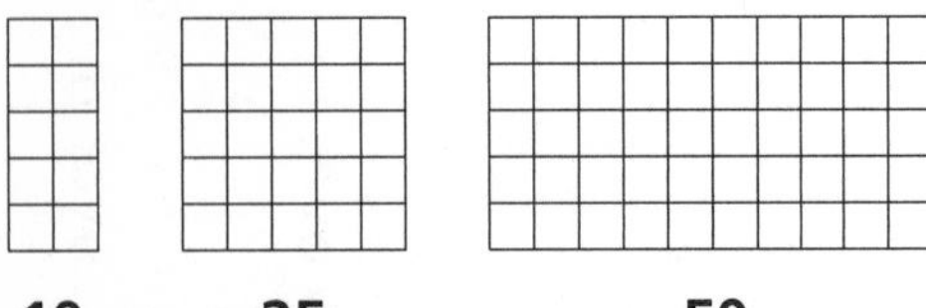

10 **25** **50**

Repeat this activity with similar examples. When the students show understanding of the estimating process, remove the benchmark numbers from students desks. Display one set of benchmark numbers on the chalkboard and have students try an exercise using only paper and pencil. Ask students to explain why they chose a particular benchmark number.

Name ______________________ Skill ______________________

Grade 4 **Skill** 3

Benchmark Numbers to 100

Benchmark numbers are useful numbers such as 10, 25, 50, and 100. They help you estimate about how much or about how many without counting.

Example A

Choose a benchmark number. Tell about how many squares will cover the figure.

Look at the figure. Choose 10. Estimate.

Think: About 3 groups of 10 will cover the figure.

10 + 10 + 10 = 30

So, about 30 squares will cover the figure.

Example B

Choose a benchmark number. Tell about how many squares will cover the figure.

Look at the figure.

Choose 25.

Estimate.

Think: About 3 groups of 25 will cover the figure.

25 + 25 + 25 = 75

So, about 75 squares will cover the figure.

Try These

Use the benchmark number to estimate. Tell about how many squares will cover the figure.

1. 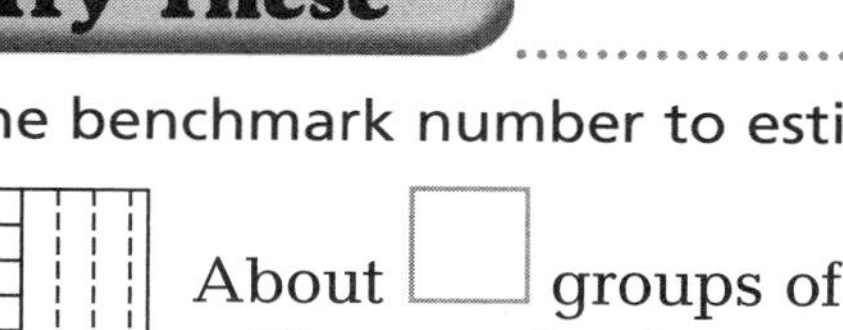

About ☐ groups of 10 will cover the figure.

About ☐ squares will cover the figure.

2. 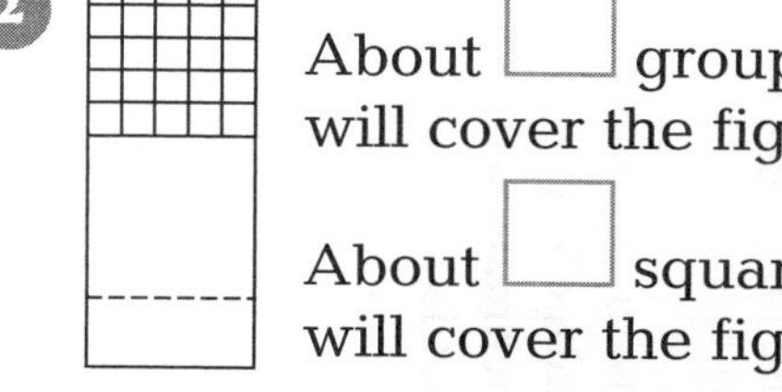

About ☐ groups of 25 will cover the figure.

About ☐ squares will cover the figure.

3. 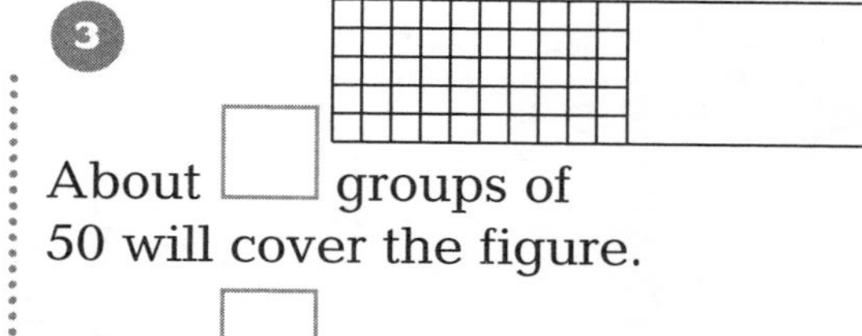

About ☐ groups of 50 will cover the figure.

About ☐ squares will cover the figure.

Go to the next side.

Name ______________________ Skill ______________________

Practice on Your Own

Skill 

About 4 groups of 100 will cover the figure.

100 + 100 + 100 + 100 = 400

So, about 400 squares will cover the figure.

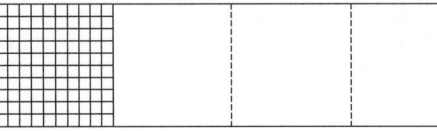

Think: Use benchmark numbers to estimate.

Use benchmark numbers to estimate. Tell about how many squares cover the figure.

 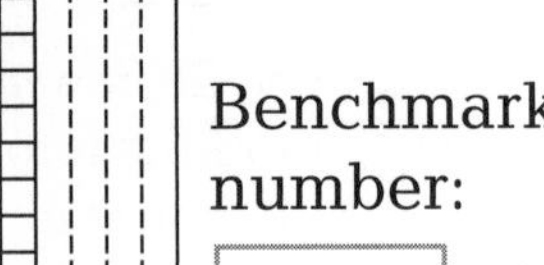

Benchmark number: ☐

Estimate: ☐ squares

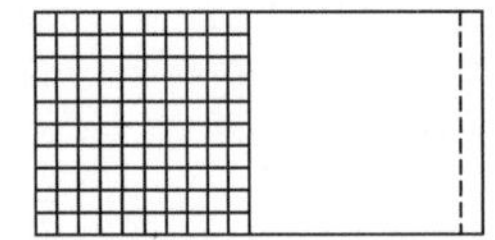

Benchmark number: ☐

Estimate: ☐ squares

Benchmark number: ☐

Estimate: ☐ squares

 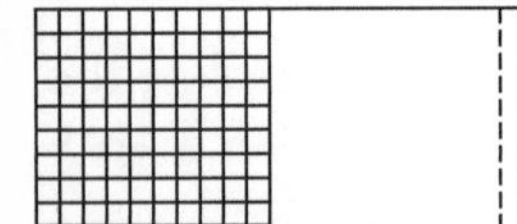

Benchmark number: 

Estimate: ☐ squares

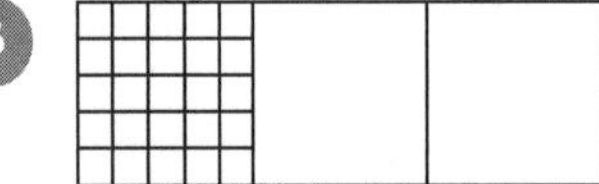

Benchmark number: ☐

Estimate: ☐ squares

Benchmark number: ☐

Estimate: ☐ squares

Check

Use benchmark numbers to estimate. Tell about how many squares cover the figure.

 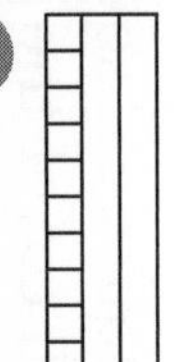

Benchmark number: ☐

Estimate: ☐ squares

Benchmark number: ☐

Estimate: ☐ squares

 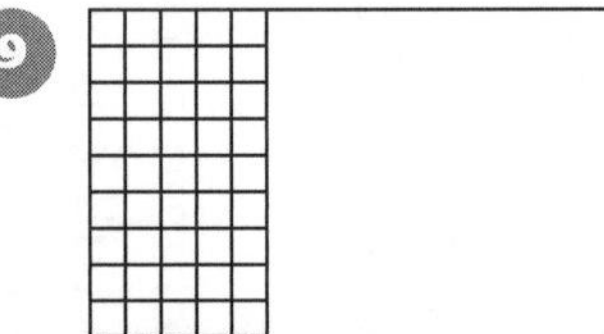

Benchmark number: ☐

Estimate: ☐ squares

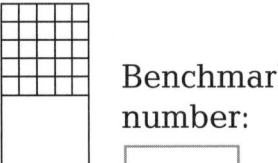

Read and Write Numbers to Thousands

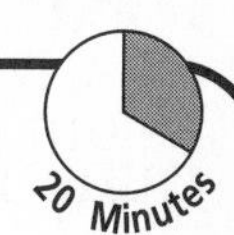

OBJECTIVE Read and write numbers to thousands

MATERIALS place-value chart

You may wish to review how using a place-value chart can help students recognize the value of each digit in a number.

Begin the skill by reminding students that a number can be expressed in different ways. The number in the place value chart can be written in expanded form, in standard form, and in words.

Say: **Look at the place-value chart. What is the value of each digit in the number 451?** 4 hundreds, 5 tens, 1 one or 400, 50, 1

Continue: **Once you know the value of the digits you can read the number by using the word form. How do you read the number 451 using words?** four hundred fifty-one **How do you write the number in standard form?** 451

Since the value of each digit is determined by its position in the number, it is important that students understand how place value affects the value of a digit.

Ask similar questions for Example 2. Remind students that a comma is used to separate the thousands place from the hundreds place.

TRY THESE Exercises 1–2 give students an opportunity to express numbers in different ways.

- **Exercise 1** 3-digit number.
- **Exercise 2** 4-digit number.

PRACTICE ON YOUR OWN Review the example with students. Remind them to look at the position of each digit in the number.

CHECK Determine if students know the value of a digit in a number, and can express the number in different ways.

Success is indicated by 2 out of 2 correct responses.

Students who successfully complete the **Practice On Your Own** and **Check** are ready to move on to the next skill.

COMMON ERRORS

- Students may have difficulty writing the standard form from the expanded form.
- Some students may write a number such as one thousand, two hundred five as 125.

Students who made more than two errors in the **Practice On Your Own**, or who were not successful in the **Check** section, may also benefit from the **Alternative Teaching Strategy** on the next page.

Optional

15 Minutes

Alternative Teaching Strategy

Use Models to Read and Write Numbers

OBJECTIVE Use models to read and write numbers to thousands

MATERIALS base-ten blocks

Provide students with base-ten blocks. Have students show 2 hundreds, 9 tens, 3 ones to represent 293. Have students express the number that the model represents in different ways.

Ask: **What blocks did you use?** 2 hundreds 9 tens 3 ones **What is the number in expanded form?** 200 + 90 + 3

Explain that the expanded form can help them write the number in standard form and in words. To write the number in standard form, students can use mental math to add.

What is the number in standard form? 293 **How would you write the expanded form as words?** two hundred ninety-three

Repeat the activity several times, having students read and write numbers for hundreds and thousands. Remind the students to use a comma to separate the thousands place from the hundreds place.

When students show an understanding of reading and writing numbers to thousands, have them read and write numbers without using the expanded form and the base-ten blocks.

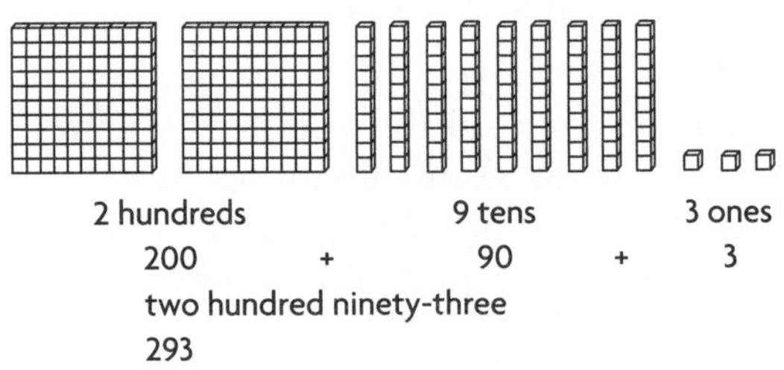

Name ____________________ Skill ____________________

Read and Write Numbers to Thousands

Grade 4 Skill 4

Model 1 Read and write the number.

Hundreds	Tens	Ones
4	5	1

Expanded Form:
400 + 50 + 1

Read: four hundred fifty-one

Write: 451

Model 2 Read and write the number.

Thousands	Hundreds	Tens	Ones
1	6	0	4

Expanded Form:
1,000 + 600 + 0 + 4

Read: one thousand, six hundred four

Write: 1,604

Try These

Complete.

1.

Hundreds	Tens	Ones
6	8	1

6 hundreds 8 tens 1 ones

600 + 80 + 1

Read: six hundred eighty–one

Write: ______

2.

Thousands	Hundreds	Tens	Ones
6	7	3	2

☐ thousands ☐ hundreds ☐ tens ☐ ones

______ + ______ + ______ + ______

Read: six thousand, seven hundred thirty-two

Write: ______

Name ____________________ Skill ____________________

Practice on Your Own

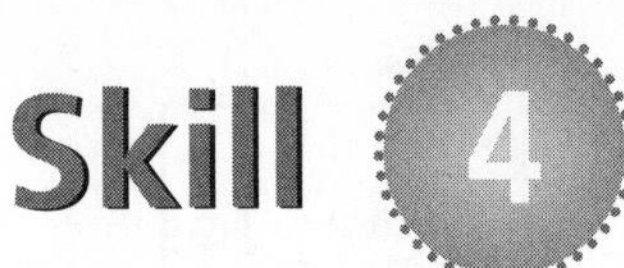

Example:

Thousands	Hundreds	Tens	Ones
3	0	8	5

Think:
3 thousands 0 hundreds 8 tens 5 ones
3,000 + 0 + 80 + 5

Read: three thousand, eighty–five
Write: 3,085

Complete.

1.

Hundreds	Tens	Ones
4	7	9

☐ hundreds ☐ tens ☐ ones

_____ + _____ + _____

Read: four hundred seventy–nine

Write: ____________________

2.

Thousands	Hundreds	Tens	Ones
7	0	6	0

☐ thousands

☐ hundreds

☐ tens ☐ ones

_____ + _____ + _____ + _____

Read: seven thousand, sixty

Write: ____________________

3. Read: nine hundred twenty-six

☐ hundreds ☐ tens ☐ ones

_____ + _____ + _____

Write: ____________________

4. Eight hundred ninety–one

Write: ____________________

5. Two thousand, six hundred fifty

Write: ____________________

Check

Write the number.

6. nine hundred seven

____

7. five thousand, two hundred twenty-five

____

Skill 5 Grade 4

Compare Numbers to Hundreds

OBJECTIVE Compare numbers to hundreds using >, <, and =

MATERIALS number lines, place value charts

Display a number line labeled 0 to 10 and ask students to mark 2 and 8. Have them tell you which number is greater. 8 Repeat the exercise for tens and hundreds with appropriately labeled number lines. Point out that the greater number is always to the right on the number line.

Review the meaning of the inequality symbols. Then draw students' attention to the first example. Remind students that the numbers on a number line are greater from left to right. Have students touch the number line to show the location of 427 and 465.

Ask: **Is 465 to the right of 427?** Yes
Which number is greater? 465

Tell students they can also use a place value chart to compare numbers. Focus students' attention on the second example. Recall that to compare numbers, begin with the digits in the greatest place value position. Guide students to understand that in this example, since both are 3-digit numbers, the greatest place value is the hundreds place.

Ask: **How many hundreds are in 465?** 4
How many hundreds are in 427? 4

Point out that the hundreds digits are the same. Tell students that when the digits in the greatest place value position are the same, they compare the digits in the next place value position to the right. Help them compare the tens and then identify the greater number. Read aloud and discuss the inequality.

TRY THESE Exercises 1–4 provide students practice in comparing numbers to hundreds.

- **Exercises 1–2** Compare on number line.
- **Exercises 3–4** Compare in place value chart.

PRACTICE ON YOUR OWN Review the example at the top of the page. In Exercises 1–2, students compare 3-digit numbers using number lines as visual cues. In Exercises 3–4, students compare 3-digit numbers using place value charts as visual cues. In Exercises 5–12, students compare 3-digit numbers without visual cues.

CHECK Determine if students correctly compare numbers using inequality symbols. Success is determined by 3 out of 4 correct responses.

Students who successfully complete the **Practice on Your Own** and **Check** are ready to move on to the next skill.

COMMON ERRORS

- Students may confuse inequality symbols when comparing numbers.
- Students may compare numbers from right to left.

Students who made more than three errors in the **Practice on Your Own,** or who were not successful in the **Check** section, may benefit from the **Alternative Teaching Strategy** on the next page.

Optional

Alternative Teaching Strategy

20 Minutes

Model Numbers to Compare to Hundreds

OBJECTIVE Use base-ten blocks to compare numbers to hundreds

MATERIALS base-ten blocks, paper

Distribute base-ten blocks to students. Have them model the numbers 42 and 48 with the blocks. Ask students to place the model of 42 on the left and the model of 48 on the right.

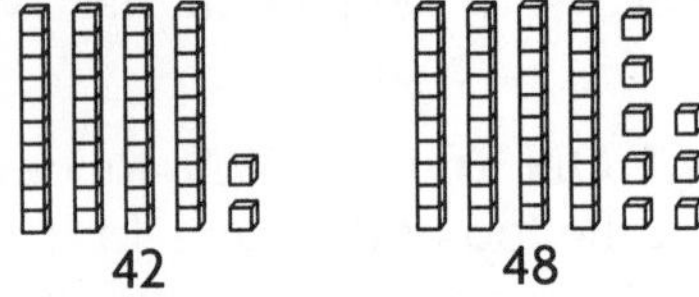

Ask: **How many tens rods are there in 42?** 4 **How many tens rods in 48?** 4 **Are the number of tens rods in 42 and 48 the same?** Yes

Next have students tell you how many ones cubes there are in 42 and 48.

Ask: **Are the number of ones cubes in 42 and 48 the same?** No **Which number has the fewest ones cubes?** 42 **Is 42 less than or greater than 48?** less than

Remind students that the > symbol, means *is greater than* and < means *is less than*.

Ask: **Which symbol would you use to compare 42 and 48?** <

Have students record the inequality, using words as well as symbols.

Continue by having students use base-ten blocks to model 375 and 365. Be sure students place the model of 375 on the left and the model of 365 on the right. Guide them to compare the numbers and record the inequality.

Repeat the activity with other 3-digit numbers. Vary the place value position of the digits students compare to determine which number is greater or less.

When students have demonstrated understanding, have them try the exercise using only paper and pencil.

Compare Numbers to Hundreds

Grade 4 Skill 5

Compare 465 and 427.

Number Line

Numbers are greater as you go from left to right.

← lesser   greater →

400   427   450   465   500

465 is to the right of 427.

So, $465 > 427$.

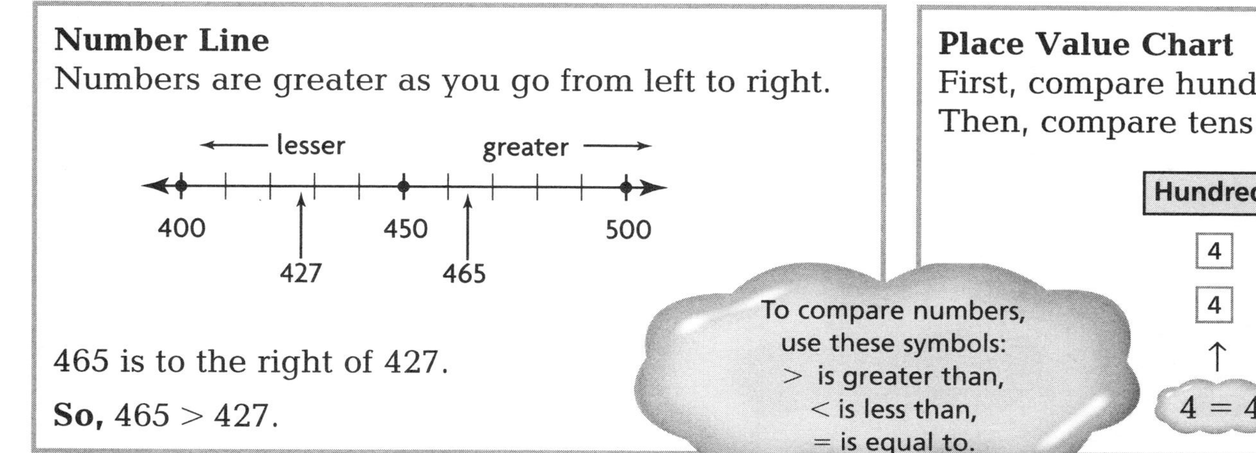

To compare numbers,
use these symbols:
> is greater than,
< is less than,
= is equal to.

Place Value Chart

First, compare hundreds. $4 = 4$.
Then, compare tens. $6 > 2$.

Hundreds	Tens	Ones
4	6	5
4	2	7
↑ $4 = 4$	↑ $6 > 2$	

So, $465 > 427$.

Try These

Compare. Write $>$, $<$, or $=$ for each.

1. Number line: 150, 170, 200, 230, 250

170 ◯ 230

2. Number line: 350, 351, 354, 355

354 ◯ 351

3. 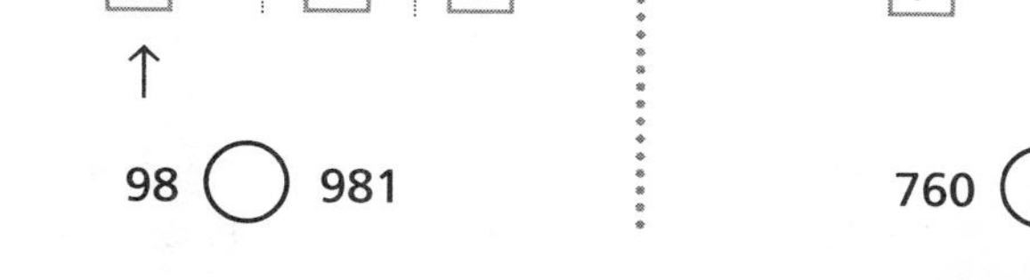

Hundreds	Tens	Ones
	9	8
9	8	1
↑		

98 ◯ 981

4.

Hundreds	Tens	Ones
7	6	0
7	6	5
		↑

760 ◯ 765

Go to the next side.

Name ______________________ Skill ______________________

Practice on Your Own

Skill 5

Compare 703 and 784.

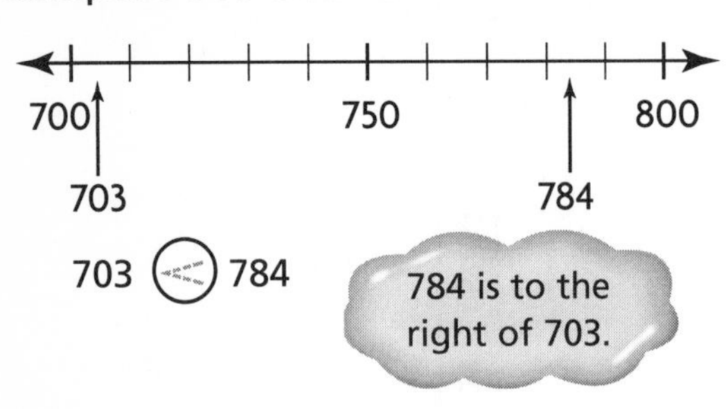

703 (<) 784

784 is to the right of 703.

Hundreds	Tens	Ones
7	0	3
7	8	4
↑	↑	

7 = 7

8 > 0

703 (<) 784

Compare. Use >, <, or =.

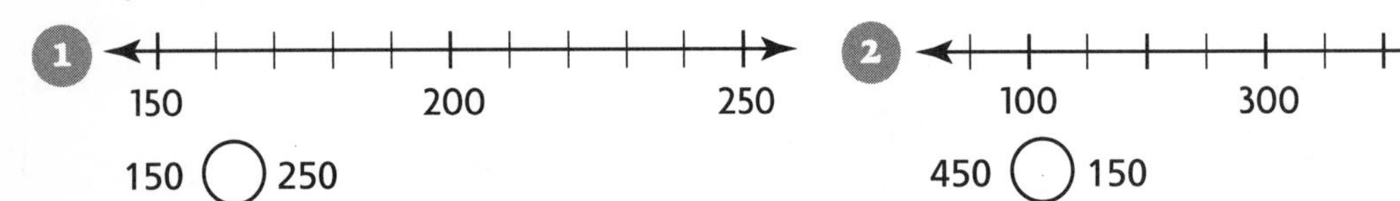

1. 150 ◯ 250

2. 450 ◯ 150

3.

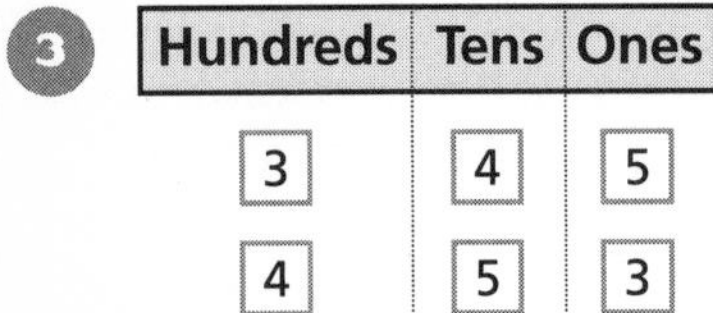

Hundreds	Tens	Ones
3	4	5
4	5	3

345 ◯ 453

4.

Hundreds	Tens	Ones
	9	0
1	6	0

90 ◯ 160

5. 349 ◯ 315
6. 129 ◯ 290
7. 690 ◯ 96
8. 38 ◯ 380

9. 442 ◯ 444
10. 609 ◯ 69
11. 724 ◯ 742
12. 962 ◯ 967

Check

Compare. Use >, <, or =.

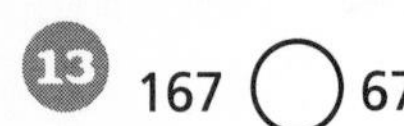

13. 167 ◯ 67
14. 941 ◯ 914
15. 807 ◯ 87
16. 22 ◯ 202

Skill 6 Grade 4

Compare Numbers to Thousands

10 Minutes

OBJECTIVE Compare numbers to thousands using >, <, and =.

Begin the lesson by discussing the first example. Point out the value of each column in the place value chart—thousands, hundreds, tens, ones. Stress that each place is 10 times the value of the one on its left. Review the meaning of the signs <, =, and >.

Students need to understand that when comparing numbers, they must start at the left, or the greatest place.

Ask: **What does the 8 stand for?** 8,000

Repeat for the other places. Point out that the tens place is the first place in which the numbers differ.

Stress that on a number line, any number to the right is greater than a number to the left. Thus, any number to the left of another number on the number line is less.

Show the students a number line similar to the one below.

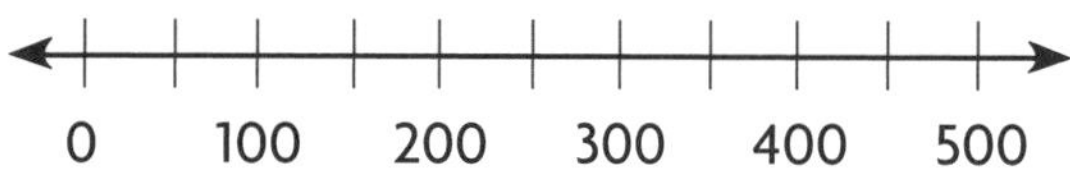

Ask questions such as: **Which is greater, 200 or 300?** 300 **Which is less, 500 or 400?** 400 **How can you tell?** **300 is to the right of 200; 400 is to the left of 500**

TRY THESE In Exercises 1–4, students compare numbers to thousands.

- **Exercise 1** Use the number line to compare numbers in which the thousands digits are different.

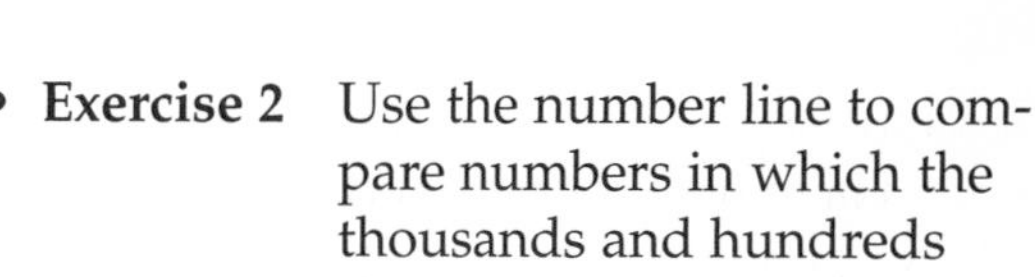

- **Exercise 2** Use the number line to compare numbers in which the thousands and hundreds digits are the same.
- **Exercise 3** Use a place value chart to compare numbers with different numbers of places.
- **Exercise 4** Use a place value chart to compare numbers with the same number of places.

PRACTICE ON YOUR OWN Review the examples at the top of the page. Have students explain how they know which number is greater or less.

CHECK Success is determined by 4 out of 4 correct responses.

Students who successfully complete the **Practice On Your Own** and **Check** are ready to move on to the next skill.

COMMON ERRORS

- Students often confuse < and >.
- When the first digit of a 3-place number is greater than the first digit of a 4-place number, students may think that the 3-place number is greater. For example, they may think that 925 is greater than 3,100.

Students who made more than three errors in the **Practice On Your Own** or who were not successful in the **Check** section may benefit from the **Alternative Teaching Strategy** on the next page.

Optional

Alternative Teaching Strategy

10 Minutes

Model Numbers to Compare to Thousands

OBJECTIVE Use place-value blocks to compare numbers

MATERIALS place-value blocks, index cards

Distribute place-value blocks to pairs of students. Have one partner model the number 1,452 with the blocks. Have the other partner model the number 2,372. Then ask the students to compare the numbers.

Ask: **Which number has more thousands?** 2,375 **Which number is greater?** 2,375 **Why?** Because 2,000 is greater than 1,000.

Write 1,452 _____ 2,375 on the board or on a sheet of paper and have one partner fill in the blank.

Repeat with 3,945 and 3,250.

Say: **Compare the thousands. What do you notice?** They are the same. **What should you do next?** Compare the hundreds. **What do you notice?** They are different. **Which number has more hundreds?** 3,945 **Which number is greater?** 3,945 **Why?** Because 900 is greater than 200. So, 3,945 > 3,250.

Continue with other pairs of numbers until you feel that the students are proficient in comparing.

Be sure students understand that if two numbers have a different number of places, the one with more places is greater. In comparing numbers such as 3,495 and 987, students may find it helpful to write the numbers in a place value chart. Then they would see that for 987 the thousands place does not have a digit.

Check to be sure that students are not confusing < and >. Tell them that each symbol always points to the smaller number.

Name ______________________ Skill ______________________

Compare Numbers to Thousands

Grade 4 Skill 6

Compare 8,571 and 8,522.

Number Line

Numbers are greater as you move from left to right.

← lesser   greater →

8,500   8,522   8,550   8,571   8,600

8,571 is to the right of 8,522.

So, 8,571 > 8,522.

Remember
> is greater than,
< is less than,
= is equal to.

Place Value Chart

- First, compare thousands. They are the same. 8 = 8.
- Then, compare hundreds. They are the same. 5 = 5.
- Now compare tens. 7 > 2.

So, 8,571 > 8,522.

TH	H	T	O
8,	5	7	1
8,	5	2	2
↑	↑	↑	
same	same	different	

Try These

Compare. Choose >, <, or =.

1. 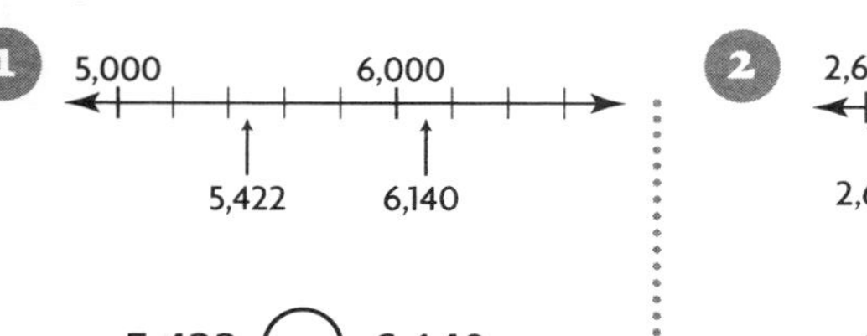

5,422 ◯ 6,140

2. 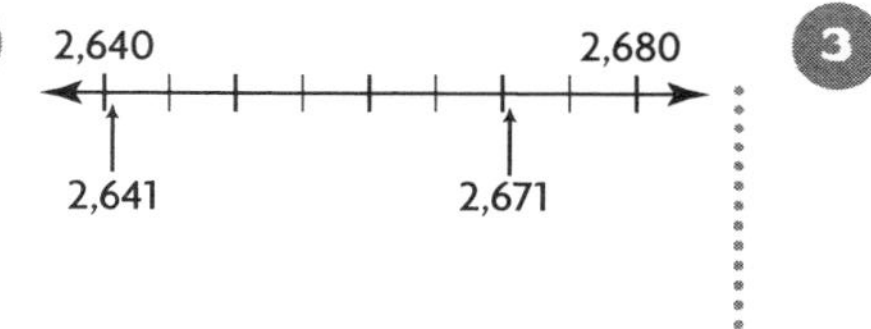

2,671 ◯ 2,641

3. 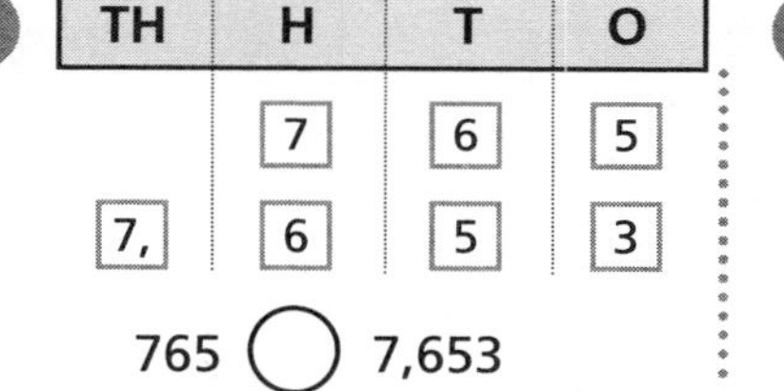

TH	H	T	O
	7	6	5
7,	6	5	3

765 ◯ 7,653

4. 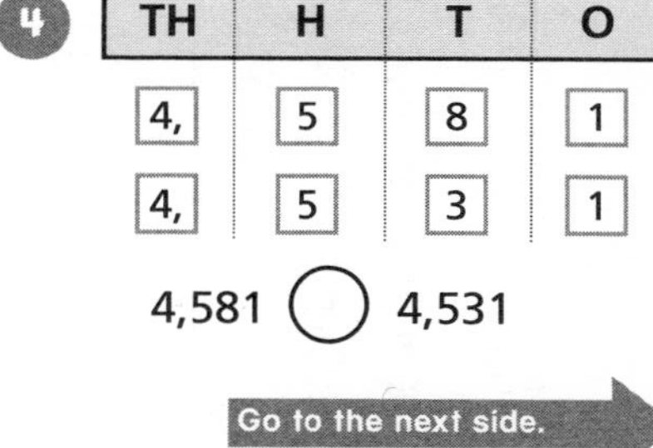

TH	H	T	O
4,	5	8	1
4,	5	3	1

4,581 ◯ 4,531

Go to the next side.

Name ______________________ Skill ______________________

Practice on Your Own

Skill 6

Compare 5,452 and 5,412.

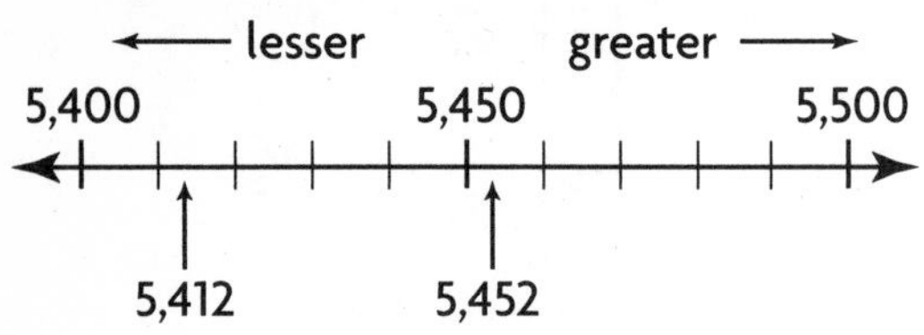

5,452 is to the the right of 5,412.

5,452 (>) 5,412

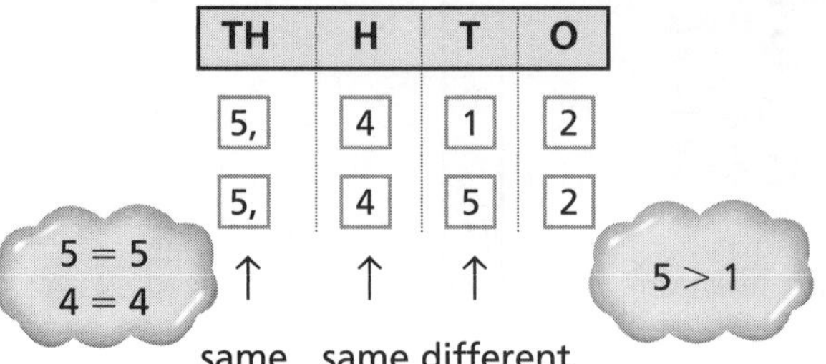

Compare. Choose >, <, or =.

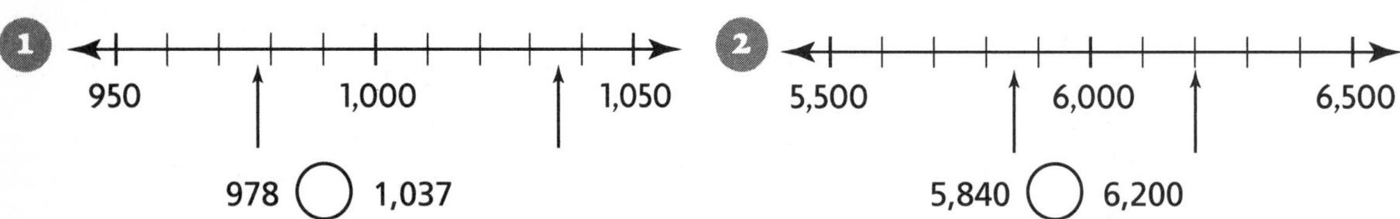

1 978 ◯ 1,037

2 5,840 ◯ 6,200

3

TH	H	T	O
	7	8	5
1,	0	5	5

785 ◯ 1,055

4

TH	H	T	O
8,	9	0	6
8,	7	9	0

8,906 ◯ 8,790

5 1,981 ◯ 9,812 **6** 4,105 ◯ 1,504 **7** 960 ◯ 6,900 **8** 3,050 ◯ 350

9 6,420 ◯ 6,440 **10** 5,090 ◯ 5,900 **11** 724 ◯ 2,047 **12** 5,600 ◯ 562

Check

Compare. Choose >, <, or =.

13 1,670 ◯ 2,009 **14** 4,871 ◯ 4,876 **15** 905 ◯ 9,001 **16** 1,420 ◯ 420

Skill 7 Grade 4 Round to Tens and Hundreds

OBJECTIVE Round to tens and hundreds

MATERIALS number line

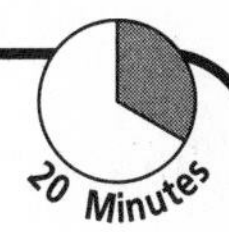

You may wish to begin by reviewing how to round 2-digit numbers on a number line.

Remind students that they have learned to round numbers two ways. They can use a number line or they can use the rounding rules.

Say: **You are rounding 1,463 to the nearest ten on the number line. 1,463 is between which two tens?** 1,460 and 1,470 **What number is halfway between 1,460 and 1,470?** 1,465 **Is 1,463 closer to 1,460 or 1,470?** 1,460 So, 1,463 rounded to the nearest ten is 1,460.

Ask similar questions for rounding 1,463 to the nearest hundred.

As you work through the example with the students using rounding rules, you may wish to suggest that students underline the digit to be rounded and circle the digit to the right. Marking in this way will help the students focus on the correct digits.

TRY THESE Exercises 1 and 2 give students the opportunity to round both ways.

- **Exercise 1** Round up to the nearest ten.
- **Exercise 2** Round down to the nearest hundred.

PRACTICE ON YOUR OWN Review the example with the students. Remind them to read the directions carefully, so they will know to which place they are rounding.

CHECK Students may choose to use either a number line or the rounding method to complete the **Check.** Success is indicated by 3 out of 4 correct responses.

Students who successfully complete the **Practice on Your Own** and **Check** are ready to move on to the next skill.

COMMON ERRORS

- Some students may look at a digit in the wrong place; for example to round to the nearest 10 they may use the digit in the tens place instead of the ones place.
- Some students may round to the nearest ten instead of the nearest hundred and vice versa.

Students who made more than three errors in the **Practice on Your Own,** or who were not successful in the **Check** section, may benefit from the **Alternative Teaching Strategy** on the next page.

Optional

Alternative Teaching Strategy

Round to Tens and Hundreds

OBJECTIVE Use a place value chart to round to tens and hundreds

MATERIALS place-value chart, number cards

Provide students with place-value charts or have them make the charts. Display a card showing 62. Then explain to the students that they are to round 62 to the nearest ten. Have them tell you the two tens that 62 is between. **60 and 70**

Display a card showing 60 and a card showing 70. Place 62 between 60 and 70 and ask the students to decide whether to round 62 up to 70, or round it down to 60. Have them write 62 in the place-value chart in preparation for rounding.

Have them underline and label the digit in the tens place to show that it's the rounding place. Then have them circle the digit in the ones place.

Hundreds	Tens	Ones
	6 rounding place	2 ← Think: digit to the right

Say: **This is the digit to the right of the rounding place. You use this digit to decide how to round.**

Have students use the rounding rules to determine that since $2 < 5$, they need to round down. So, 62 rounded to the nearest ten is 60.

Repeat the activity several times, having students round two digit numbers to the nearest ten. Then move on to rounding numbers to the nearest hundred.

When students show an understanding of the rounding process, have them round numbers without the place-value chart.

Name ______________________ Skill ______________________

Round to Tens and Hundreds

Grade 4 **Skill 7**

Round 1,463 to the nearest ten using a number line.

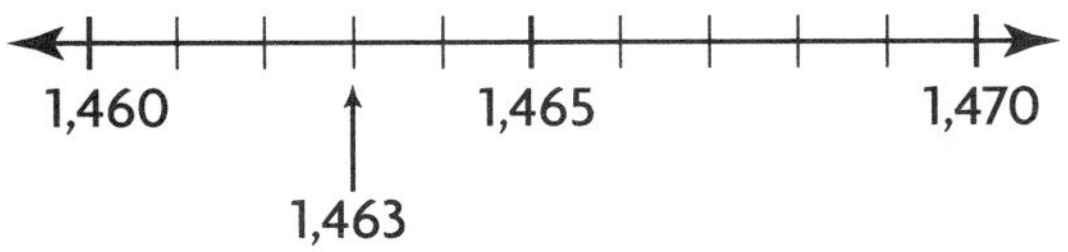

1,463 is closer to 1,460 than to 1,470.

1,463 rounded to the nearest ten is 1,460.

Round 1,463 to the nearest hundred.

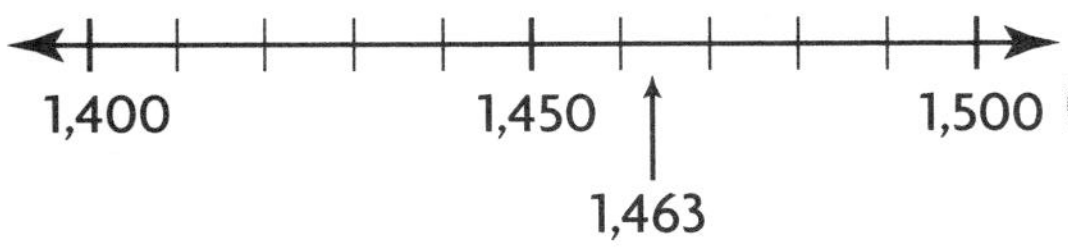

1,463 is closer to 1,500 than to 1,400.

1,463 rounded to the nearest hundred is 1,500.

To round to the nearest **ten,** look at the digit in the ones place. If the digit is less than 5, the digit to be rounded stays the same.

digit to be rounded → 1 , 4 <u>6</u> **3** ← ones place

3 < 5

1,463 rounded to the nearest ten is 1,460.

To round to the nearest **hundred,** look at the digit in the tens place. If the digit is 5 or greater, the digit to be rounded increases by 1.

digit to be rounded → 1 , <u>4</u> **6** 3 ← tens place

6 > 5

1,463 rounded to the nearest hundred is 1,500.

Try These

Complete.

1. Use a number line. Round 126 to the nearest ten.

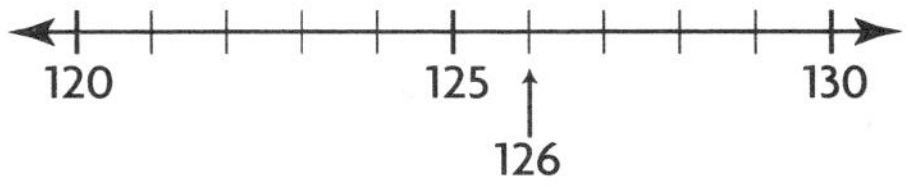

126 is closer to ☐ than to ☐.

126 rounded to the nearest ten is ☐.

2. Use rounding rules. Round 1,617 to the nearest hundred.

The digit in the tens place is ☐.

1, <u>6</u> 1 7 The digit to be rounded is ☐.

1,617 rounded to the nearest hundred is ☐.

Go to the next side.

Name ______________________ Skill ______________________

Practice on Your Own

Skill 7

Round 259 to the nearest hundred.

Number Line

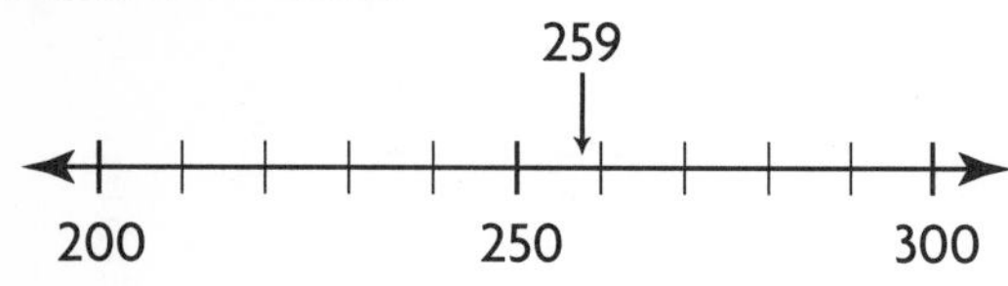

Look at the number line.

259 is closer to 300 than to 200.

Rounding Rules

To round to the nearest hundred, look at the tens digit.

tens digit

$\underline{2}$ **5** 9

If the digit is 5 or greater, the digit to be rounded increases by 1.
If the digit is less than 5, the digit to be rounded stays the same.

So, 259 rounded to the nearest hundred is 300.

Round to the nearest ten.

1

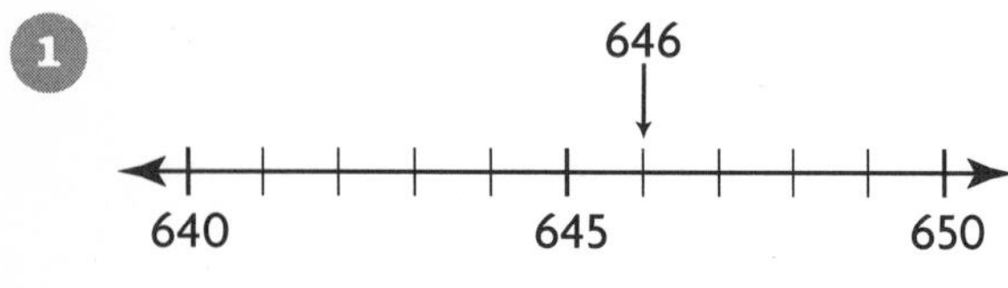

646 to the nearest ten is ☐.

2

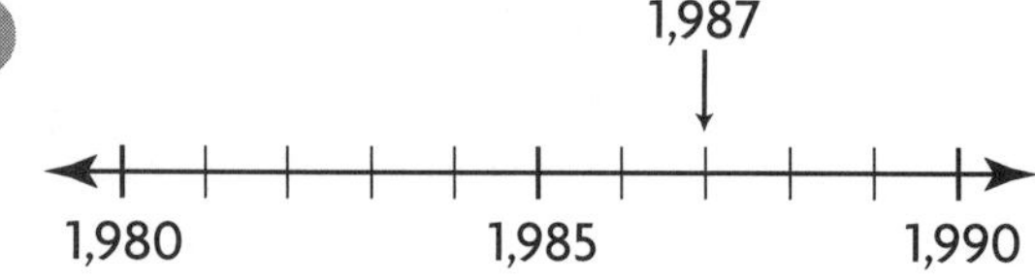

1,987 to the nearest ten is ☐.

Use rounding rules to round the numbers to the nearest hundred.

3 380

↓

$\underline{3}$ **8** 0

380 rounded to the nearest hundred is ☐.

4 1,794

↓

1, $\underline{7}$ 9 4

1,794 rounded to the nearest hundred is ☐.

Round to the place of the underlined digit.

5 $4\underline{6}2$ ____________

6 $6{,}9\underline{2}7$ ____________

7 $\underline{2}65$ ____________

8 $3{,}\underline{8}78$ ____________

Check

Round to the place of the underlined digit.

9 $5\underline{8}6$ ____________

10 $9{,}0\underline{3}8$ ____________

11 $\underline{6}30$ ____________

12 $5{,}\underline{6}83$ ____________

Answer Card

Place Value

Grade 4

SKILL 1

TRY THESE

1. 4, 400; 3, 30; 2, 2; 4, 3, 2
2. 8, 800; 0, 0; 9, 9; 8, 0, 9

PRACTICE

1. 9, 900; 5, 50; 6, 6; 9, 5, 6
2. 6, 600; 0, 0; 0, 0; 6, 0, 0
3. 7, 700; 2, 20; 0, 0; 7, 2, 0
4. 3, 300; 9, 90; 4, 4; 3, 9, 4

CHECK

5. 1 ten or 10
6. 0 ones or 0
7. 9 hundreds or 900

SKILL 2

TRY THESE

1. 7, 7,000; 5, 500; 6, 60; 9, 9; 7, 5, 6, 9
2. 6, 6,000; 4, 400; 0, 0; 3, 3; 6, 4, 0, 3

PRACTICE

1. 3, 3,000; 0, 0; 7, 70; 9, 9
2. 7, 7,000; 5, 500; 0, 0; 3, 3
3. 8, 8,000; 2, 200; 9, 90; 0, 0
4. 6, 6,000; 1, 100; 0, 0; 0, 0

CHECK

5. 4 thousands or 4,000
6. 7 tens or 70

SKILL 3

TRY THESE

1. 4, 40
2. 2, 50
3. 2, 100

PRACTICE

1. 10, 50
2. 100, 200
3. 50, 150
4. 100, 200
5. 25, 75
6. 10, 20

CHECK

7. 10, 30
8. 25, 50
9. 50, 150

SKILL 4

TRY THESE

1. 681
2. 6, 7, 3, 2; 6,000, 700, 30, 2; 6,732

PRACTICE

1. 4, 7, 9; 400, 70, 9; 479
2. 7, 0, 6, 0; 7,000, 0, 60, 0; 7,060
3. 9, 2, 6; 900, 20, 6; 926
4. 891
5. 2,650

CHECK

6. 907
7. 5,225

Answer Card

Place Value

Grade 4

SKILL 5

TRY THESE

1. <
2. >
3. <
4. <

PRACTICE

1. <
2. >
3. <
4. <
5. >
6. <
7. >
8. <
9. <
10. >
11. <
12. <

CHECK

13. >
14. >
15. >
16. <

SKILL 6

TRY THESE

1. <
2. >
3. <
4. >

PRACTICE

1. <
2. <
3. <
4. >
5. <
6. >
7. <
8. >
9. <
10. <
11. <
12. >

CHECK

13. <
14. <
15. <
16. >

SKILL 7

TRY THESE

1. 130, 120, 130
2. 1, 6, 1,600

PRACTICE

1. 650
2. 1,990
3. 400
4. 1,800
5. 460
6. 6,930
7. 300
8. 3,900

CHECK

9. 590
10. 9,040
11. 600
12. 5,700

Number Sense

Whole Number Addition

2-Digit Addition: Regrouping Ones as Tens

15 Minutes

OBJECTIVE Add 2-digit numbers, regrouping ones as tens

Explain to students that in this skill they will be finding the sum of 2-digit numbers. Point out the place-value labels and suggest that students use them to remember the value of each digit as they add.

Draw students' attention to Step 1. Have them add the ones.

Ask: **You have more than 9 ones. What can you do next?** Regroup the ones.

Work through the regrouping with students. Have them look at Step 2.

Ask: **How do you regroup the ones?** There are 13 ones, I regroup 10 ones as 1 ten 3 ones.

Emphasize that the regrouped digit is placed in the tens column.

Ask: **How do you show the regrouping?** I write 3 in the ones place of the sum, and write the 1 ten in the tens column.

As students complete the addition in Step 3, remind them to add the regrouped digit when they add the tens.

TRY THESE Exercises 1–4 prepare students for the regrouping exercises they will find on the **Practice on Your Own** page.

- **Exercise 1** Regroup 14 ones.
- **Exercise 2** Regroup 10 ones.
- **Exercise 3** Regroup 11 ones.
- **Exercise 4** Regroup 12 ones.

PRACTICE ON YOUR OWN Review the example at the top of the page. Ask students to explain how they use the place value labels to help them in the regrouping process. Encourage students to use terms such as *ones, tens, regroup, place,* and *sum.*

CHECK Determine if students can regroup when they add 2-digit numbers. Success is indicated by 3 out of 4 correct responses.

Students who successfully complete the **Practice on Your Own** and **Check** are ready to move on to the next skill.

COMMON ERRORS

- Students may forget to add the regrouped digit.
- Students may place the regrouped digit in the wrong column.
- Students may not know their addition facts.

Students who made more than four errors in the **Practice on Your Own**, or who were not successful in the **Check** section, may benefit from the **Alternative Teaching Strategy** on the next page.

Optional

Alternative Teaching Strategy

15 Minutes

Model 2-Digit Addition: Regrouping Ones as Tens

OBJECTIVE Use base-ten blocks to model addition of 2-digit numbers, regrouping ones as tens

MATERIALS base-ten blocks, paper

Explain to students that they will be reviewing the regrouping process for adding 2-digit numbers.

First, review regrouping ones. Have students take 18 ones. Explain that since there are more than 9 ones, they can regroup 10 ones as 1 ten. Note that now they have 1 ten 8 ones. Help students recognize that 18 ones and 1 ten 8 ones represent 18.

Next, have students work in pairs. One partner models with the blocks, the other records the addition on paper. Present this addition example.

tens	ones
2	4
+3	7
□	□

Explain that the place-value labels can help students remember the value of the digits as they add. The grid helps them to write the digits in the correct place. Have the students show the addends 24 and 37 with the blocks.

Say: **Begin by adding ones. Group the ones blocks together. How many ones do you have in all?** 11 ones **You have more than 9 ones. What do you do next?** Regroup 11 ones as 1 ten 1 one. **Where do you put the regrouped ten?** with the other tens

Remind students to write the regrouped digit above the 2 in tens column.

When you add the tens, what must you remember to do? Add the regrouped digit. **What is the sum?** 61

Repeat the activity with similar examples. When students show understanding of the regrouping process, encourage them to add using only paper and pencil.

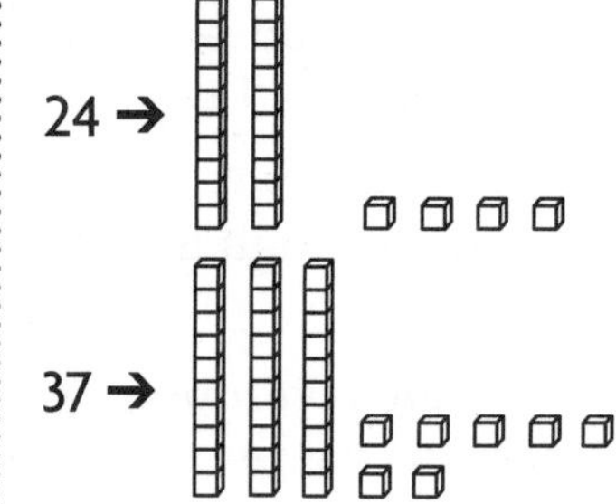

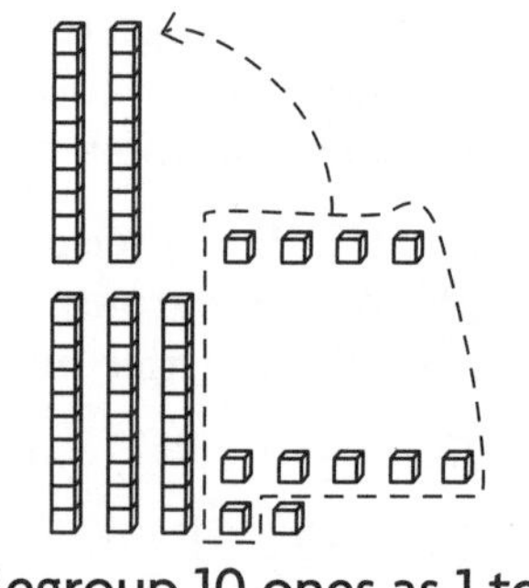

Regroup 10 ones as 1 ten

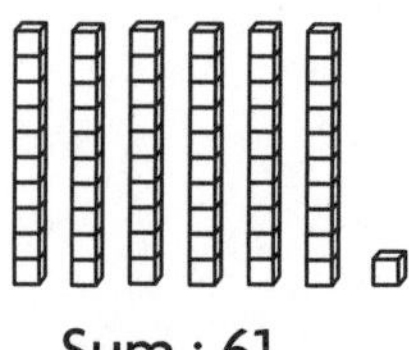

Sum : 61

Name ______________________ Skill ______________________

Grade 4 Skill 8

2-Digit Addition: Regrouping Ones as Tens

Find 36 + 27 = ■.

Step 1 Add the ones.

	Tens	Ones
	☐	
	3	6
+	2	7
	☐	☐

6 + 7 = 13 ones

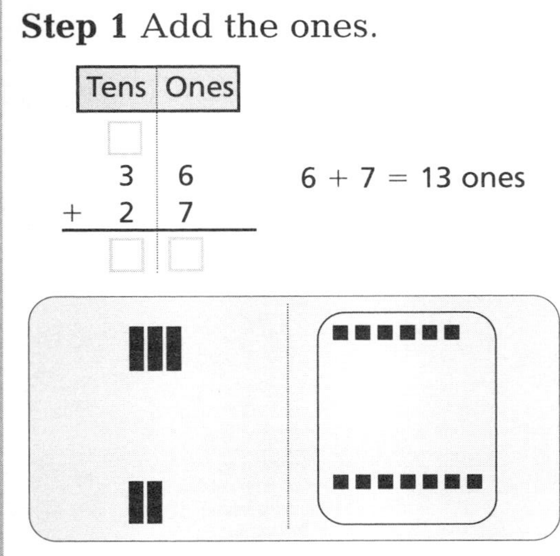

Step 2 Regroup.

	Tens	Ones
	1	
	3	6
+	2	7
	☐	3

Regroup 13 ones as 1 ten 3 ones.

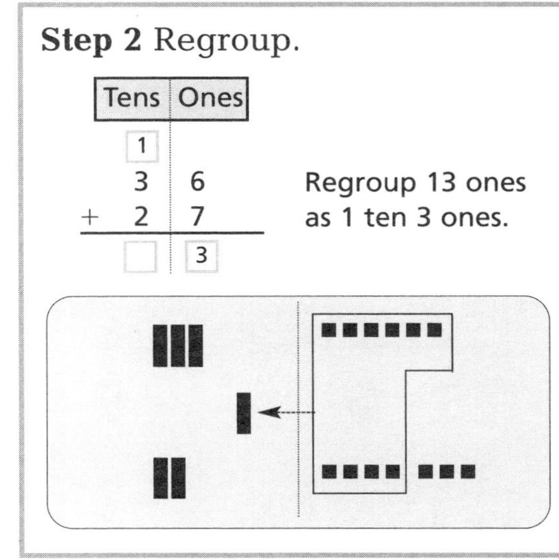

Step 3 Add the tens.

	Tens	Ones
	1	
	3	6
+	2	7
	6	3

1 + 3 + 2 = 6 tens

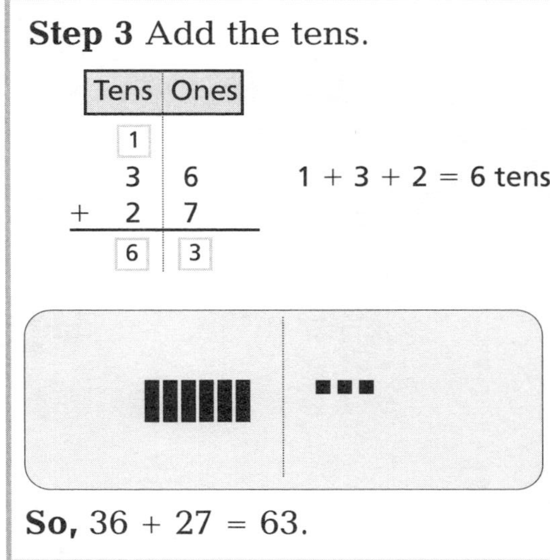

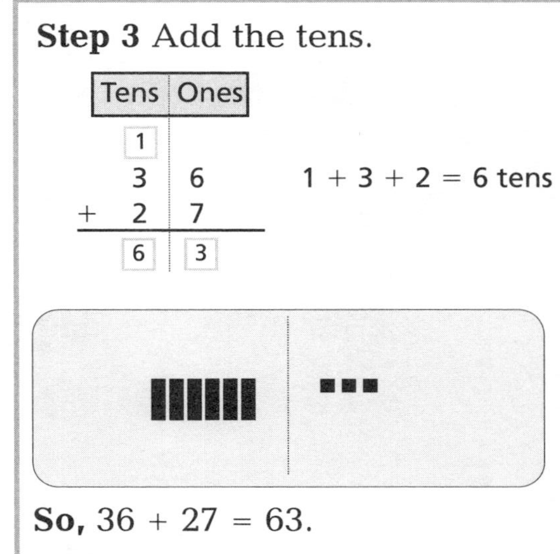

So, 36 + 27 = 63.

Try These

Find the sum.

1.

	Tens	Ones
	☐	
	4	8
+		6
	☐	☐

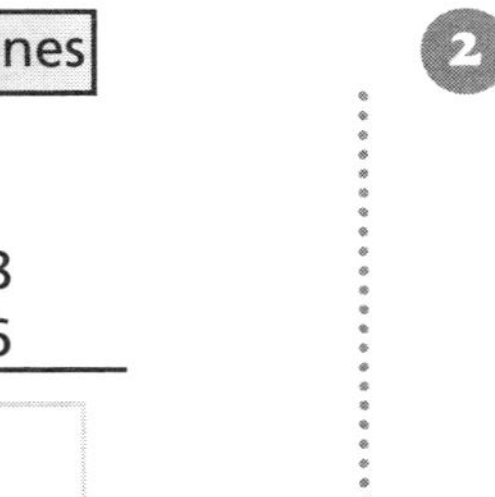

2.

	Tens	Ones
	☐	
	5	9
+	3	1
	☐	☐

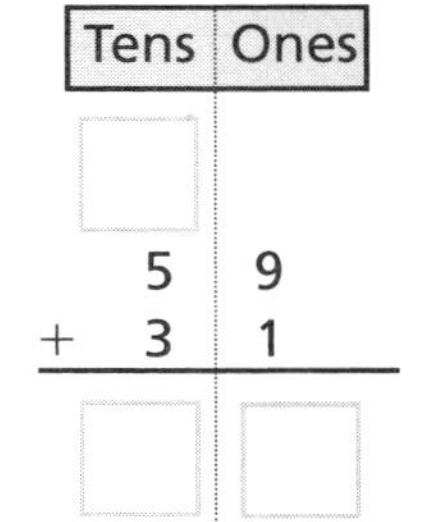

3.

	Tens	Ones
	☐	
	3	3
+	4	8
	☐	☐

4.

	Tens	Ones
	☐	
	1	6
+	7	6
	☐	☐

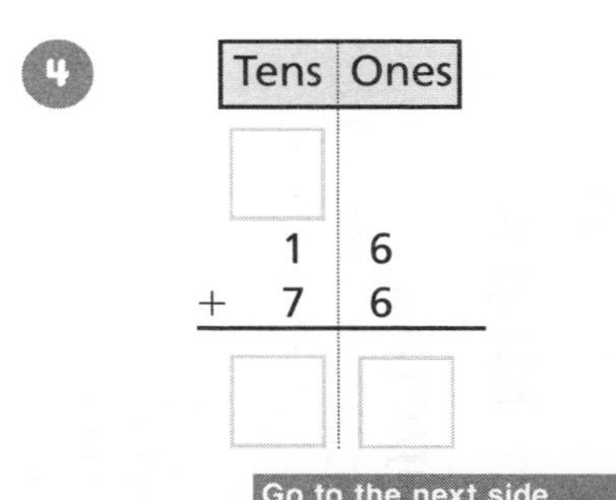

Go to the next side.

Name ________________ Skill ________________

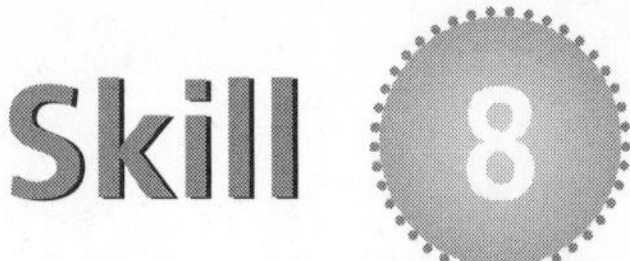

Practice on Your Own

Skill 8

Think:
Regroup ones when the sum of the digits is 10 or greater.

Tens	Ones
1	
5	5
+ 2	6
8	1

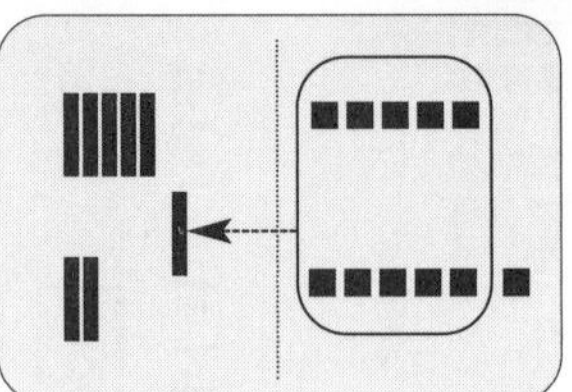

Find the sum.

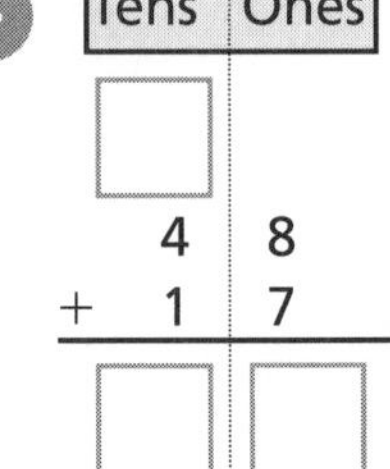

1.

Tens	Ones
☐	
1	7
+ 1	3
☐	☐

2.

Tens	Ones
☐	
3	6
+ 2	8
☐	☐

3.

Tens	Ones
☐	
4	5
+ 4	5
☐	☐

4.

Tens	Ones
☐	
4	8
+ 1	7
☐	☐

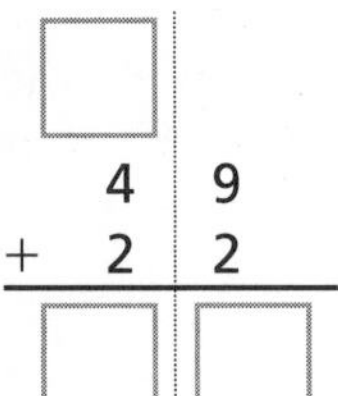
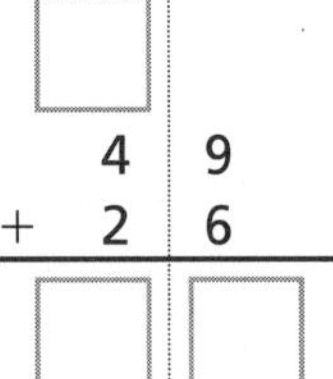
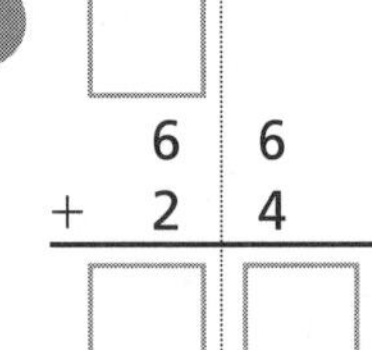

5. ☐ 49 + 22 = ☐☐

6. ☐ 49 + 26 = ☐☐

7. ☐ 66 + 24 = ☐☐

8. ☐ 18 + 35 = ☐☐

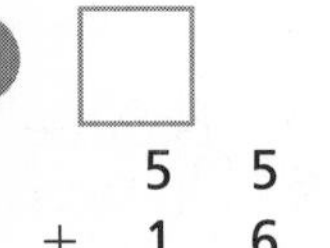
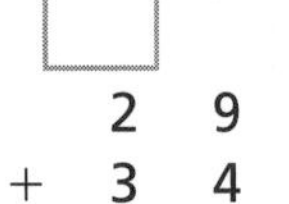
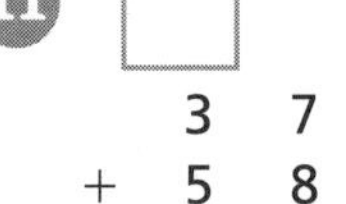

9. ☐ 55 + 16

10. ☐ 29 + 34

11. ☐ 37 + 58

12. ☐ 14 + 57

13. 11 + 19

14. 65 + 29

15. 27 + 33

16. 36 + 16

Check

Find the sum.

17. 25 + 25

18. 37 + 48

19. 56 + 34

20. 79 + 19

Skill 9 Grade 4

2-Digit Addition: Regrouping Tens as Hundreds

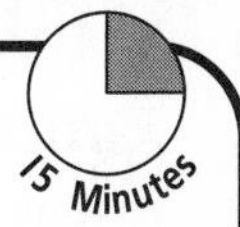

OBJECTIVE Add 2-digit numbers, regrouping tens as hundreds

Some students may benefit from practicing basic addition facts before beginning the lesson. Use flash cards or an addition table as a warm-up.

Point to the place-value labels in Step 1. Review place-value names and the value of the digits in the addends.

Suggest that students refer to the models of the base-ten blocks as they add.

Ask: **How many ones are there in all?** 9 **Can you regroup the ones?** No **How do you know?** There are fewer than 10 ones; 10 ones are needed to make a ten. **How many tens are there in all?** 11 tens **Can you regroup the tens?** Yes **Explain how you will regroup them.** Regroup 11 tens as 1 hundred 1 ten

Point out that the hundred is recorded in the hundreds place of the sum.

TRY THESE Exercises 1–4 provide practice in 2-digit addition with regrouping tens.

- **Exercise 1** 11 tens as 1 hundred 1 ten.
- **Exercise 2** 11 tens as 1 hundred 1 ten.
- **Exercise 3** 11 tens as 1 hundred 1 ten.
- **Exercise 4** 13 tens as 1 hundred 3 tens.

PRACTICE ON YOUR OWN Review the example at the top of the page. As students work through the example, ask them to explain how they would regroup the tens.

In Exercises 1–8, students add 2-digit numbers, regrouping tens as hundreds, with place-value labels and answer boxes provided as visual cues. In Exercises 9–16, students add 2-digit numbers, regrouping tens as hundreds.

CHECK Determine if students can add 2-digit numbers, regrouping tens as hundreds. Success is determined by 3 out of 4 correct responses.

Students who successfully complete the **Practice on Your Own** and **Check** are ready to move on to the next skill.

COMMON ERRORS

- Students may forget to write the regrouped hundred in the sum when they regroup the tens.
- Students may add incorrectly because they do not know basic addition facts.

Students who made more than three errors in the **Practice on Your Own,** or who were not successful in the **Check** section, may benefit from the **Alternative Teaching Strategy** on the next page.

Optional

Alternative Teaching Strategy

Model 2-Digit Addition: Regrouping Tens as Hundreds

OBJECTIVE Model 2-digit addition, regrouping tens as hundreds

MATERIALS base-ten blocks

Distribute base-ten blocks. Give students 14 tens rods. Explain to students that they have enough tens rods to regroup 10 tens as 1 hundred. Have students exchange 10 tens rods for 1 hundreds block. Point out that although they have regrouped the tens as hundreds, the total value of the blocks remains the same.

14 tens = 1 hundred 4 tens

Present students with the following 2-digit addition.

	hundreds	tens	ones
		8	2
+		4	1

Ask: **Can you regroup the ones?** No, there are fewer than 10 ones; 10 ones are needed to make 1 ten.

Can you regroup the tens? Yes, there are more than 10 tens; 12 tens is regrouped as 1 hundred 2 tens.

Where will you write the hundred? In the hundreds place of the sum.

Repeat the activity several times with other examples. When students show understanding of the regrouping process, remove the base-ten blocks and have them try some similar examples, using only paper and pencil.

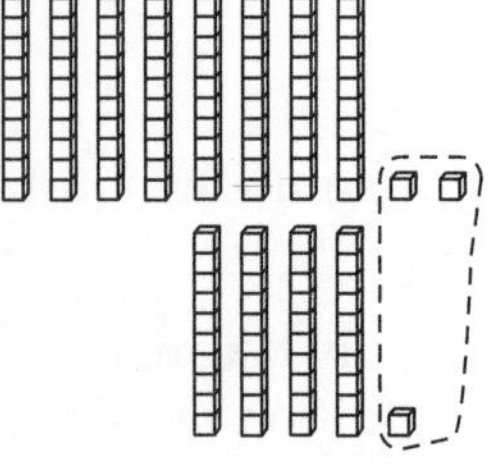

Add the ones.

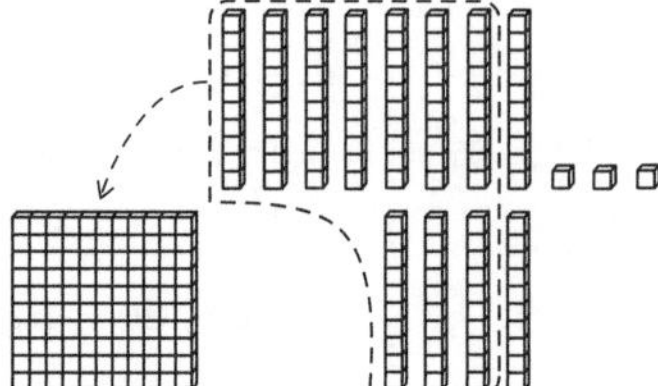

Regroup the tens.

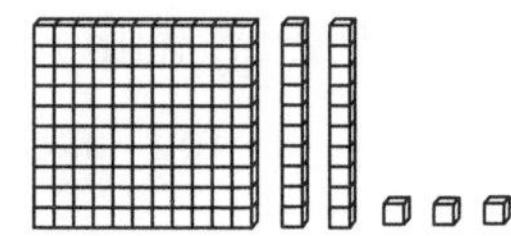

82 + 41 = 123

Name ______________________ Skill ______________________

Grade 4 Skill 9

2-Digit Addition: Regrouping Tens as Hundreds

Find 52 + 67 = ■.

Step 1 Add the ones.

H	T	O
☐		
	5	2
+	6	7
		9

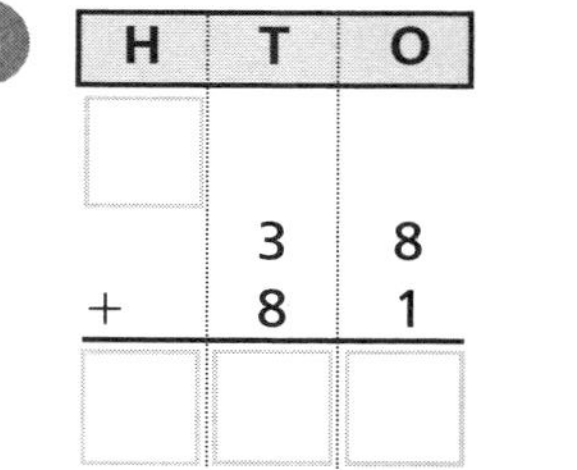

Step 2 Add the tens.
Regroup 11 tens as 1 ten 1 hundred.

H	T	O
1		
	5	2
+	6	7
1	1	9

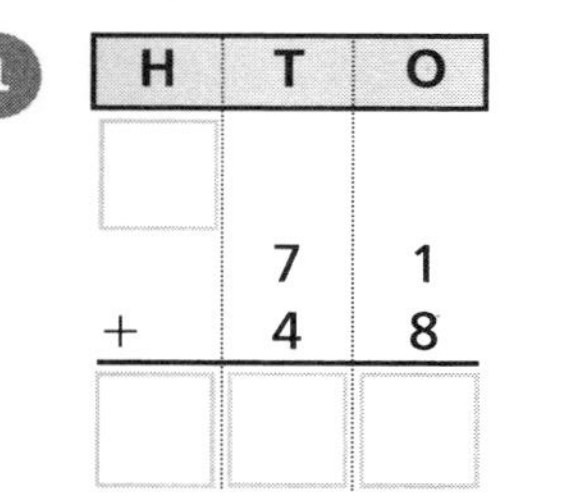

5 tens + 6 tens
= 11 tens

So, 52 + 67 = 119.

Try These

Find the sum.

1

H	T	O
☐		
	7	1
+	4	8
☐	☐	☐

2

H	T	O
☐		
	3	8
+	8	1
☐	☐	☐

3

H	T	O
☐		
	5	5
+	6	2
☐	☐	☐

4

H	T	O
☐		
	9	5
+	3	4
☐	☐	☐

Go to the next side.

Name ______________________ Skill ______________________

Practice on Your Own

Skill 9

Example:

Add the ones.

H	T	O
	7	3
+	5	3
		6

Then add the tens.

H	T	O
1	7	3
+	5	3
1	2	6

Think
7 tens + 5 tens = 12 tens
12 tens = 1 hundred 2 tens

Find the sum.

1.

H	T	O
	3	3
+	7	6

2.

H	T	O
	4	4
+	8	5

3.

H	T	O
	5	7
+	7	1

4.

H	T	O
	5	6
+	7	3

5.

H	T	O
	9	0
+	3	2

6.

H	T	O
	2	2
+	8	4

7.

H	T	O
	8	6
+	5	0

8.

H	T	O
	5	6
+	5	1

9. $92 + 35$

10. $73 + 74$

11. $31 + 82$

12. $96 + 82$

13. $56 + 92$

14. $43 + 82$

15. $74 + 84$

16. $90 + 92$

Check

Find the sum.

17. $61 + 95$

18. $94 + 71$

19. $86 + 93$

20. $94 + 95$

Skill 10 Grade 4 — 3-Digit Addition: Regrouping Ones as Tens

OBJECTIVE Add two 3-digit numbers, regrouping ones as tens

MATERIALS base-ten blocks

You may wish to work through Steps 1–3 with the students using base-ten blocks. Introduce the skill by explaining to the students that they will be adding 3-digit numbers. Have them look at Step 1 and notice the place-value labels.

Ask: **What is the sum of the digits in the ones column?** 14 **Can you regroup the ones?** yes **Why?** There are more than 9 ones, so I regroup 14 ones as 1 ten 4 ones. **Where do you write the regrouped ten?** Above the digit 4 in the tens column.

In Step 2, emphasize adding the regrouped ten to the sum of 4 and 2. Then for Step 3, note that the sum of these hundreds digits is recorded in the hundreds column.

TRY THESE Exercises 1–4 model the type of exercises students will find in the **Practice on Your Own** page.

- **Exercise 1** Regroup 13 ones 1 ten 3 ones.
- **Exercise 2 and 4** Regroup 10 as 1 ten 0 ones.
- **Exercise 3** Regroup 12 ones as 1 ten 2 ones.

PRACTICE ON YOUR OWN Work through the example at the top of the page. Remind students as they complete the page, that to find the correct sum, they should place the regrouped digit carefully at the top of the tens column. Then they will remember to add the regrouped ten when they add the tens.

CHECK Determine if students know the basic facts, and can add two 3-digit numbers, regrouping ones as tens correctly. Success is indicated by 3 out of 4 correct responses.

Students who successfully complete the **Practice on Your Own** and **Check** are ready to move on to the next skill.

COMMON ERRORS

- Students may not know basic facts for addition.
- Students may write the regrouped digit in the wrong place, or forget to write it.
- Students may forget to add the regrouped digit when adding the tens.

Students who made more than three errors in the **Practice on Your Own,** or who were not successful in the **Check** section, may benefit from the **Alternative Teaching Strategy** on the next page.

Optional

Alternative Teaching Strategy

15 Minutes

Model 3-Digit Addition: Regrouping Ones as Tens

OBJECTIVE Use base-ten blocks to add two 3-digit numbers regrouping ones as tens

MATERIALS base-ten blocks

Have students work in groups of two or three. Suggest they take turns modeling the addition with the base-ten blocks, and recording the addition with pencil and paper.

Present the following example and have the students model the addition and then record it.

Hundreds	Tens	Ones
3	0	9
+ 1	2	3

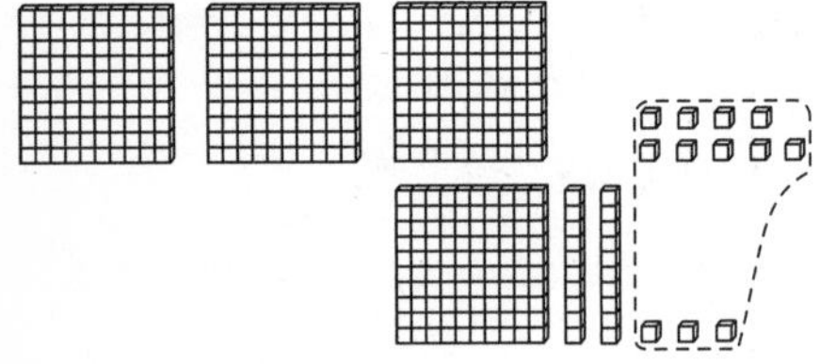

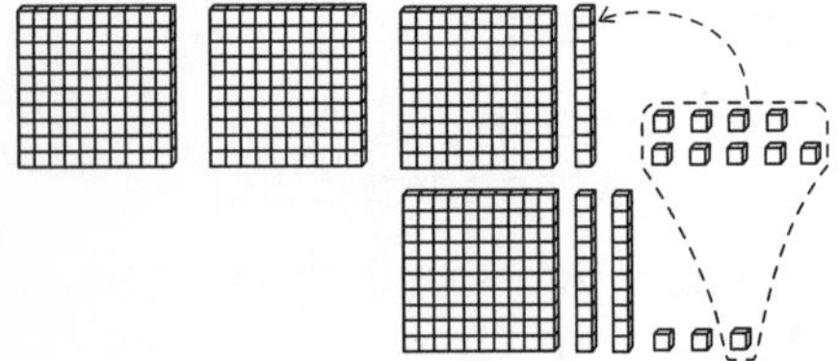

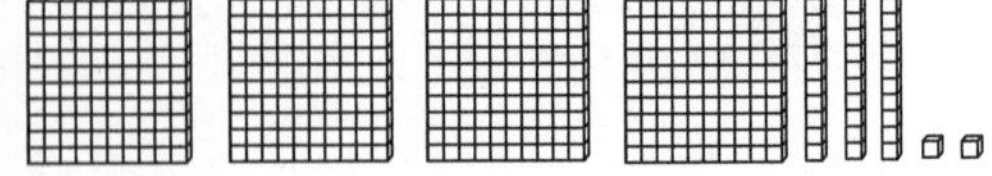

Say: **Begin with the ones. What is the sum of the digits in the ones column?** 12 **Can you regroup?** yes **How do you know?** There are more than 9 ones, so I regroup 12 ones as 1 ten 2 ones. **Where do you write the regrouped digit?** Above the digit 0 in the tens column.

Discuss the importance of recording the regrouped digit, as well as placing it at the top of the correct column.

What is the sum of the digits in the tens column? 3 **Add the hundreds. What is the sum?** 432

Repeat the activity with similar examples until the students understand the regrouping process. Then have them do several exercises without using the base-ten blocks.

Name ____________________ Skill ____________________

Grade 4 **Skill 10**

3-Digit Addition: Regrouping Ones as Tens

Find 145 + 229 = ■.

Step 1 Add the ones.
5 ones + 9 ones = 14 ones
Regroup 14 ones as 1 ten 4 ones.

	H	T	O
		1	
	1	4	5
+	2	2	9
			4

Step 2 Add the tens.
1 ten + 4 tens + 2 tens = 7 tens

	H	T	O
		1	
	1	4	5
+	2	2	9
		7	4

Step 3 Add the hundreds.
1 hundred + 2 hundreds
= 3 hundreds.

	H	T	O
		1	
	1	4	5
+	2	2	9
	3	7	4

So, 145 + 229 = 374.

Try These

Find the sum.

1.

	H	T	O
		☐	
	1	3	4
+	3	1	9
	☐	☐	☐

2.

	H	T	O
		☐	
	3	0	8
+	2	5	2
	☐	☐	☐

3.

	H	T	O
		☐	
	2	2	9
+		2	3
	☐	☐	☐

4.

	H	T	O
		☐	
	6	1	2
+	1	6	8
	☐	☐	☐

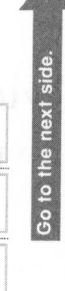

Go to the next side.

Name ______________________ Skill ______________________

Practice on Your Own

Example:
Add the ones.
Do you need to regroup?
Add the tens.
Then add the hundreds.

	H	T	O
		1	
	3	6	7
+	2	1	5
	5	8	2

7 ones + 5 ones = 12 ones
12 ones = 1 ten, 2 ones

Find the sum.

1. 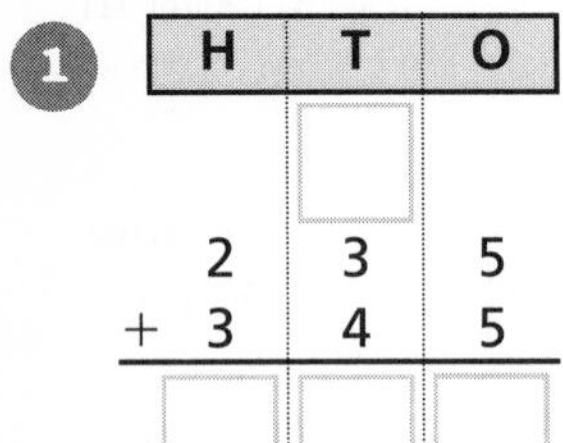

	H	T	O
	2	3	5
+	3	4	5

2. 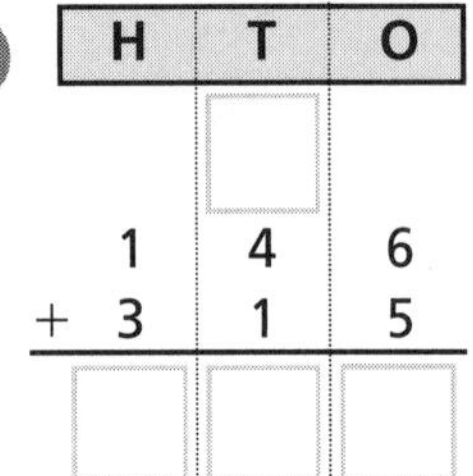

	H	T	O
	1	4	6
+	3	1	5

3. 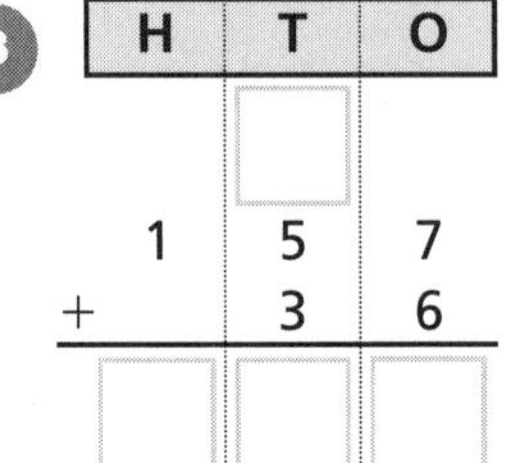

	H	T	O
	1	5	7
+		3	6

4.

	H	T	O
	2	5	6
+		2	9

5. 171 + 19

6. 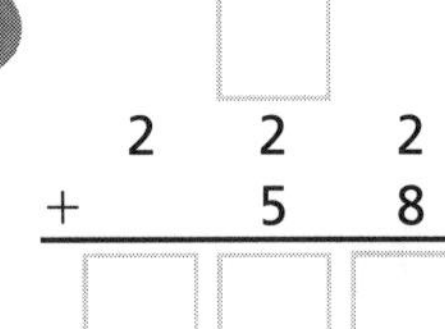 222 + 58

7. 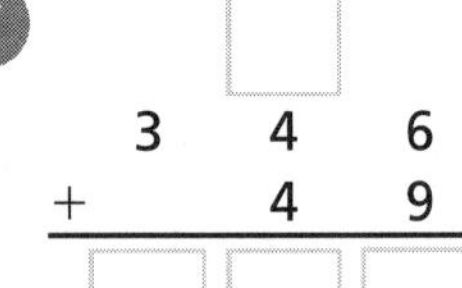 346 + 49

8. 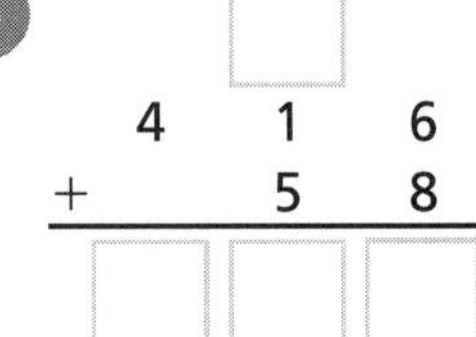 416 + 58

9. 238 + 135

10. 473 + 119

11. 631 + 129

12. 186 + 305

13. 356 + 218

14. 445 + 126

15. 375 + 209

16. 189 + 102

Check

Find the sum.

17. 569 + 223

18. 184 + 308

19. 746 + 135

20. 179 + 206

Skill 11 Grade 4

3-Digit Addition: Regrouping Tens as Hundreds

OBJECTIVE Add two 3-digit numbers, regrouping tens as hundreds

Begin by pointing to the place value table, recalling that "H" represents hundreds, "T" tens, and "O" ones. Remind students to use the labels to help them remember the value of the digits as they add. Explain that they will be adding two 3-digit numbers, regrouping tens as hundreds. Have students look at Step 1.

Say: **Always begin adding in the ones place.**

Continue: **What is 5 ones plus 4 ones?** 9 ones **Can you regroup?** no **Why not?** There are 9 ones, I can only regroup if there are 10 or more ones.

Look at Step 2, What is 4 tens plus 8 tens? 12 tens **Can you regroup the tens?** Yes, there are 12 tens; 12 tens = 1 hundred 2 tens. **How do you record 1 hundred 2 tens?** Write the 2 tens in the tens column in the sum; write the 1 hundred above the 4 in the hundreds column. **What is 4 hundreds plus 1 hundred?** 5 hundreds **What do you do with the regrouped hundred?** Add it to the 5 hundreds: 5 hundreds + 1 hundred = 6 hundreds. **When you add, how do you know when you can regroup?** when there are ten or more in any place

TRY THESE Exercises 1–3 model the type of exercises students will find on the **Practice on Your Own** page.

- **Exercises 1 and 2** Regroup 10 tens as 1 hundred.
- **Exercise 3** Regroup 11 tens as 1 hundred, 1 ten.

PRACTICE ON YOUR OWN Work through the example at the top of the page. Ask the students to explain the adding and regrouping process, using the terms *ones, tens,* and *hundreds.* Caution students to align the digits and to remember to add the regrouped digit when adding in the hundreds place.

CHECK Determine if students know how to add two 3-digit numbers, recognizing when to regroup the tens as hundreds. Success is indicated by 3 out of 4 correct responses.

Students who successfully complete the **Practice on Your Own** and **Check** are ready to move on to the next skill.

COMMON ERRORS

- Students may write the regrouped digit in the wrong place or may forget to record the regrouping.
- Students may forget to add the regrouped digit when adding the hundreds.
- Students may not know their addition facts.

Students who made more than four errors in the **Practice on Your Own,** or were not successful in the **Check** section, may benefit from the **Alternative Teaching Strategy** on the next page.

Optional

Alternative Teaching Strategy

20 Minutes

Model 3-Digit Addition: Regrouping Tens

OBJECTIVE Use base-ten blocks to model addition of two 3-digit numbers, regrouping tens as hundreds.

MATERIALS base-ten blocks, paper

You may wish to have the students work in pairs. One student models the addition with the base-ten blocks, while the partner records the addition process using paper and pencil.

Distribute the base-ten blocks. Have the recording students write the following addition example in a place-value grid leaving enough room above the top addend to show regrouping with the appropriate labels. The partner models each number using the blocks.

Hundreds	Tens	Ones
2	6	3
+1	8	2

Have students begin adding with the ones. Students will note that the sum is 5 and no regrouping is needed. Have them explain why. The answer should state that since there are fewer than 10 ones, there is no need to regroup.

As they add the tens, students should recognize regrouping is necessary. Have them explain that since there are more than 9 tens, they can regroup the 14 tens as 1 hundred 4 tens.

Discuss where to place the regrouped digit and why it should be carefully aligned with the other hundreds. As students add the hundreds, remind them to add the regrouped hundred also.

Repeat the activity with similar examples and have partners change roles. You may also wish to provide several examples in which students only determine if regrouping is involved. Have them examine the digits in each place to tell if the sum will be ten or more. Have students also decide where the regrouped digits should be recorded.

When the students understand the regrouping process, have them do several additions without using the base-ten blocks.

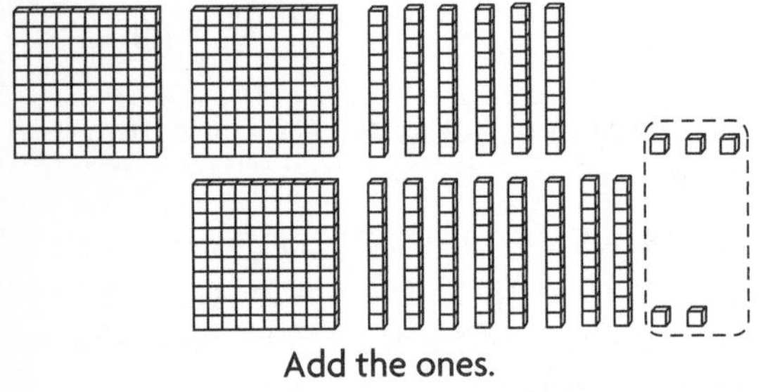
Add the ones.

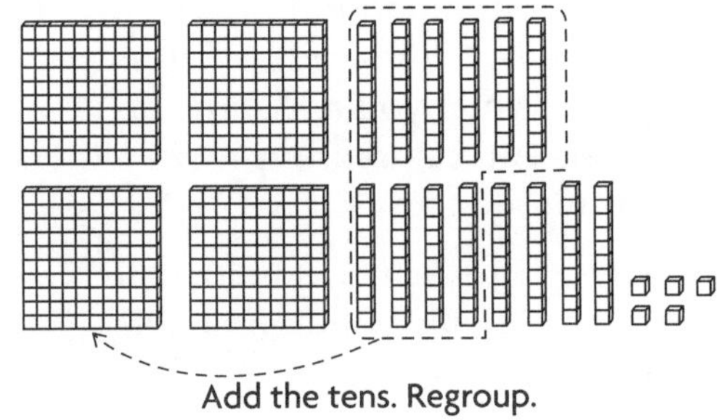
Add the tens. Regroup.

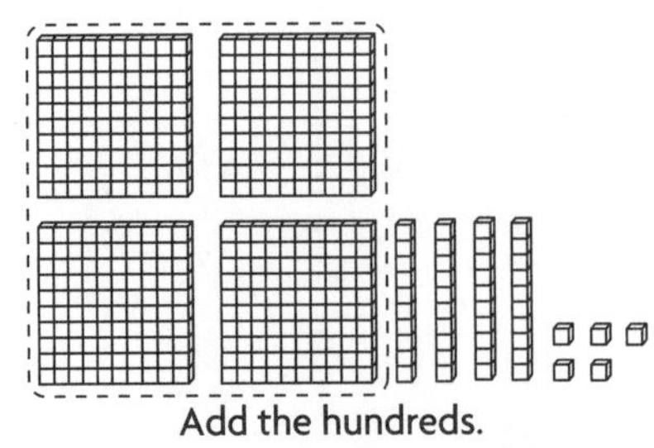
Add the hundreds.

Name ______________________ Skill ______________________

Grade 4 **Skill 11**

3-Digit Addition: Regrouping Tens as Hundreds

Find 445 + 184 = ■.

Step 1 Add the ones.
5 ones + 4 ones = 9 ones

	H	T	O
	4	4	5
+	1	8	4
			9

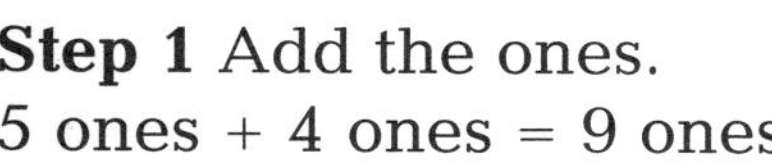

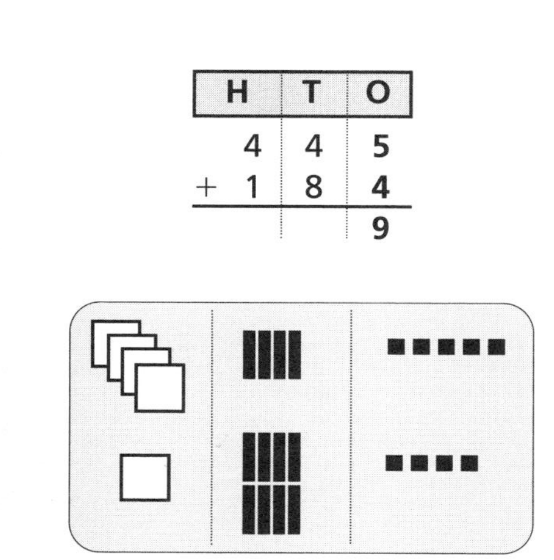

Step 2 Add the tens.
4 tens + 8 tens = 12 tens
Regroup 12 tens as 1 hundred 2 tens.

	H	T	O
	1		
	4	**4**	5
+	1	**8**	4
		2	9

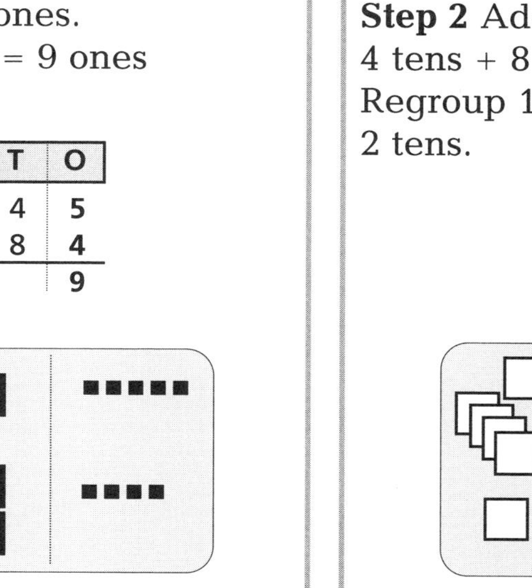

Step 3 Add the hundreds.
Write the sum.

	H	T	O
	1		
	4	4	5
+	**1**	8	4
	6	2	9

So, 445 + 184 = 629.

Try These

Add.

1.

	H	T	O
	☐		
	6	9	1
+	2	1	0
	☐	☐	☐

2.

	H	T	O
	☐		
	3	8	0
+	5	2	0
	☐	☐	☐

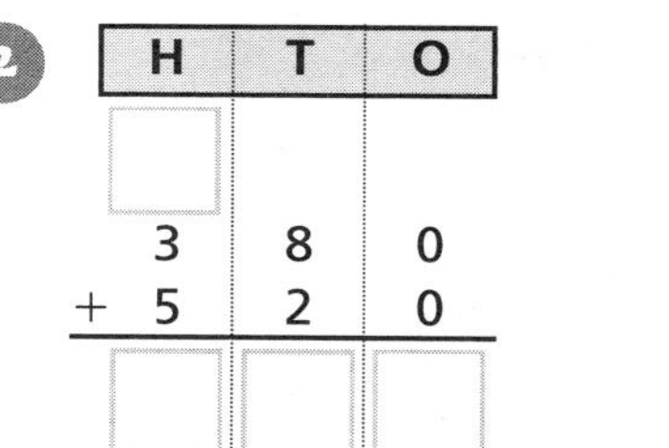

3.

	H	T	O
	☐		
	4	3	8
+	3	8	1
	☐	☐	☐

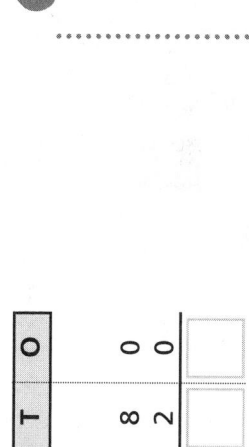

Go to the next side.

Name ______________________ Skill ______________________

Practice on Your Own

Skill 11

Find 765 + 143 = ■.

Add the ones.
Then add the tens.
Do you need to regroup?
Add the hundreds.

H	T	O
1		
7	6	5
+ 1	4	3
9	0	8

6 tens + 4 tens = 10 tens
10 tens = 1 hundred 0 tens

Find the sum.

1. 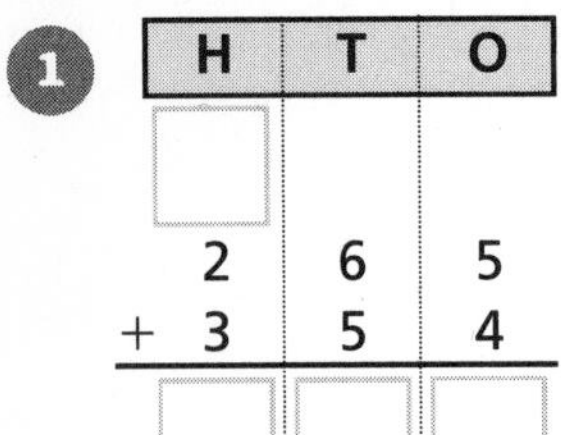

H	T	O
2	6	5
+ 3	5	4

2. 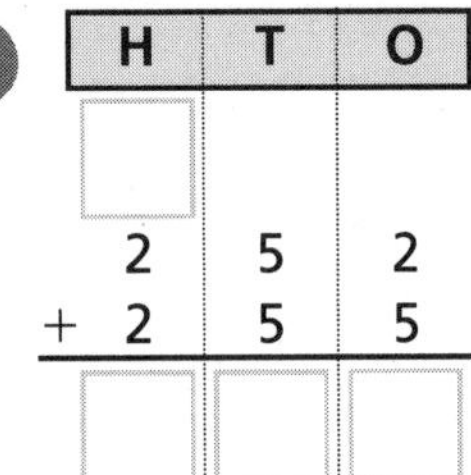

H	T	O
2	5	2
+ 2	5	5

3. 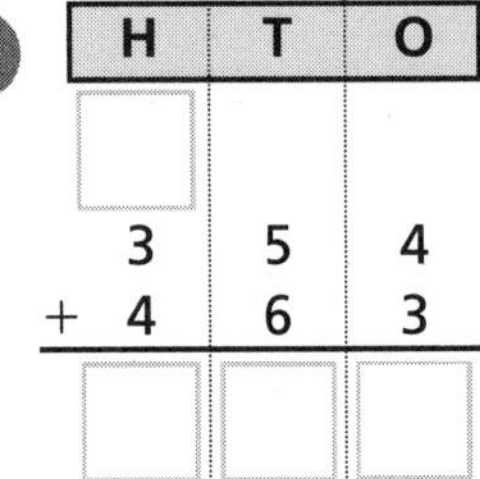

H	T	O
3	5	4
+ 4	6	3

4.

H	T	O
2	7	0
+ 6	3	6

5. 163 + 392

6. 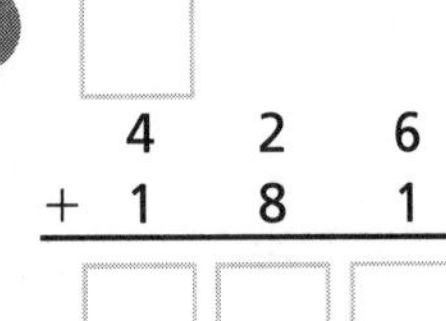
426 + 181

7. 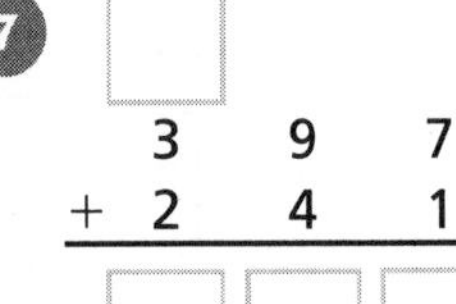
397 + 241

8. 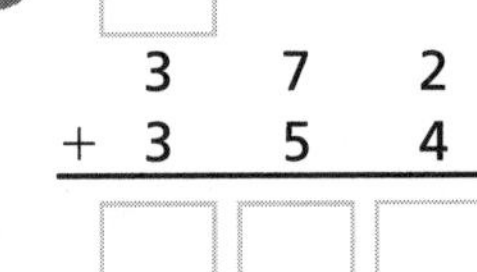
372 + 354

9. 490 + 468

10. 343 + 273

11. 435 + 393

12. 552 + 291

13. 346 + 462

14. 371 + 394

15. 482 + 485

16. 340 + 399

Check

Find the sum.

17. 242 + 662

18. 592 + 386

19. 474 + 480

20. 498 + 391

Skill 12 Grade 4

3-Digit Addition: Regrouping Hundreds as Thousands

OBJECTIVE Add two 3-digit numbers, regrouping to thousands

MATERIALS base-ten blocks

You may wish to have students use base-ten blocks to model Steps 1–4. Draw students' attention to Step 1.

Ask: **What is the sum of the digits in the ones column?** 10 **Can you regroup the ones?** Yes **So there is 1 ten and how many ones?** zero ones **What do you write in the sum under the ones column?** zero **Where do you record the regrouped ten?** above the tens column.

Continue in a similar way for Steps 2, 3 and 4.

TRY THESE Exercises 1–4 provide practice for regrouping ones as tens, tens as hundreds, and hundreds as thousands.

- **Exercise 1** Regroup once: ones as tens.
- **Exercise 2** Regroup twice: tens as hundreds, hundreds as thousands.
- **Exercise 3** Regroup three times: ones as tens, tens as hundreds, hundreds as thousands.

PRACTICE ON YOUR OWN Review the example at the top of the page. Exercises 1–3 provide practice regrouping once, with place-value tables, regrouping boxes and answer boxes as visual cues. Exercises 4–6 provide practice regrouping once with regrouping boxes and answer boxes as visual cues. Exercises 7–9 provide practice regrouping twice.

CHECK Determine if students know the basic facts; can regroup ones, tens, and hundreds; and can record the sum correctly by aligning the digits in the appropriate columns. Success is indicated by 3 out of 4 correct responses.

Students who successfully complete the **Practice on Your Own** and **Check** are ready to move on to the next skill.

COMMON ERRORS

- Students may record a regrouped digit in the wrong column.
- Students may forget to record or add a regrouped digit.

Students who made more than four errors in the **Practice on Your Own,** or who were not successful in the **Check** section, may benefit from the **Alternative Teaching Strategy** on the next page.

Optional

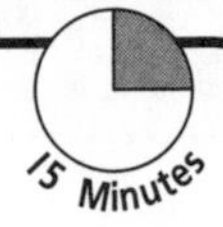

Alternative Teaching Strategy
Add with Play Money

OBJECTIVE Use play money to understand regrouping in addition

MATERIALS play money, paper

Have students count and exchange coins and bills before they begin to add money amounts. Have them verbalize that 10 pennies are regrouped or exchanged for 1 dime, and 10 dimes are regrouped as 1 dollar.

For this activity, work through several examples aloud without having the students record the steps symbolically. When the students show confidence with the process, have them start recording each result.

Have the students model \$2.45 + \$6.87. Explain that they will use the bills and coins to find the sum. Guide the students to join the 5 pennies with the 7 pennies and regroup as 1 dime and 2 pennies. Next, join the 4 dimes and 8 dimes and regroup as 1 dollar and 2 dimes. Then join the dollars.

Ask the students to identify the total amount by counting all of the regrouped money, starting with the bills. **\$9.32**

Repeat the activity with other examples. Have the students try an example that includes recording the results. When students have demonstrated understanding, have them add money amounts using paper and pencil only.

Name ______________________ Skill ______________________

Grade 4 Skill 12

3-Digit Addition: Regrouping Hundreds as Thousands

Find 681 + 439 = ■.

Step 1 Add the ones. Regroup.

	thousands	hundreds	tens	ones
			1	
		6	8	1
+		4	3	9
				0

Think:
1 one + 9 ones = 10 ones
10 ones = 1 ten 0 ones

Step 2 Add the tens. Regroup.

	thousands	hundreds	tens	ones
		1	1	
		6	8	1
+		4	3	9
			2	0

Think:
1 ten + 8 tens
+ 3 tens = 12 tens
12 tens =
1 hundred 2 tens

Step 3 Add the hundreds. Regroup.

	thousands	hundreds	tens	ones
	1	1	1	
		6	8	1
+		4	3	9
		1	2	0

Think: 1 hundred
+ 6 hundreds + 4 hundreds
= 11 hundreds
11 hundreds =
1 thousand 1 hundred

Step 4 Add the thousands.

	thousands	hundreds	tens	ones
	1	1	1	
		6	8	1
+		4	3	9
	1,	1	2	0

Think:
Place a comma between the thousands and hundreds places

So, 681 + 439 = 1, 120.

Try These

Find the sum.

1.

	thousands	hundreds	tens	ones
	☐			
		6	5	5
+		7	1	4
	☐	☐	☐	☐

2.

	thousands	hundreds	tens	ones
	☐	☐		
		7	3	5
+		4	8	4
	☐	☐	☐	☐

3.

	thousands	hundreds	tens	ones
	☐	☐	☐	
		8	3	6
+		2	7	5
	☐	☐	☐	☐

Go to the next side.

Name ______________________ Skill ______________________

Practice on Your Own

Skill 12

Find 543 + 678 = ■.

Think:
Add the ones. Regroup. Add the tens. Regroup. Add the hundreds. Regroup. Write the thousands.

	thousands	hundreds	tens	ones
	1	1	1	
		5	4	3
+		6	7	8
	1,	2	2	1

Find the sum.

	TH	H	T	O
	☐			
		8	4	6
+		5	5	2
	☐,	☐	☐	☐

	TH	H	T	O
	☐			
		8	1	3
+		5	7	4
	☐,	☐	☐	☐

	TH	H	T	O
	☐			
		2	5	4
+		8	4	3
	☐,	☐	☐	☐

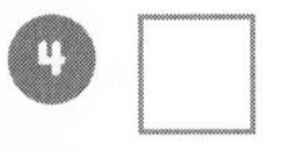

  864
+ 324

  813
+ 545

6.

  175
+ 923

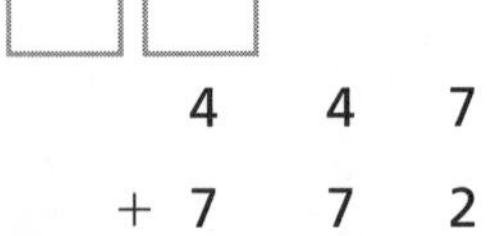

  447
+ 772

  256
+ 973

9.

  763
+ 545

Check

Find the sum.

10. 269 + 950

11. 227 + 781

12. 614 + 898

13. 177 + 924

Order and Zero Properties of Addition

OBJECTIVE Understand the Order and Zero Properties of Addition

Direct students' attention to the Order Property of Addition.

Ask: **What is an addend?** The numbers being added together **What is the sum of 6 + 5?** 11 **What is the sum of 5 + 6?** 11 **What is the difference between the two additions?** The order of the addends has been changed. **Do both ways of writing the addends have the same sum?** Yes

Continue: **So, you can change the order of addends, and the sum is always the same.**

As you work through the example for the Zero Property of Addition, draw attention to the addition table. Have students notice the pattern in the sums and addends. Summarize by stating the Zero Property of Addition.

Say: **When you add 0 to any number, the sum is that number.**

TRY THESE Exercises 1–4 model the type of exercises students will find on the **Practice on Your Own** page.

- **Exercises 1 and 3** Order Property of Addition.
- **Exercises 2 and 4** Zero Property of Addition.

PRACTICE ON YOUR OWN Review the example at the top of the page. As they work through the exercise, remind students to identify whether the order of addends is changed, or whether one of the addends is zero.

CHECK Determine if the students know that changing the order of addends does not change the sum, and adding zero to a number gives a sum that is the number. Success is determined by 4 out of 4 correct responses.

Students who successfully complete the **Practice on Your Own** and **Check** are ready to move on to the next skill.

COMMON ERRORS

- Students may find different sums when the order of addends is changed.
- Students may write a zero next to an addend; for example, for 2 + 0, students might write the sum as 20.

Students who made more than four errors in the **Practice on Your Own**, or who were not successful in the **Check** section, may benefit from the **Alternative Teaching** Strategy on the next page.

Optional

Alternative Teaching Strategy

15 Minutes

Model Order and Zero Properties of Addition

OBJECTIVE Use dominoes to model the Order and Zero Properties of Addition

MATERIALS dominoes

Distribute four dominoes to each student. Guide them to position one domino one way, then reposition it. Students should then record the two addition sentences.

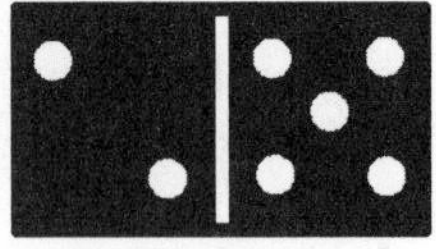

2 + 5 = 7

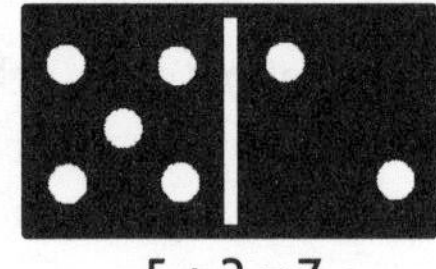

5 + 2 = 7

Ask: **What is the sum of 2 + 5?** 7 **5 + 2?** 7 **Did changing the order of the addends change the sums?** No

Have students continue with the three other dominoes.

Display a domino with a blank side. Have students write the addition sentence.

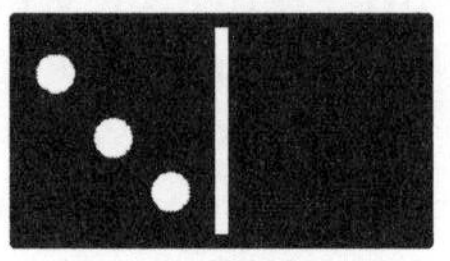

3 + 0 = 3

Repeat this activity with similar examples. When students show understanding, have them try an exercise using only paper and pencil.

Name ______________________ Skill ______________________

Grade 4 **Skill 13**

Order and Zero Properties of Addition

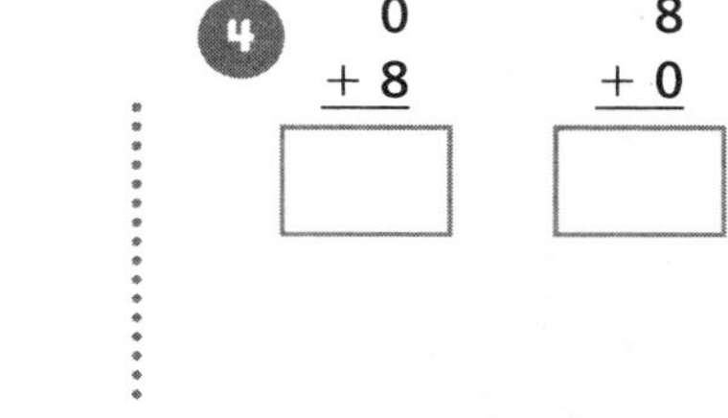

Order Property of Addition
You can change the order of addends, and the sum is always the same.

6 + 5 = 11   5 + 6 = 11

$$\begin{array}{r} 3 \\ +\,4 \\ \hline 7 \end{array} \qquad \begin{array}{r} 4 \\ +\,3 \\ \hline 7 \end{array}$$

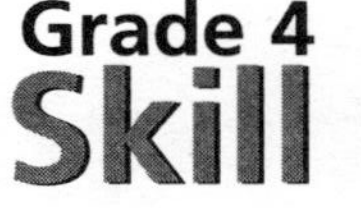

Zero Property of Addition
When you add 0 to any number, the sum is that number.

0 + 2 = 2   2 + 0 = 2

This addition table shows the Zero Property of Addition.

+	0	1	2	3	4	5
0	0	1	2	3	4	5

Try These

Find the sum.

1. 8 + 4 = ☐
   4 + 8 = ☐

2. 6 + 0 = ☐
   0 + 6 = ☐

3. $\begin{array}{r} 9 \\ +\,2 \\ \hline \end{array}$ ☐ $\begin{array}{r} 2 \\ +\,9 \\ \hline \end{array}$ ☐

4. $\begin{array}{r} 0 \\ +\,8 \\ \hline \end{array}$ ☐ $\begin{array}{r} 8 \\ +\,0 \\ \hline \end{array}$ ☐

Go to the next side.

Name ______________________ Skill ______________________

Practice on Your Own

Skill 13

Ask yourself:
Is the order changed?
Is zero involved?

Order Property
10 + 8 = 18
8 + 10 = ■

Think:
The order is changed.
The sum is the same.

So, 8 + 10 = 18.

Zero Property
32 + 0 = ■

Think: 0 is one of the addends. The sum is the other addend.

So, 32 + 0 = 32.

Find the sum.

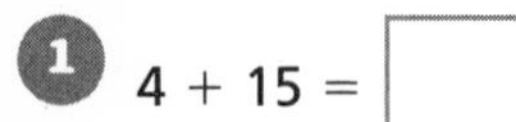
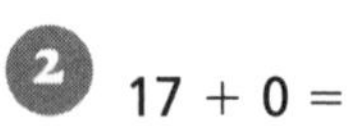
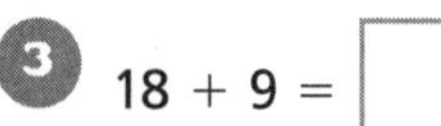

1. 4 + 15 = ☐
   15 + 4 = ☐
2. 17 + 0 = ☐
   0 + 17 = ☐
3. 18 + 9 = ☐
   9 + 18 = ☐
4. 12 + 24 = ☐
   24 + 12 = ☐

5. 10 + 25 = ☐ ; 25 + 10 = ☐
6. 26 + 0 = ☐ ; 0 + 26 = ☐
7. 21 + 34 = ☐ ; 34 + 21 = ☐
8. 45 + 12 = ☐ ; 12 + 45 = ☐

Complete using the Order Property.

9. 12 + 0 = ☐
   ☐ + 12 = ☐
10. 14 + 15 = ☐
    15 + ☐ = ☐
11. 8 + 14 = ☐
    ☐ + ☐ = ☐
12. 21 + 17 = ☐
    ☐ + ☐ = ☐

Check

Complete using the Order Property.

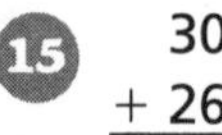

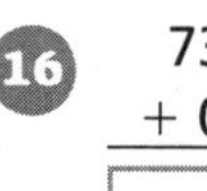
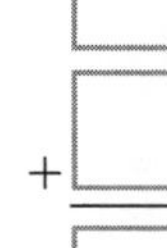

13. 7 + 19 = ☐
    ☐ + ☐ = ☐
14. 25 + 0 = ☐
    ☐ + ☐ = ☐
15. 30 + 26 = ☐
    ☐ + ☐ = ☐
16. 73 + 0 = ☐
    ☐ + ☐ = ☐

Skill 14 Grade 4

Grouping Property of Addition

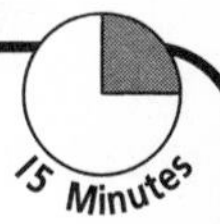

OBJECTIVE Use the Grouping Property of Addition to add three or more addends

Begin by recalling that numbers that are added are called *addends*. The result is called the *sum*.

Direct students' attention to Step 1.

Ask: **How many addends are there?** 3

Help students recognize that they can start at the top to group 1 and 9 or they can group 1 and 9 from the bottom. Either way, they can make a ten.

Can you regroup the ones? Yes, 1 ten 4 ones.

Continue to ask similar questions as you work through Step 2.

TRY THESE Exercises 1–4 model the type of exercises students will find in the **Practice on Your Own** page.

- **Exercises 1 and 3** Group ones from the top or bottom to make a 10.
- **Exercises 2 and 4** Grouping ones from the top to make a double to make a 10.

PRACTICE ON YOUR OWN Review the example at the top of the page. Ask students to explain how they would group the digits to add.

CHECK Determine if the students can regroup to form tens.

Success is determined by 3 out of 4 correct responses.

Students who successfully complete the **Practice on Your Own** and **Check** are ready to move on to the next skill.

COMMON ERRORS

- Students may forget to regroup ones as tens.
- Students may make a ten but forget to add the third digit.

Students who made more than three errors in the **Practice on Your Own,** or who were not successful in the **Check** section, may benefit from the **Alternative Teaching Strategy** on the next page.

CALIFORNIA STANDARDS G3 NS 2.1 Find the sum or difference of two whole numbers between 0 and 10,000.

Optional

Alternative Teaching Strategy

15 Minutes

Model the Grouping Property of Addition

OBJECTIVE Use base-ten blocks to model adding 3 or more addends using the Grouping Property of Addition

MATERIALS base-ten blocks

You may wish to have the students work in pairs. One student models the addition with the base-ten blocks while the other student records each step.

Distribute the base-ten blocks.

Say: **Use your blocks to model the exercise**

13 + 27 + 13.

You may wish to write the addition on the board in a vertical format for the students to copy.

Ask: **What is the result if you group the ones digits starting at the top?** It will make a 10. **What is the result if you group from the bottom?** It will also make a 10.

Have students model the grouping and then combine all the ones.

Say: **Exchange 10 ones for 1 ten, to regroup.**

Then have the students place the regrouped ten with the other tens. Have them record the sum of the ones digits.

Then work through the grouping of the tens digits. Have a student note that the sum of the tens is 4 plus the 1 regrouped ten. Then ask for the sum of the three numbers. 53

Repeat this activity with similar examples. When the students show they understand how to group digits, remove the base-ten blocks and have them try an exercise using only paper and pencil. Ask students to explain how they group digits and how they regroup.

Name ______________________ Skill ______________________

Grade 4 **Skill 14**

Grouping Property of Addition

The Grouping Property of Addition states that you can group addends in different ways. The sum is always the same. Use the Grouping Property of Addition to find the sum of three or more addends. Find 21 + 84 + 59 = ■.

Step 1 Add the ones. Group the addends differently to make a ten.

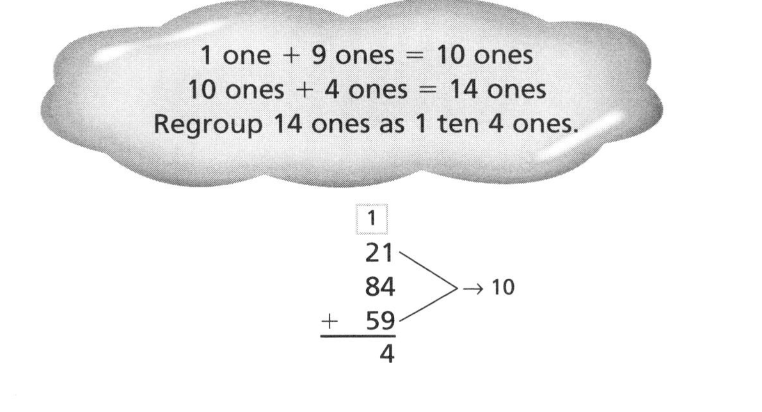

1 one + 9 ones = 10 ones
10 ones + 4 ones = 14 ones
Regroup 14 ones as 1 ten 4 ones.

$$\begin{array}{r} \boxed{1} \\ 21 \\ 84 \\ +\ 59 \\ \hline 4 \end{array} \quad 1 + 9 \rightarrow 10$$

Step 2 Add the tens. Write the sum.

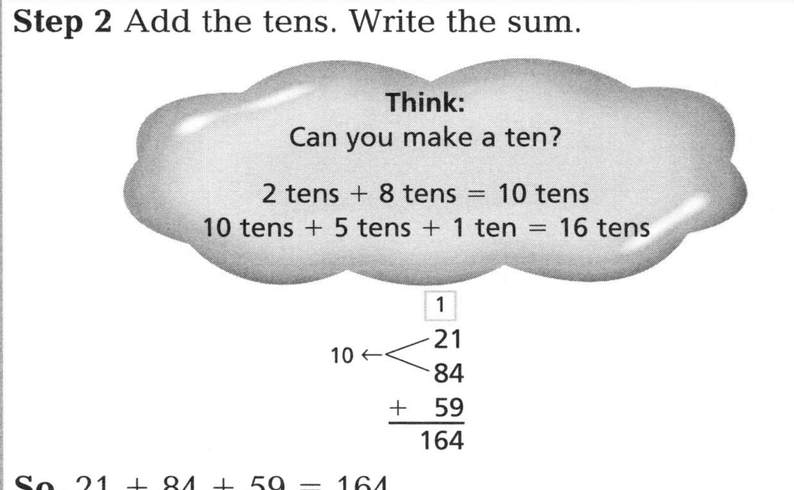

$$10 \leftarrow (2 + 8) \quad \begin{array}{r} \boxed{1} \\ 21 \\ 84 \\ +\ 59 \\ \hline 164 \end{array}$$

So, 21 + 84 + 59 = 164.

Try These

Find the sum.

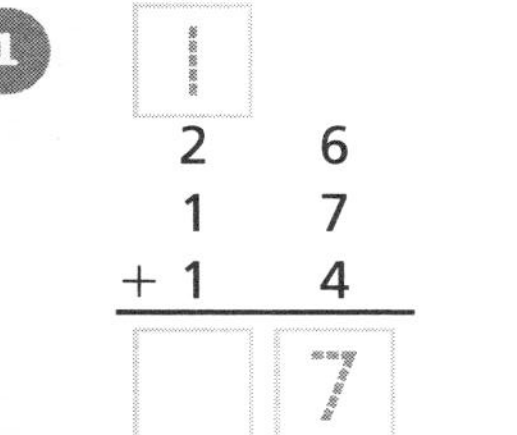

1.
$$\begin{array}{r} 26 \\ 17 \\ +\ 14 \\ \hline \end{array}$$

2.
$$\begin{array}{r} 82 \\ 53 \\ +\ 12 \\ \hline \end{array}$$

3.
$$\begin{array}{r} 62 \\ 18 \\ +\ 39 \\ \hline \end{array}$$

4.
$$\begin{array}{r} 43 \\ 15 \\ +\ 63 \\ \hline \end{array}$$

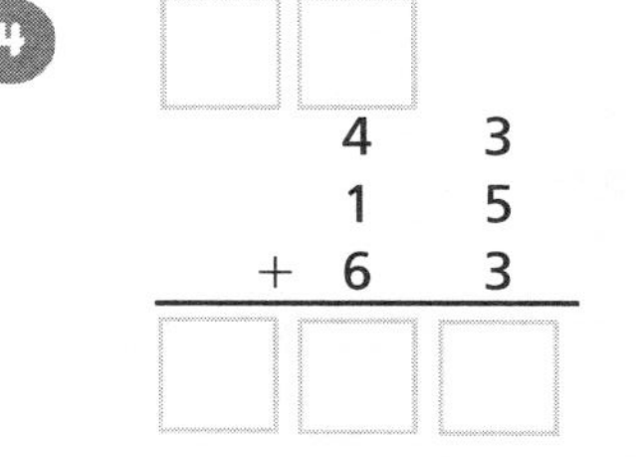

Go to the next side.

Name ______________________ Skill ______________________

Practice on Your Own

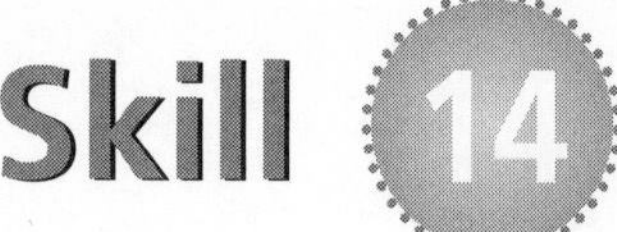

Make a ten.

6 + 4 = 10

Add doubles.

Add the tens.

```
   32          1          1
   26         32         32
   42         26         26
 + 54         42         42
            + 54       + 54
              4         154
```

Think: Can you make a ten? Can you add doubles?

10 + 4 = 14

3 + 2 = 5
1 + 4 + 5 = 10
5 + 10 = 15

Find the sum.

1. □ 32 + 25 + 35
2. □ 62 + 28 + 35
3. □ 37 + 11 + 87
4. □ 75 + 25 + 86

5. □ 11 + 28 + 10 + 49
6. □ 16 + 34 + 58 + 92
7. □ 17 + 23 + 51 + 79
8. □ 57 + 32 + 57 + 23

9. 65 + 35 + 87
10. 59 + 11 + 48
11. 16 + 83 + 53
12. 54 + 85 + 16

Check

Find the sum.

13. 41 + 16 + 29
14. 37 + 17 + 99
15. 38 + 72 + 13 + 15
16. 47 + 25 + 63 + 45

Skill 15 Grade 4 Fact Families (Addition and Subtraction) 1–10

OBJECTIVE Recall fact families for addition and subtraction sums through 10

You may wish to review that addition and subtraction are inverse operations. Have students use the cubes to represent Steps 1 and 2.

Ask students to look at the four number sentences in Steps 1 and 2.

Ask: **How are they alike?** **All four sentences use the same numbers.** **How are the number sentences different?** **Step 1 is addition. Step 2 is subtraction. The order of the numbers in Step 1 is different from the order of numbers in Step 2.** **How do fact families in Step 3 show inverse operations?** **Addition and subtraction are opposite operations. They undo each other.**

TRY THESE In Exercises 1–3, students are asked to use addition and subtraction fact families.

- **Exercise 1** Fact family for 3, 5, 8.
- **Exercise 2** Fact family for 2, 5, 7.
- **Exercise 3** Fact family for 4, 6, 10.

MATERIALS colored cubes and white cubes

PRACTICE ON YOUR OWN Review the examples at the top of the page with students. Have them identify the inverse operations. **addition and subtraction**

CHECK Determine if students can use the basic addition and subtraction facts in the fact family.

Success is indicated by 3 out of 3 correct responses.

Students who successfully complete the **Practice on Your Own** and **Check** are ready to move on to the next skill.

COMMON ERRORS

- Some students may reverse the order of numbers when writing subtraction sentences.
- Some students may add and subtract incorrectly.
- Students may write one number twice.

Students who made more than three errors in the **Practice on Your Own**, or who were not successful in the **Check** section, may benefit from the **Alternative Teaching Strategy** on the next page.

Optional

Alternative Teaching Strategy

15 Minutes

Model Fact Families

OBJECTIVE Use counters to model a fact family

MATERIALS two-sided counters and paper

Provide students with a group of two-sided counters and a piece of paper. Ask students to fold the paper to divide it into four sections.

Tell students to show 6 counters of one color and 2 counters of the other color.

Ask: **How many counters are in each group?** 6 and 2

Have students write the number of counters under each group. 6 and 2 Together generate a complete fact family. Write four number sentences using the two numbers 6 and 2.

Repeat this activity several times. Use a separate section on the paper for each new fact family.

●●●●●● ●●   6 2   6 + 2 = 8
2 + 6 = 8
8 − 6 = 2
8 − 2 = 6

Determine if students can generate a fact family without counters. Give three numbers. Have students write the number sentences. If students still are unable to generate a fact family, provide numeral cards and minus cards.

Have students arrange the cards to show one addition sentence, record it, then show the next addition sentence. Repeat for subtraction.

Name ______________________ Skill ______________________

Fact Families (Addition and Subtraction) 1–10

Grade 4 Skill 15

Write the addition and subtraction facts in the fact family 3, 4, and 7.

Step 1 Count the shaded cubes. Then count the white cubes. Write the two addition facts.

4 + 3 = 7
3 + 4 = 7

Fact families for addition and subtraction use the same numbers.

Step 2 Remove the shaded cubes from the 7 cubes. Then remove the white cubes from the 7 cubes. Write the two subtraction facts.

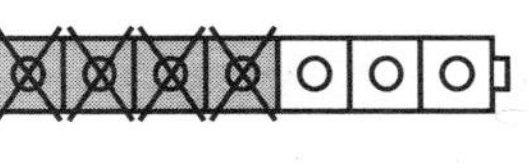

7 − 4 = 3
7 − 3 = 4

Step 3 Write the addition and subtraction facts for the fact family.

Fact families use opposite, or inverse, operations.

3 + 4 = 7
4 + 3 = 7
7 − 3 = 4
7 − 4 = 3

Try These

Write the addition and subtraction facts in the fact family.

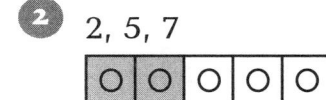

1. 3, 5, 8

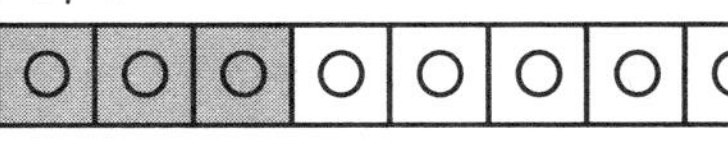

3 + [5] = 8 8 − [3] = 5

5 + [ ] = 8 8 − [ ] = 3

2. 2, 5, 7

[ ] + 5 = 7 7 − [ ] = 2

[ ] + 2 = 7 7 − [ ] = 5

3. 4, 6, 10

[ ] + [ ] = 10 10 − [ ] = [ ]

[ ] + [ ] = 10 10 − [ ] = [ ]

Go to the next side.

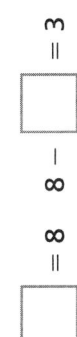

Name ______________________ Skill ______________________

Practice on Your Own

Skill 15

Write the addition and subtraction facts in the fact family 4, 5, 9.

5 + 4 = 9

4 + 5 = 9

9 − 5 = 4

9 − 4 = 5

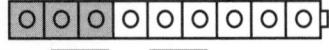

Fact families use opposite, or inverse, operations.

Write the addition and subtraction facts for each fact family.

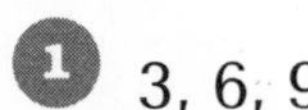

1. 3, 6, 9

☐ + ☐ = 9

☐ + ☐ = 9

9 − ☐ = ☐

9 − ☐ = ☐

2. 2, 6, 8

☐ + ☐ = 8

☐ + ☐ = 8

8 − ☐ = ☐

8 − ☐ = ☐

3. 1, 3, 4

☐ + ☐ = ☐

☐ + ☐ = ☐

☐ − ☐ = ☐

☐ − ☐ = ☐

4. 2, 4, 6

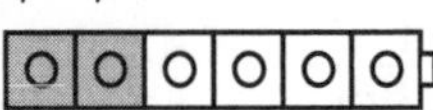

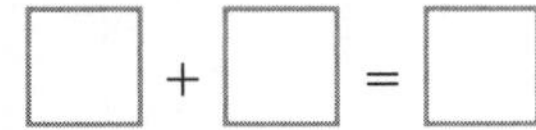

☐ + ☐ = ☐

☐ + ☐ = ☐

☐ − ☐ = ☐

☐ − ☐ = ☐

5. 2, 3, 5

____

____

____

____

6. 1, 8, 9

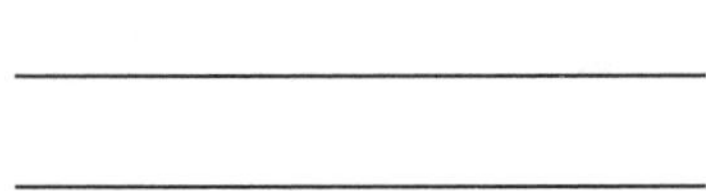

____

____

____

____

Check

Write the addition and subtraction facts for each fact family.

7. 1, 7, 8

____

____

____

____

8. 5, 5, 10

____

____

____

____

9. 2, 7, 9

____

____

____

____

Skill 16 Grade 4

Fact Families (Addition and Subtraction) 11–20

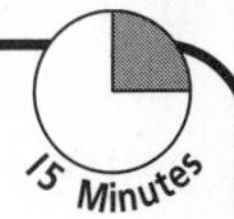

OBJECTIVE Recall fact families for addition and subtraction sums 11–20

Begin by having students describe what a fact family is. Have them recall that fact families have two addition facts and two subtraction facts. Tell students that there are three numbers in a fact family, for example: 5, 8, and 13.

Draw students' attention to Step 1 of Skill 16.

Say: **The cube trains represent two addition facts. What two addition facts do the trains show?** 5 + 8 = 13 and 8 + 5 = 13 **What 3 numbers are in the addition facts?** 5, 8, 13

Call attention to Step 2.

Ask: **What two subtraction facts do the cubes show?** 13 − 5 = 8, 13 − 8 = 5 **What three numbers are in the subtraction facts?** 5, 8, 13

Explain that in Step 3 the addition and subtraction facts are written together to form a fact family. Point out that the same three numbers are used in each fact.

TRY THESE In Exercises 1–3, students are given some of the numbers in the fact families and are asked to complete the number sentences.

- **Exercise 1** Fact family for 3, 9, and 12.
- **Exercise 2** Fact family for 7, 8, and 15.
- **Exercise 3** Fact family for 6, 10, and 16.

PRACTICE ON YOUR OWN Review the fact family for 5, 10, and 15. Have students count the cubes. Remind students that there are two addition facts and two subtraction facts in a fact family.

CHECK Determine if students know how to write fact families using the correct numbers. Success is indicated by correctly completing 3 out of 3 fact families.

Students who successfully complete the **Practice on Your Own** and **Check** are ready to move on to the next skill.

COMMON ERRORS

- Students may not reverse the digits to write the second addition or subtraction fact.
- Students may not know their addition and subtraction facts.
- Students may not begin the subtraction sentences with the greatest number.

Students who made more than three errors in the **Practice on Your Own,** or who were not successful in the **Check** section, may benefit from the **Alternative Teaching Strategy** on the next page.

Optional

20 Minutes

Alternative Teaching Strategy
Model Fact Families

OBJECTIVE Use counters to model a fact family

MATERIALS 2-color counters, pencil, paper

Distribute 11 counters to students and have students count them. Turn two of the counters over so that they are the second color. Have students count the light counters, the dark counters, and the total. Tell students that these are the three numbers they will use to write a fact family.

Have students count the number of light counters and the number of dark counters and add the numbers together.

Ask: **What addition sentence do the counters show?** 2 + 9 = 11

Have students write it down. Students can check the addition by counting the total number of counters. Continue with the fact: 9 + 2 = 11.

Ask: **What is the same about these addition sentences?** They use the same numbers. The total is the same.

○○●●●●●●●●● ○○○○○○○○○⊗⊗

2 + 9 = 11 11 − 2 = 9

●●●●●●●●●○○ ○○⊗⊗⊗⊗⊗⊗⊗⊗⊗

9 + 2 = 11 11 − 9 = 2

Then, count the total number of counters with the students, and move all of the light counters away from the dark counters.

Ask: **How many counters did you have at the start?** 11 **How many did you take away?** 2 **How many are left?** 9 **What subtraction sentence do the counters show?** 11 − 2 = 9

Have students write it down.

Continue with the fact: 11 − 9 = 2.

Ask: **What do these subtraction sentences have that are the same?** They are written with the same three numbers. The beginning number is the same.

Provide additional counters and have students create and write their own fact families. The goal is for the students to recognize that fact families are made up of two addition and two subtraction sentences, all using the same three numbers.

When students show understanding of the concept, have them write fact families without using the counters.

Name ____________________ Skill ____________________

Grade 4 Skill 16

Fact Families (Addition and Subtraction) 11–20

Write the addition and subtraction facts in the fact family 5, 8, and 13.

Step 1 Count the shaded cubes. Then count the white cubes. Write the two addition facts.

5 + 8 = 13
8 + 5 = 13

Step 2 Remove the shaded cubes from the 13 cubes. Then remove the white cubes from the 13 cubes. Write the two subtraction facts.

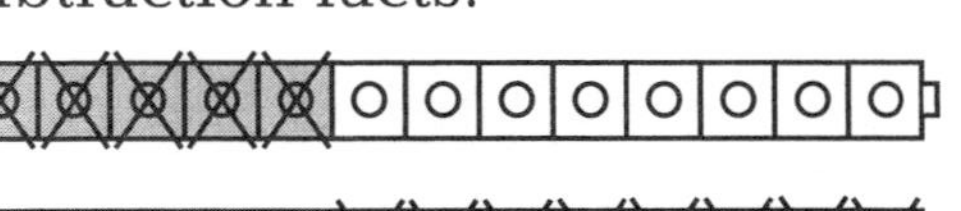

13 − 5 = 8
13 − 8 = 5

Step 3 Write the addition and subtraction facts for the fact family.

Fact families for addition and subtraction use opposite, or inverse, operations.

5 + 8 = 13
8 + 5 = 13
13 − 5 = 8
13 − 8 = 5

Try These

Write the addition and subtraction facts in the fact family.

1. 3, 9, 12

3 + ☐ = 12　　12 − ☐ = 9
9 + ☐ = 12　　12 − ☐ = 3

2. 7, 8, 15

☐ + 8 = 15　　15 − ☐ = 7
☐ + 7 = 15　　15 − ☐ = 8

3. 6, 10, 16

☐ + ☐ = 16　　16 − ☐ = ☐
☐ + ☐ = 16　　16 − ☐ = ☐

Go to the next side.

Name ______________________ Skill ______________________

Practice on Your Own

Skill 16

Write the addition and subtraction facts in the fact family 5, 10, 15.

5 + 10 = 15
10 + 5 = 15

15 − 10 = 5
15 − 5 = 10

Fact families use opposite, or inverse, operations.

Write the addition and subtraction facts for each fact family.

1 1, 10, 11

☐ + ☐ = 11
☐ + ☐ = 11
11 − ☐ = ☐
11 − ☐ = ☐

2 8, 10, 18

☐ + ☐ = 18
☐ + ☐ = 18
18 − ☐ = ☐
18 − ☐ = ☐

3 4, 7, 11

☐ + ☐ = ☐
☐ + ☐ = ☐
☐ − ☐ = ☐
☐ − ☐ = ☐

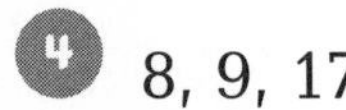

4 8, 9, 17

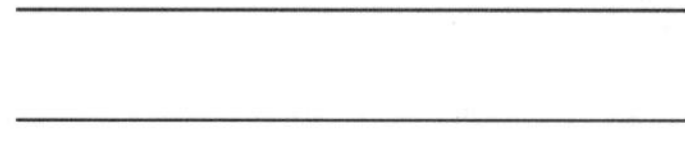

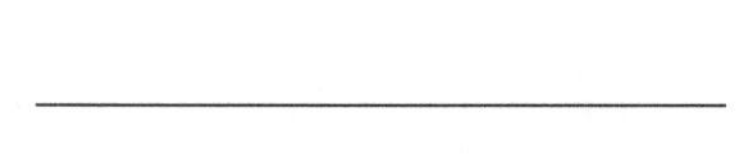

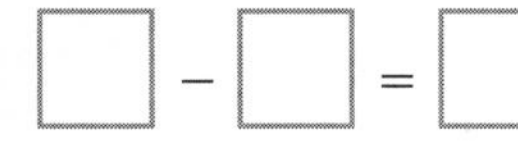

☐ + ☐ = ☐
☐ + ☐ = ☐
☐ − ☐ = ☐
☐ − ☐ = ☐

5 3, 10, 13

____
____
____
____

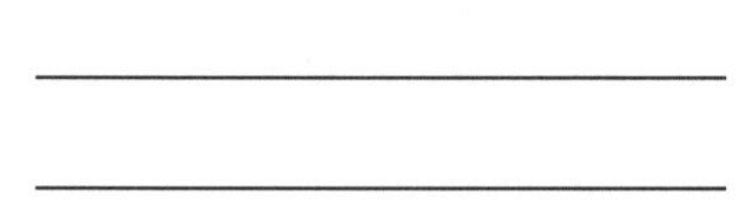

6 5, 9, 14

____
____
____
____

Check

Write the addition and subtraction facts for each fact family.

7 3, 8, 11

____
____
____
____

8 5, 7, 12

____
____
____
____

9 6, 8, 14

____
____
____
____

Answer Card
Addition
Grade 4

SKILL 8

TRY THESE

1. 1; 54
2. 1; 90
3. 1; 81
4. 1; 92

PRACTICE

1. 1; 30
2. 1; 64
3. 1; 90
4. 1; 65
5. 1; 71
6. 1; 75
7. 1; 90
8. 1; 53
9. 1; 71
10. 1; 63
11. 1; 95
12. 1; 71
13. 30
14. 94
15. 60
16. 52

CHECK

17. 50
18. 85
19. 90
20. 98

SKILL 9

TRY THESE

1. 1; 119
2. 1; 119
3. 1; 117
4. 1; 129

PRACTICE

1. 109
2. 129
3. 128
4. 129
5. 122
6. 106
7. 136
8. 107
9. 127
10. 147
11. 113
12. 178
13. 148
14. 125
15. 158
16. 182

CHECK

17. 156
18. 165
19. 179
20. 189

SKILL 10

TRY THESE

1. 1; 453
2. 1; 560
3. 1; 252
4. 1; 780

PRACTICE

1. 1; 580
2. 1; 461
3. 1; 193
4. 1; 285
5. 1; 190
6. 1; 280
7. 1; 395
8. 1; 474
9. 1; 373
10. 1; 592
11. 1; 760
12. 1; 491
13. 574
14. 571
15. 584
16. 291

CHECK

17. 792
18. 492
19. 881
20. 385

SKILL 11

TRY THESE

1. 901
2. 1; 900
3. 1; 819

PRACTICE

1. 1; 619
2. 1; 507
3. 1; 817
4. 1; 906
5. 1; 555
6. 1; 607
7. 1; 638
8. 1; 726
9. 1; 958
10. 1; 616
11. 1; 828
12. 1; 843
13. 808
14. 765
15. 967
16. 739

CHECK

17. 904
18. 978
19. 954
20. 889

Answer Card

Addition

Grade 4

SKILL 12

TRY THESE

1. 1; 1,369
2. 1; 1; 1,219
3. 1; 1; 1; 1,111

PRACTICE

1. 1; 1,398
2. 1; 1,387
3. 1; 1,097
4. 1; 1; 1,188
5. 1; 1; 1,358
6. 1; 1,098
7. 1; 1; 1,219
8. 1; 1; 1,229
9. 1; 1; 1,308

CHECK

10. 1,219
11. 1,008
12. 1,512
13. 1,101

SKILL 13

TRY THESE

1. 12; 12
2. 6; 6
3. 11; 11
4. 8; 8

PRACTICE

1. 19; 19
2. 17; 17
3. 27; 27
4. 36; 36
5. 35; 35
6. 26; 26
7. 55; 55
8. 57; 57
9. 12; 0; 12
10. 29; 14; 29
11. 22; 14; 8; 22
12. 38; 17; 21; 38

CHECK

13. 26; 19; 7; 26
14. 25; 0; 25; 25
15. 56; $\begin{array}{r} 26 \\ +\ 30 \\ \hline 56 \end{array}$
16. 73; $\begin{array}{r} 0 \\ +\ 73 \\ \hline 73 \end{array}$

SKILL 14

TRY THESE

1. 1; 57
2. 1; 147
3. 1; 1; 119
4. 1; 1; 121

PRACTICE

1. 1; 92
2. 1; 125
3. 1; 135
4. 1; 186
5. 1; 98
6. 2; 200
7. 2; 170
8. 1; 169
9. 187
10. 118
11. 152
12. 155

CHECK

13. 86
14. 153
15. 138
16. 180

Answer Card
Addition
Grade 4

SKILL 15

TRY THESE

1. 5, 3, 3, 5
2. 2, 5, 5, 2
3. 4, 6; 6, 4; 6, 4; 4, 6 or 6, 4; 4, 6; 4, 6; 6, 4

PRACTICE

In Exercises 1-9, the order of the answers may vary.

1. 3 + 6 = 9, 6 + 3 = 9, 9 − 3 = 6, 9 − 6 = 3
2. 2 + 6 = 8, 6 + 2 = 8, 8 − 6 = 2, 8 − 2 = 6
3. 3 + 1 = 4, 1 + 3 = 4, 4 − 1 = 3, 4 − 3 = 1
4. 4 + 2 = 6, 2 + 4 = 6, 6 − 4 = 2, 6 − 2 = 4

SKILL 15

(continued)

5. 2 + 3 = 5, 3 + 2 = 5, 5 − 2 = 3, 5 − 3 = 2
6. 8 + 1 = 9, 1 + 8 = 9, 9 − 1 = 8, 9 − 8 = 1

CHECK

7. 1 + 7 = 8, 7 + 1 = 8, 8 − 1 = 7, 8 − 7 = 1
8. 5 + 5 = 10, 10 − 5 = 5
9. 2 + 7 = 9, 7 + 2 = 9, 9 − 2 = 7, 9 − 7 = 2

SKILL 16

TRY THESE

1. 9, 3, 3, 9
2. 7, 8, 8, 7
3. 6 + 10 = 16, 10 + 6 = 16, 16 − 10 = 6, 16 − 6 = 10

PRACTICE

In Exercises 1-9, the order of the answers may vary.

1. 1 + 10 = 11, 10 + 1 = 11, 11 − 10 = 1, 11 − 1 = 10
2. 8 + 10 = 18, 10 + 8 = 18, 18 − 10 = 8, 18 − 8 = 10
3. 7 + 4 = 11, 4 + 7 = 11, 11 − 7 = 4, 11 − 4 = 7

SKILL 16

(continued)

4. 9 + 8 = 17, 8 + 9 = 17, 17 − 8 = 9, 17 − 9 = 8
5. 10 + 3 = 13, 3 + 10 = 13, 13 − 10 = 3, 13 − 3 = 10
6. 5 + 9 = 14, 9 + 5 = 14, 14 − 5 = 9, 14 − 9 = 5

CHECK

7. 3 + 8 = 11, 8 + 3 = 11, 11 − 3 = 8, 11 − 8 = 3
8. 7 + 5 = 12, 5 + 7 = 12, 12 − 7 = 5, 12 − 5 = 7
9. 8 + 6 = 14, 6 + 8 = 14, 14 − 8 = 6, 14 − 6 = 8

Number Sense

Whole Number Subtraction

2-Digit Subtraction: Regrouping Tens as Ones

OBJECTIVE Subtract 2-digit numbers, regrouping tens as ones

Begin by pointing to the place value labels on the subtraction exercises. Suggest that students use the labels to help them remember the value of the digits when they regroup. Explain that in this activity students are asked to subtract two 2-digit numbers.

Direct students' attention to Step 1.

Ask: **Do you have enough ones to subtract?** No, 2 ones are less than 9 ones. **What can you do?** Regroup 4 tens as 3 tens 10 ones.

Continue with Step 2.

Ask: **How many ones do you have after you regroup?** 12 ones **How many tens do you have after you regroup?** 3 tens

Look at Step 3.

Ask: **How many tens are left after you subtract?** 2

TRY THESE In Exercises 1–4, students regroup 1 ten as 10 ones.

- **Exercise 1** Regroup 2 tens as 1 ten 10 ones.
- **Exercise 2** Regroup 3 tens as 2 tens 10 ones.
- **Exercise 3** Regroup 2 tens as 1 ten 10 ones.
- **Exercise 4** Regroup 3 tens as 2 tens 10 ones.

PRACTICE ON YOUR OWN Review the example at the top of the page. Ask students to explain the subtraction and regrouping process, using the terms *ones* and *tens*.

CHECK Determine if students know when to regroup tens as ones before they subtract the ones.

Success is indicated by 3 out of 4 correct responses.

Students who successfully complete the **Practice on Your Own** and **Check** are ready to move on to the next skill.

COMMON ERRORS

- Students may subtract the top digit from the bottom digit instead of regrouping.
- Students may forget that they have regrouped a ten when they subtract the tens, and subtract from the original number of tens.

Students who made more than four errors in the **Practice on Your Own**, or who were not successful in the **Check** section, may benefit from the **Alternative Teaching Strategy** on the next page.

Optional

Alternative Teaching Strategy

15 Minutes

Model 3-Digit Subtraction: Regrouping Tens as Ones

OBJECTIVE Use base-ten blocks to model subtracting 2-digit numbers, regrouping tens as ones

MATERIALS base-ten blocks, place-value grid

Distribute base-ten blocks. Provide an example for students to complete using only the blocks. For example, have students find 53 − 26. Display the exercise, then ask students to model 53 with the blocks.

Ask: **What do you subtract first?** The ones, or 6 ones **Can you remove 6 ones from 3 ones?** No **What can you do?** Regroup one of the tens rods as 10 ones **What do you have now?** 4 tens 13 ones

Have students remove the ones, then remove 2 tens to complete the subtraction.

Continue with another example having students verbalize the steps in the regrouping and subtraction process.

Next, have students record 42 − 19 in a place-value grid. As they regroup and subtract with the models, have students record the result on the place-value grid.

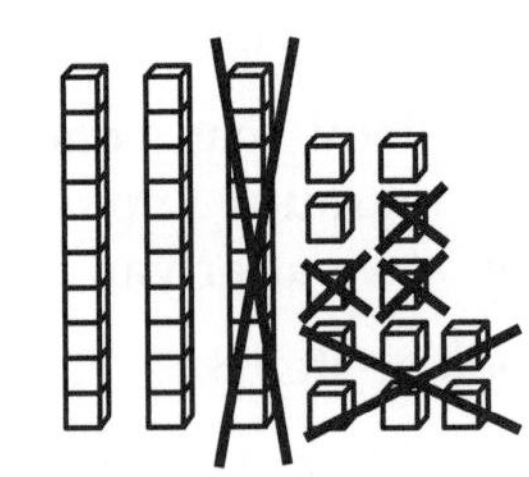

Name ______________________ Skill ______________________

Grade 4
Skill
17

2-Digit Subtraction: Regrouping Tens as Ones

Find 42 − 19 = ■.

Step 1 Show 42 as 4 tens and 2 ones. Since 9 > 2, regroup 42 as 3 tens and 12 ones.

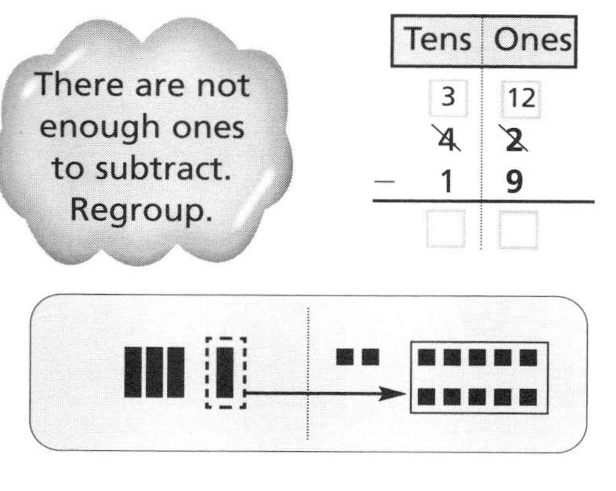
There are not enough ones to subtract. Regroup.

	Tens	Ones
	3	12
	~~4~~	~~2~~
−	1	9

Step 2 Subtract the ones.

12 ones − 9 ones = 3 ones

	Tens	Ones
	3	12
	~~4~~	~~2~~
−	1	9
		3

Step 3 Subtract the tens.

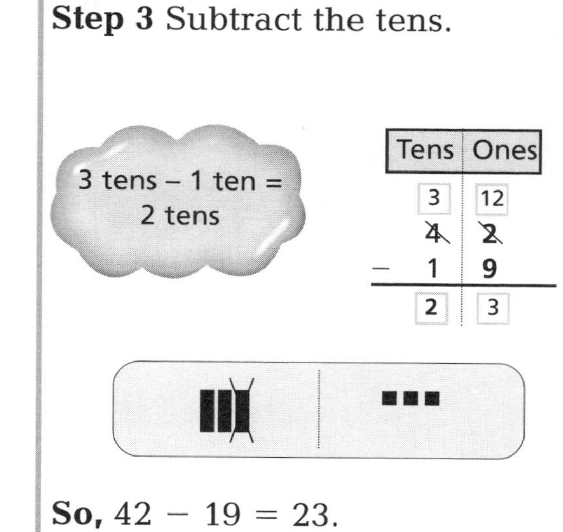
3 tens − 1 ten = 2 tens

	Tens	Ones
	3	12
	~~4~~	~~2~~
−	1	9
	2	3

So, 42 − 19 = 23.

Try These

Find the difference.

1.

	Tens	Ones
	1	12
	~~2~~	2
−	1	4

2.

	Tens	Ones
	3	1
−	1	6

3.

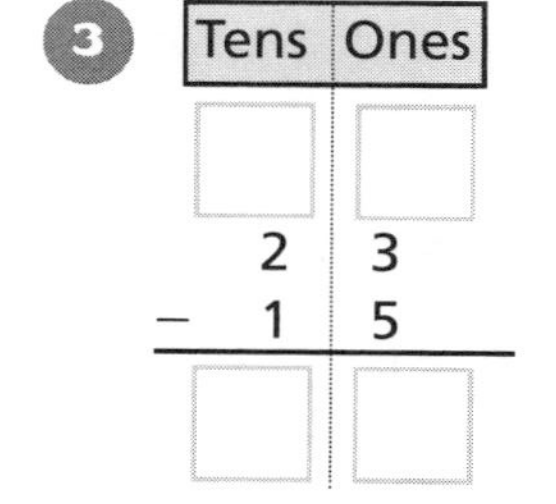

	Tens	Ones
	2	3
−	1	5

4.

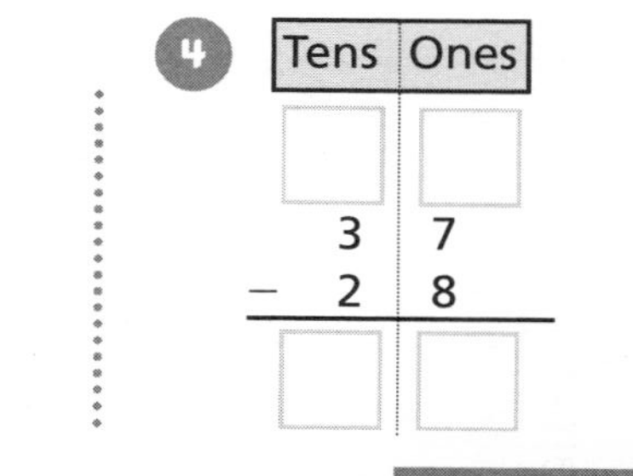

	Tens	Ones
	3	7
−	2	8

Go to the next side.

Name ____________________ Skill ____________________

Practice on Your Own

Skill 17

Think: Do you need to regroup?

Tens	Ones
6	14
~~7~~	~~4~~
− 4	5
☐	9

Regroup 1 ten as 10 ones.
Subtract the ones.

Tens	Ones
6	14
~~7~~	~~4~~
− 4	5
2	9

Subtract the tens.

Find the difference.

1. 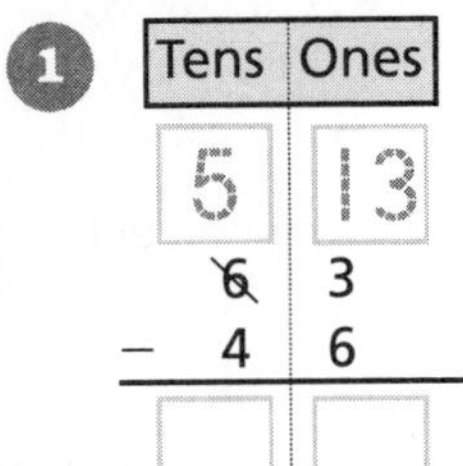

Tens	Ones
5	13
~~6~~	3
− 4	6
☐	☐

2. 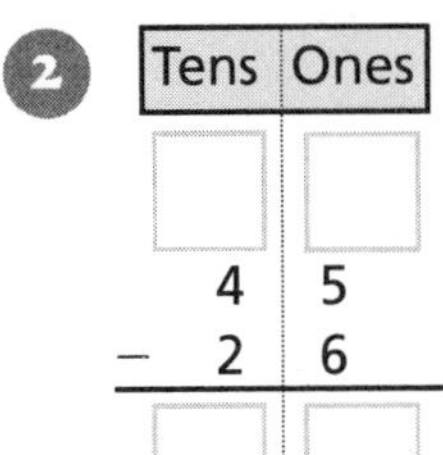

Tens	Ones
☐	☐
4	5
− 2	6
☐	☐

3. 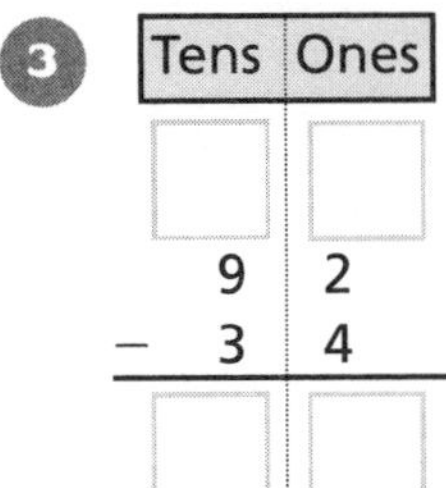

Tens	Ones
☐	☐
9	2
− 3	4
☐	☐

4. 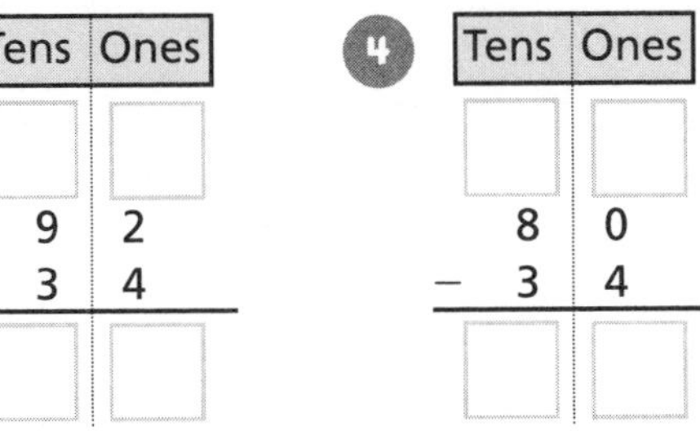

Tens	Ones
☐	☐
8	0
− 3	4
☐	☐

5. 54 − 8 = ☐☐

6. 63 − 27 = ☐☐

7. 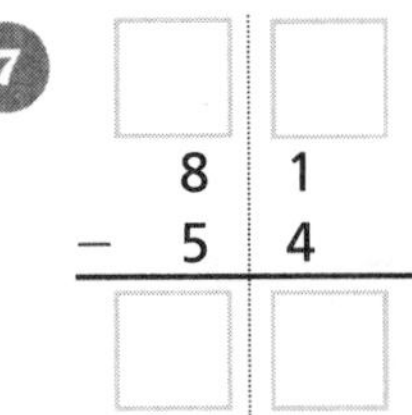
81 − 54 = ☐☐

8. 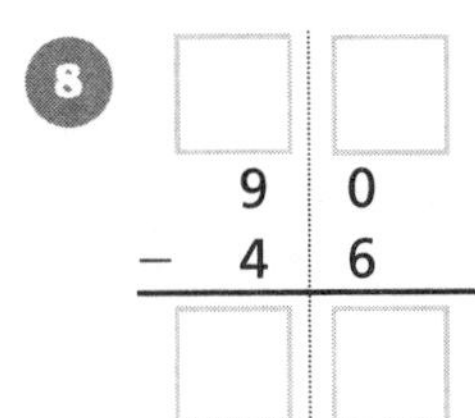
90 − 46 = ☐☐

9. 31 − 9

10. 57 − 48

11. 50 − 29

12. 71 − 34

13. 81 − 5

14. 47 − 18

15. 64 − 35

16. 90 − 35

Check

Find the difference.

17. 25 − 9

18. 43 − 16

19. 72 − 38

20. 70 − 59

3-Digit Subtraction: Regrouping Tens as Ones

OBJECTIVE Subtract two 3-digit numbers, regrouping tens as ones

Begin by pointing to the place-value labels on the addition tables. Suggest that the students use the labels to help them remember the value of the digits when they regroup. Explain that they are asked to subtract two 3-digit numbers.

Direct students' attention to Step 1.

Ask: **Do you have enough ones to subtract?** No. 4 ones is less than 5 ones. **What should you do before subtracting?** Regroup 1 ten as 10 ones. **How many ones do you have after you regroup?** 14 **How many tens?** 3 **Do you need to regroup the tens?** No, because 3 tens are greater than 2 tens.

Continue to ask similar questions as you work through Step 2 and Step 3. Refer to the models of base-ten blocks to illustrate each step.

TRY THESE Exercises 1–4 model the type of exercises students will find on the **Practice on Your Own** page.

- **Exercise 1** 2-digit difference.
- **Exercises 2–4** 3-digit difference.

PRACTICE ON YOUR OWN Review the example at the top of the page. Ask students to explain the subtraction and regrouping process, using the terms *ones, tens,* and *hundreds.*

CHECK Determine if students know when to regroup in the tens place before they subtract the ones.

Success is indicated by 4 out of 4 correct responses.

Students who successfully complete the **Practice on Your Own** and **Check** are ready to move on to the next skill.

COMMON ERRORS

- Students may forget to cross out the digit in the tens column when regrouping and then subtract from the original number of tens.
- Students may regroup in the tens or ones column when it is not necessary.

Students who made more than three errors in the **Practice on Your Own**, or who were not successful in the **Check** section, may benefit from the **Alternative Teaching Strategy** on the next page.

Optional

Alternative Teaching Strategy

15 Minutes

Model 3-Digit Subtraction: Regrouping Tens as Ones

OBJECTIVE Use base-ten blocks to model two 3-digit numbers, regrouping tens as ones

MATERIALS base-ten blocks

You may wish to have students work in pairs. One student models the subtraction with the base-ten blocks while the other student records each step with paper and pencil.

Distribute the base-ten blocks. Remind students to write the subtraction in a place-value grid as shown.

Hundreds	Tens	Ones
2	8	1
− 1	2	5

Have the partner use the base-ten blocks to model the number 281.

Say: **Begin by subtracting the ones.**

Ask: **Do you have enough ones to subtract?** No, 5 ones are greater than 1 one. **What should you do first?** Regroup 1 ten as 10 ones.

As the partner shows the regrouping with the blocks, the other student records the regrouping on paper.

Ask: **How many ones do you have?** 11 **How many tens?** 7 **Do you have enough ones to subtract now?** Yes

Have students subtract the tens and hundreds.

Repeat the activity with similar examples. When the students show understanding of the regrouping process, remove the base-ten blocks and have them try an exercise using only paper and pencil. Ask students to tell you each step as they complete the subtraction.

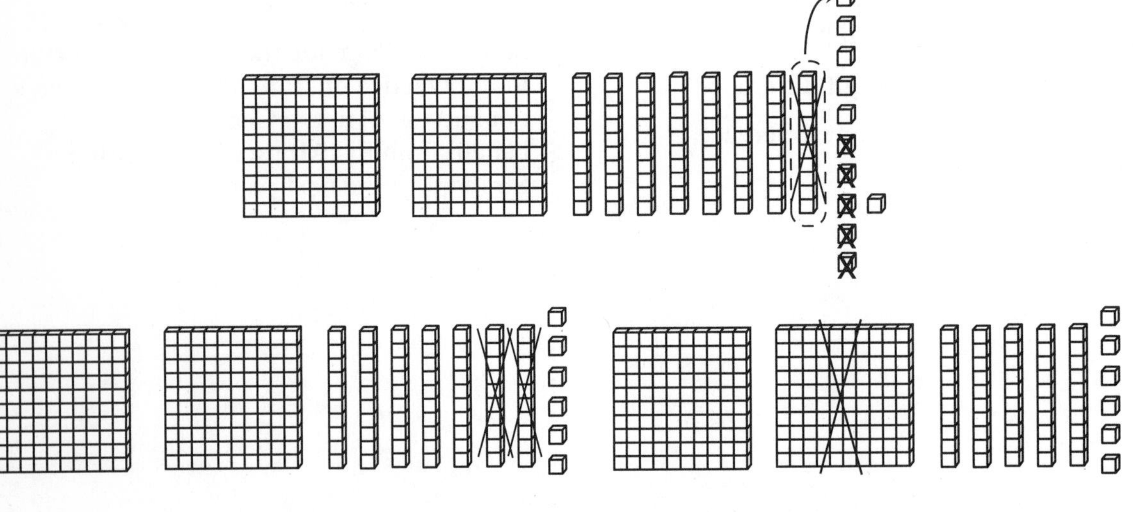

Name ____________________ Skill ____________________

3-Digit Subtraction: Regrouping Tens as Ones

Grade 4 **Skill 18**

Find 244 − 125 = ■.

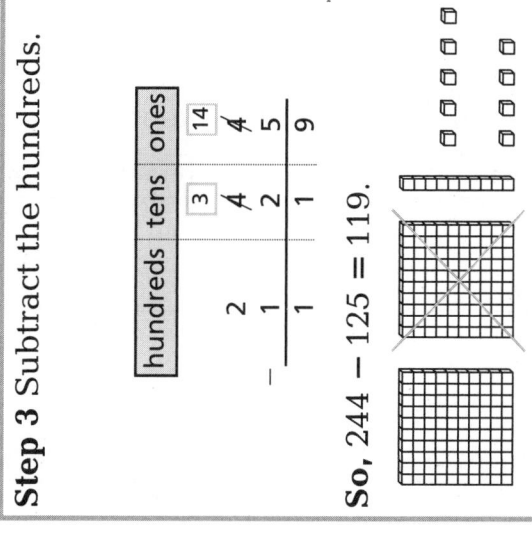

Step 1 There are not enough ones to subtract. Regroup 1 ten as 10 ones. Subtract the ones.

	hundreds	tens	ones
		3	14
	2	~~4~~	~~4~~
−	1	2	5
			9

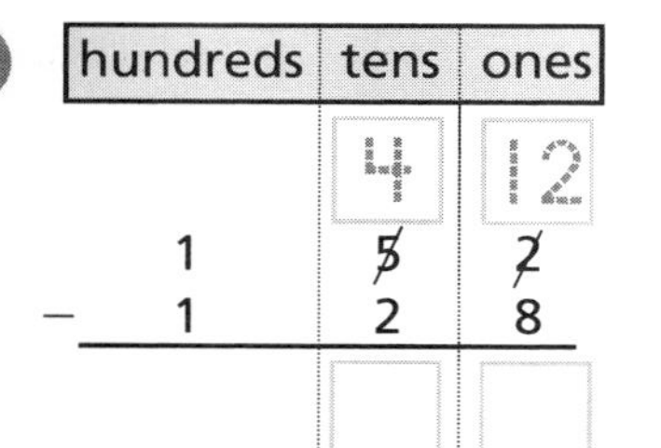

Step 2 Subtract the tens.

	hundreds	tens	ones
		3	14
	2	~~4~~	~~4~~
−	1	2	5
		1	9

Step 3 Subtract the hundreds.

	hundreds	tens	ones
		3	14
	2	~~4~~	~~4~~
−	1	2	5
	1	1	9

So, 244 − 125 = 119.

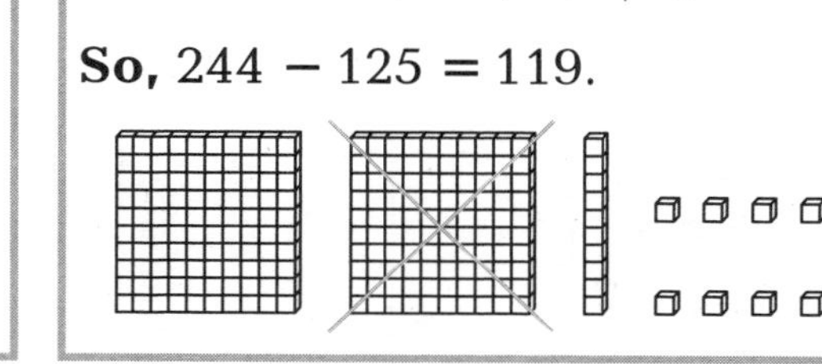

Try These

Find the difference.

1.

	hundreds	tens	ones
		4	12
	1	~~5~~	~~2~~
−	1	2	8
		☐	☐

2.

	hundreds	tens	ones
		☐	☐
	2	7	6
−	1	4	9
	☐	☐	☐

3.

	hundreds	tens	ones
		☐	☐
	3	2	4
−	2	1	6
	☐	☐	☐

4.

	hundreds	tens	ones
		☐	☐
	5	4	8
−	3	2	9
	☐	☐	☐

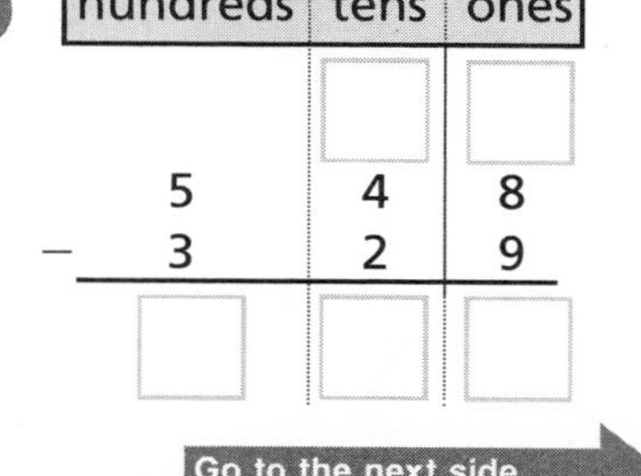

Go to the next side.

Name ______________________ Skill ______________________

Practice on Your Own

Skill 18

Think: When there are not enough ones, regroup 1 ten as 10 ones.

hundreds	tens	ones
	5	13
4	~~6~~	~~3~~
− 2	1	8
		5

hundreds	tens	ones
	5	13
4	~~6~~	~~3~~
− 2	1	8
	4	5

hundreds	tens	ones
	5	13
4	~~6~~	~~3~~
− 2	1	8
2	4	5

Find the difference. Regroup if you need to.

1.

hundreds	tens	ones
	☐	☐
3	2	2
− 1	0	7
☐	☐	☐

2.

hundreds	tens	ones
	☐	☐
2	8	4
− 1	2	6
☐	☐	☐

3.

hundreds	tens	ones
	☐	☐
4	5	7
− 2	2	9
☐	☐	☐

4. 425 − 216 = ☐☐☐

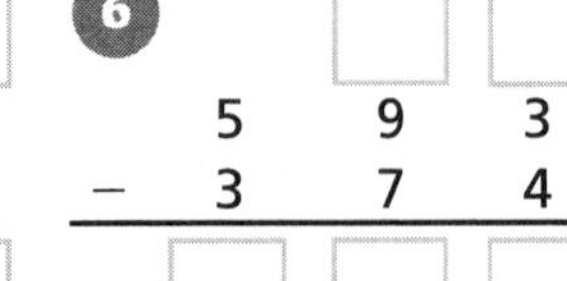

5. 587 − 368 = ☐☐☐

6. 593 − 374 = ☐☐☐

7. 692 − 163 = ☐☐☐

8. 148 − 29 = ☐☐☐

9. 252 − 136 = ☐☐☐

10. 487 − 268 = ☐☐☐

11. 854 − 547 = ☐☐☐

12. 275 − 157

13. 391 − 85

14. 560 − 442

15. 981 − 739

Check

Find the difference.

16. 380 − 271

17. 438 − 209

18. 595 − 68

19. 742 − 423

3-Digit Subtraction: Regrouping in More than One Place

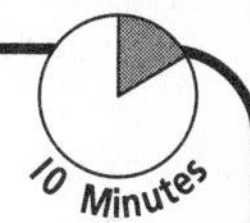

OBJECTIVE Subtract two 3-digit numbers, regrouping in more than one place

You may wish to have students state what the place-value name is for each abbreviation in the place value table.

Direct students' attention to Step 1. Have them look at the two digits in the ones place.

Ask: **Why is the 5 crossed out and 15 written above it?** There were not enough ones, so 1 ten was regrouped as 10 ones. **How many tens are there now?** 2

Work through the subtraction of the ones, then have students look at Step 2.

Ask: **Do you have enough tens to subtract?** No, 2 tens are less than 8 tens. **What can you do to have enough tens to subtract?** Regroup 1 hundred as 10 tens **How many tens do you have after you regroup?** 12 tens **How many hundreds?** 3

Complete the subtraction in Step 3. Ask the students to summarize the steps they can use to subtract with regrouping.

TRY THESE Exercises 1-4 model the type of exercises students will find on the **Practice on Your Own** page.

- **Exercise 1** Regroup tens as ones.
- **Exercise 2** Regroup hundred as tens.
- **Exercises 3–4** Regroup both tens and hundreds.

PRACTICE ON YOUR OWN Review the example at the top of the page. Ask the students to study the two numbers in the example, and without subtracting, tell you how many times they will need to regroup. twice: tens as ones, then hundreds as tens

CHECK Determine if students know when to regroup the tens and hundreds. Success is determine by 3 out of 4 correct responses.

Students who successfully complete the **Practice on Your Own** and **Check** are ready to move on to the next skill.

COMMON ERRORS

- Student may subtract the lesser digit from the greater digit when they should regroup.
- Students may regroup by adding the ten or the hundred as a one.

Students who made more than two errors in the **Practice on Your Own**, or who were not successful in the **Check** section, may benefit from the **Alternative Teaching Strategy** on the next page.

Optional

20 Minutes

Alternative Teaching Strategy
Model 3-Digit Subtraction, Regrouping in More Than One Place

OBJECTIVE Use base-ten blocks to model 3-digit subtraction with regrouping in more than one place

MATERIALS base-ten blocks

You may wish to form groups of three. One student can be in charge of exchanging the blocks when regrouping is needed. Another student can model the subtraction with the base-ten blocks. The third student can record each step on paper.

Distribute base-ten blocks. Have a student write the problem as shown.

Hundreds	Tens	Ones
2	2	5
− 1	3	7

Say: **Start by using the base-ten blocks to model the number 225.**

Ask: **What is the first step to begin subtracting?** Start by subtracting the ones. **Are there enough ones to subtract?** No, 5 ones are less than 7 ones **How do you regroup so that you can subtract the ones?** Regroup 1 ten as 10 ones.

Continue: **Exchange 1 ten for 10 ones and subtract the ones.**

Observe as students remove the ones blocks and record the step in the place-value table.

Ask: **How did you show the regrouping with paper and pencil?** Crossed off the 2 tens, wrote 1 ten above it; crossed off 5 ones, wrote 15 ones above it.

Have students continue modeling the subtraction.

Ask: **How do you regroup so that you can subtract the tens?** Regroup 1 hundred as 10 tens.

Say: **Exchange 1 hundred for 10 tens and subtract the tens.**

Have students subtract the hundreds.

Repeat this activity with similar examples. When the students show understanding of regrouping to subtract, remove the base-ten blocks and have them try an exercise using only paper and pencil. Ask students to explain each step in the subtraction process.

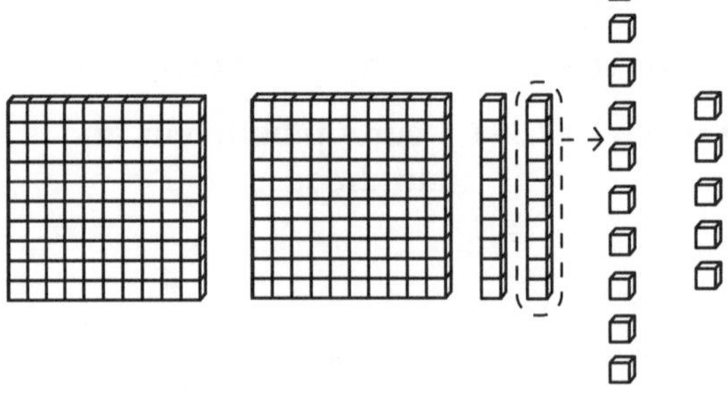

Name ______________________ Skill ______________________

© Harcourt

3-Digit Subtraction: Regrouping in More than One Place

Grade 4 Skill 19

Find 435 − 186 = ■.

Step 1 There are not enough ones to subtract. Regroup 3 tens as 2 tens 10 ones. Subtract the ones.

H	T	O
	2	15
4	~~3~~	~~5~~
− 1	8	6
		9

10 ones + 5 ones = 15 ones

15 ones − 6 ones = 9 ones

Step 2 There are not enough tens to subtract. Regroup 4 hundreds as 3 hundreds 10 tens. Subtract tens.

H	T	O
	12	
3	~~2~~	15
~~4~~	~~3~~	~~5~~
− 1	8	6
	4	9

10 tens + 2 tens = 12 tens

12 tens − 8 tens = 4 tens

Step 3 Subtract the hundreds.

H	T	O
	12	
3	~~2~~	15
~~4~~	~~3~~	~~5~~
− 1	8	6
2	4	9

3 hundreds − 1 hundred = 2 hundreds

So, 435 − 186 = 249.

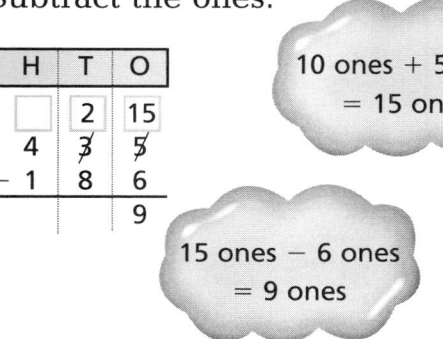

Try These

Find the difference. Regroup as needed.

1.

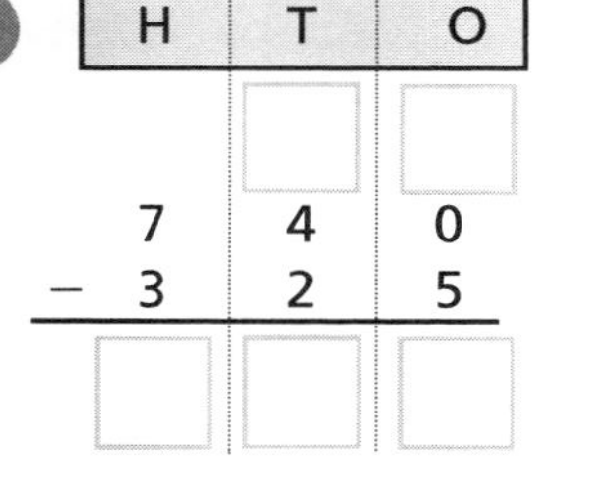

H	T	O
7	4	0
− 3	2	5

2.

H	T	O
6	5	7
− 2	8	7

3.

H	T	O
5	4	6
− 2	8	9

4.

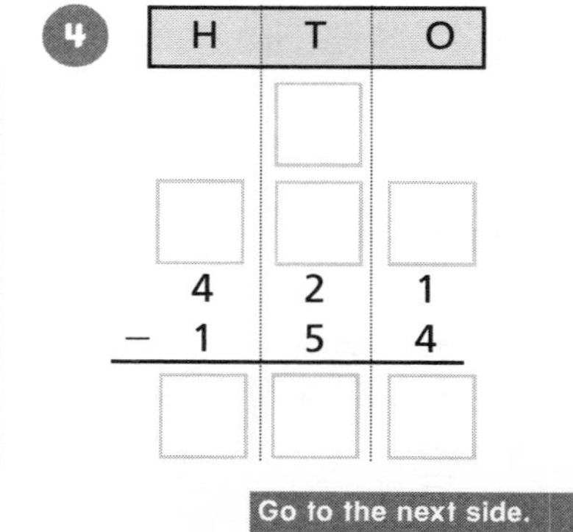

H	T	O
4	2	1
− 1	5	4

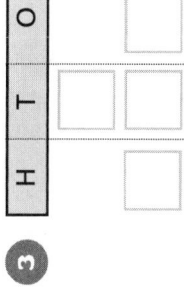

Go to the next side.

Name ____________________ Skill ____________________

Practice on Your Own

Find 310 − 126 = ■

Think: Are there enough ones to subtract? Are there enough tens?

Regroup 1 ten as 10 ones. Subtract the ones.

H	T	O
	0	10
3	~~1~~	~~0~~
− 1	2	6
		4

Regroup 3 hundreds as 2 hundreds 10 tens. Subtract the tens, then the hundreds.

H	T	O
	10	
2	~~0~~	10
~~3~~	~~1~~	~~0~~
− 1	2	6
1	8	4

Find the difference.

1.

H	T	O
2	7	3
− 1	5	6

2.

H	T	O
5	3	5
− 2	8	7

3.

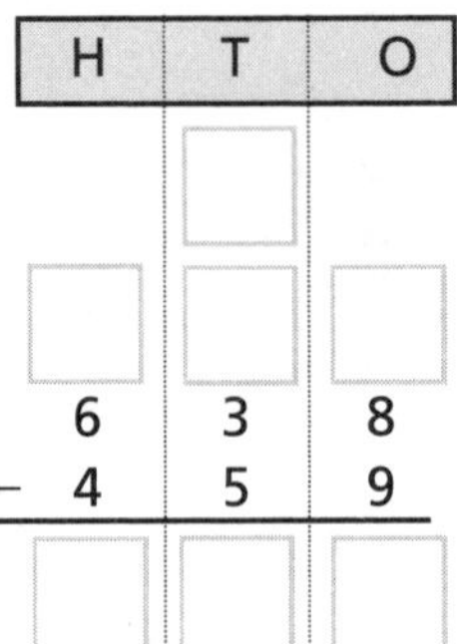

H	T	O
6	3	8
− 4	5	9

4.

H	T	O
8	2	0
− 3	4	2

5.

3	9	5
− 1	8	7

6.

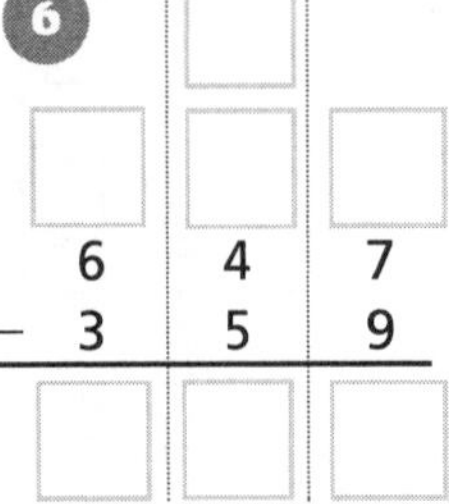

6	4	7
− 3	5	9

7.

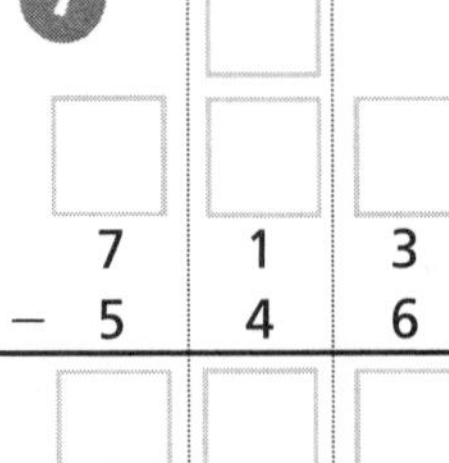

7	1	3
− 5	4	6

8.

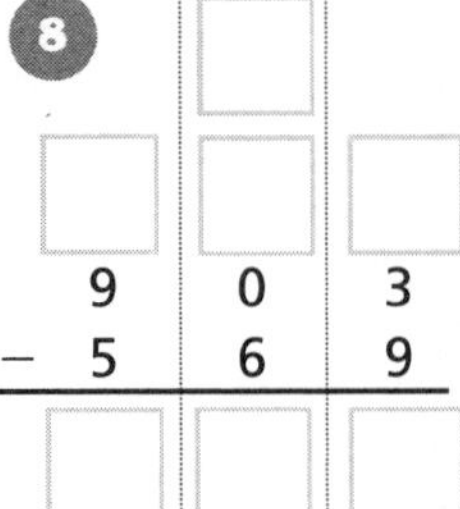

9	0	3
− 5	6	9

9. 530 − 275

10. 613 − 346

11. 325 − 187

12. 763 − 494

Check

Find the difference.

13. 807 − 629

14. 374 − 157

15. 624 − 338

16. 970 − 682

Skill 20 Grade 4

3-Digit Subtraction: Regrouping Across Zeros

15 Minutes

OBJECTIVE Subtract 3-digit numbers, regrouping across zeros

Focus students' attention on Step 1 of the example at the top of the page.

Ask: **Are there enough ones to subtract?** No **What should you do?** Go to tens to regroup tens as ones. **Are there any tens?** No

Lead students to see that they must go to hundreds to regroup 1 hundred as 10 tens.

Ask: **How many hundreds are there after you regroup?** 3

Focus on the regrouping in Step 2.

Ask: **Now are there enough ones to subtract?** no **What should you do?** Go to tens. **Now are there enough tens to regroup?** yes

Lead students to see that now they can regroup 1 ten as 10 ones.

Ask: **How many tens are there after you regroup?** 9 **How many ones?** 10

Subtract the ones.

In Step 3, subtract the tens and hundreds.

Lead students to realize that, in this example, you have to go all the way to the hundreds and regroup those first as tens before you can regroup any tens as ones.

TRY THESE Exercises 1–4 model regrouping across zeros.

- **Exercise 1** Regroup tens as ones.
- **Exercise 2** Regroup hundreds as tens.
- **Exercises 3–4** Regroup hundreds as tens; then regroup tens as ones.

PRACTICE ON YOUR OWN Review the example at the top of the page. Have students explain the regrouping using place-value names.

CHECK When there are no ones or tens, make sure students first go to hundreds to begin regrouping.

Success is indicated by 4 out of 4 correct responses.

Students who successfully complete the **Practice On Your Own** and **Check** are ready to move on to the next skill.

COMMON ERRORS

- Students may forget to mark their regrouping in the exercises, causing them to subtract from the original numbers.
- Students may regroup when it is not necessary.

Students who made more than three errors in **Practice on Your Own**, or who were not successful in **Check**, may benefit from the **Alternative Teaching Strategy** on the next page.

Optional

20 Minutes

Alternative Teaching Strategy
Model 3-Digit Subtraction, Regrouping Across Zeros

OBJECTIVE Model 3-digit subtraction, regrouping across zeros

MATERIALS base-ten blocks

Present students with the following 3-digit subtraction.

$$\begin{array}{r} 300 \\ -\ 152 \\ \hline \end{array}$$

Ask: **Are there enough ones to subtract?** No **Are there enough tens to regroup as ones?** No

Distribute base-ten blocks. Help students model the regrouping needed to subtract.

Ask students to explain the regrouping, using place-value names, and to complete the subtraction.

Repeat the activity several times with other examples. When students show understanding of the regrouping process, remove the base-ten blocks and have them try similar examples using only paper and pencil. Ask students to explain each step as they regroup and subtract.

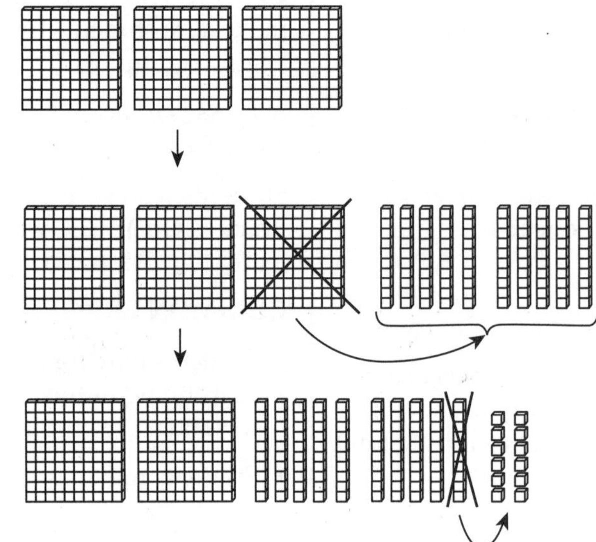

Name ____________________ Skill ____________________

Grade 4 Skill 20

3-Digit Subtraction: Regrouping Across Zeros

Find 400 − 194 = ■.

Step 1 There are not enough ones to subtract. There are no tens to regroup as ones. Regroup 4 hundreds as 3 hundreds, 10 tens.

	Hundreds	Tens	Ones
	3	10	
	~~4~~	~~0~~	0
−	1	9	4

Step 2 Since there are not enough ones to subtract, regroup 10 tens as 9 tens, 10 ones. Subtract the ones.

	Hundreds	Tens	Ones
		9	
	3	~~10~~	10
	~~4~~	~~0~~	~~0~~
−	1	9	4

Think:
10 ones −
4 ones =
6 ones

Step 3 Subtract the tens. Subtract the hundreds.

	Hundreds	Tens	Ones
		9	
	3	~~10~~	10
	~~4~~	~~0~~	~~0~~
−	1	9	4
	2	0	6

Think:
9 tens − 9 tens = 0 tens
3 hundreds − 1 hundred = 2 hundreds

Try These

Find the difference.

1.

	Hundreds	Tens	Ones
		☐	☐
	4	8	0
−	3	2	6
	☐	☐	☐

2.

	Hundreds	Tens	Ones
	☐	☐	
	6	0	3
−	2	7	2
	☐	☐	☐

3.

	Hundreds	Tens	Ones
		☐	
	☐	☐	☐
	5	0	0
−	1	1	2
	☐	☐	☐

4.

	Hundreds	Tens	Ones
		☐	
	☐	☐	☐
	5	0	0
−	4	1	2
	☐	☐	☐

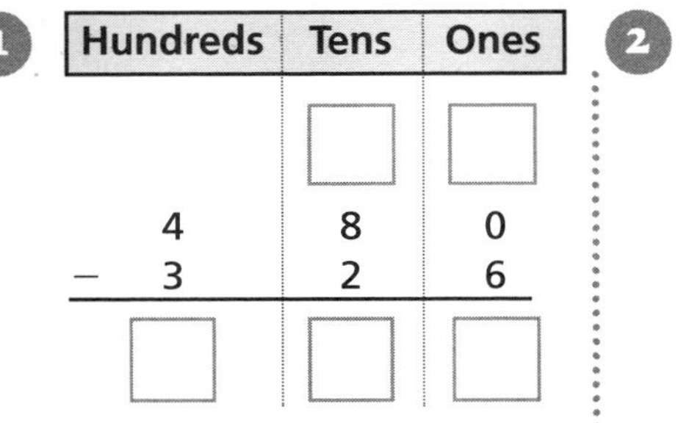

Go to the next side.

Name ______________________ Skill ______________________

Practice on Your Own

Skill 20

Find 300 − 125 = ■.

Think:
There are no ones or tens.
Regroup 3 hundreds as
2 hundreds 10 tens.
Then regroup 10 tens as
9 tens 10 ones.

Hundreds	Tens	Ones
	9	
2	~~10~~	10
~~3~~	~~0~~	~~0~~
− 1	2	5
1	7	5

So, $300 - 125 = 175$.

Find the difference.

1.

H	T	O
3	7	0
− 2	3	5

2.

H	T	O
2	0	7
− 1	4	5

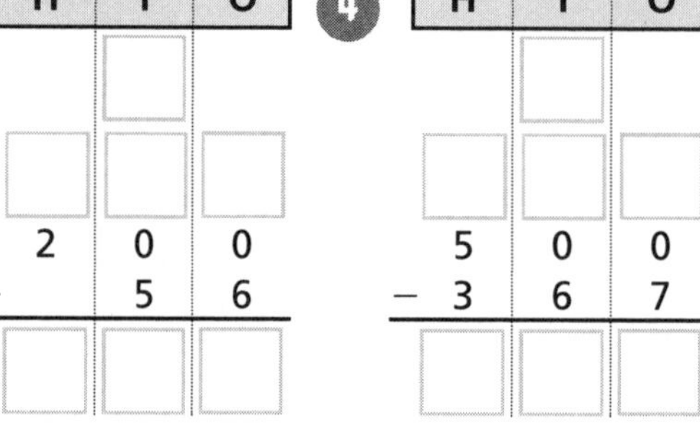

3.

H	T	O
2	0	0
−	5	6

4.

H	T	O
5	0	0
− 3	6	7

5. 450 − 122

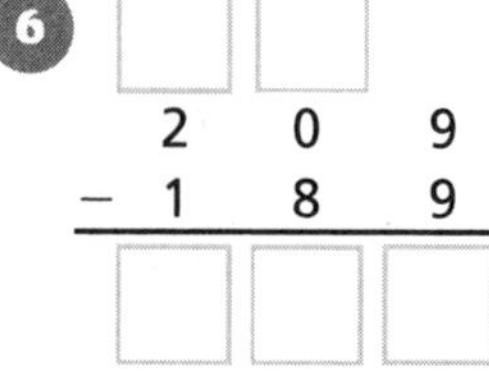

6. 209 − 189

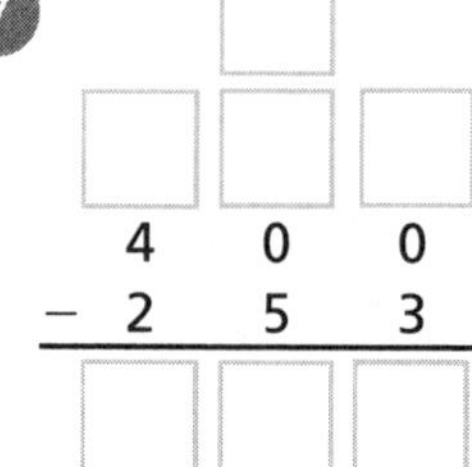

7. 400 − 253

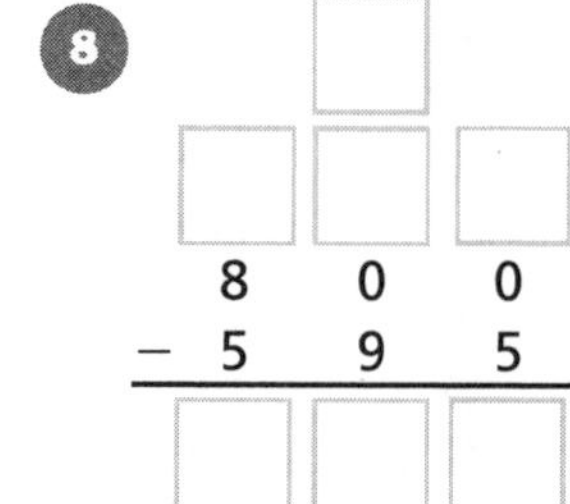

8. 800 − 595

9. 294 − 165

10. 306 − 125

11. 400 − 315

12. 500 − 338

Check

Find the difference.

13. 290 − 165

14. 509 − 367

15. 400 − 139

16. 400 − 252

Answer Card
Subtraction
Grade 4

SKILL 17

TRY THESE

1. 1, 12, 8
2. 2, 11, 15
3. 1, 13, 8
4. 2, 17, 9

PRACTICE

1. 5, 13, 17
2. 3, 15, 19
3. 8, 12, 58
4. 7, 10, 46
5. 4, 14, 46
6. 5, 13, 36
7. 7, 11, 27
8. 8, 10, 44
9. 2, 11, 22
10. 4, 17, 9
11. 4, 10, 21
12. 6, 11, 37
13. 76
14. 29
15. 29
16. 55

CHECK

17. 16
18. 27
19. 34
20. 11

SKILL 18

TRY THESE

1. 4, 12, 24
2. 6, 16, 127
3. 1, 14, 108
4. 3, 18, 219

PRACTICE

1. 1, 12, 215
2. 7, 14, 158
3. 4, 17, 228
4. 1, 15, 209
5. 7, 17, 219
6. 8, 13, 219
7. 8, 12, 529
8. 3, 18, 119
9. 4, 12, 116
10. 7, 17, 219
11. 4, 14, 307
12. 118
13. 306
14. 118
15. 242

CHECK

16. 109
17. 229
18. 527
19. 319

SKILL 19

TRY THESE

1. 3, 10, 415
2. 5, 15, 370
3. 4, 13, 3, 16, 257
4. 3, 11, 1, 11, 267

PRACTICE

1. 6, 13, 117
2. 4, 12, 2, 15, 248
3. 5, 12, 2, 18, 179
4. 7, 11, 1, 10, 478
5. 8, 15, 208
6. 5, 13, 3, 17, 288
7. 6, 10, 0, 13, 167
8. 8, 9, 10, 13, 334
9. 255
10. 267
11. 138
12. 269

CHECK

13. 178
14. 217
15. 286
16. 288

SKILL 20

TRY THESE

1. 7, 10, 154
2. 5, 10, 331
3. 4, 9, 10, 10, 388
4. 4, 9, 10, 10, 88

PRACTICE

1. 6, 10, 135
2. 1, 10, 62
3. 1, 9, 10, 10, 144
4. 4, 9, 10, 10, 133
5. 4, 10, 328
6. 1, 10, 20
7. 3, 9, 10, 10, 147
8. 7, 9, 10, 10, 205
9. 129
10. 181
11. 85
12. 162

CHECK

13. 125
14. 142
15. 261
16. 148

Number Sense

Money

Round Money Amounts

OBJECTIVE Round money amounts

MATERIALS number line

You may wish to review rounding whole numbers to the nearest ten and hundred. Then discuss rounding to the nearest ten cents and nearest dollar.

Direct students' attention to Example 1.

Ask: **What number is halfway between $0.70 and $0.80?** $0.75 **Is $0.77 closer to $0.70 or $0.80?** $0.80 **Do you increase the rounding place or leave it the same? Why?** Increased it, because $0.77 is more than halfway between $0.70 and $0.80.

Continue to ask similar questions as you work through Example 2.

TRY THESE In Exercises 1–4 students round money amounts to the nearest ten cents and nearest dollar.

- **Exercises 1–2** Round to nearest ten cents.
- **Exercises 3–4** Round to nearest dollar.

PRACTICE ON YOUR OWN Have students explain the example.

CHECK Determine if students understand that when rounding, if a number is halfway or more between two amounts, round to the greater amount. When a number is less than halfway between two amounts, round down to the lesser amount.

Success is indicated by 3 out of 3 correct responses.

Students who successfully complete the **Practice on Your Own** and **Check** are ready to move on to the next skill.

COMMON ERRORS

- Students may be confused by the decimal notation.
- Students may look at the wrong digit for rounding.

Students who made more than four errors in the **Practice on Your Own**, or who were not successful in the **Check** section, may also benefit from the **Alternative Teaching Strategy** on the next page.

Optional

Alternative Teaching Strategy

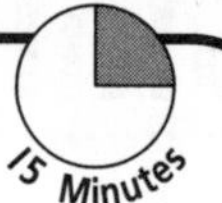

Round Money Amounts

OBJECTIVE To round money amounts

MATERIALS place-value grid for money

Provide students with place-value grids labeled with dollars, dimes, and pennies. Choose an amount of money such as $6.34 and have students write it in their place-value grid as you model rounding to the nearest ten cents.

dollars	dimes	pennies
$ 6 .	3	4

rounding digit (3) — digit to right (4)

$6.34 → $6.30

Ask: **What is the digit in the place to be rounded?** 3 **What is the digit to its right?** 4 **Is the digit to the right greater or less than 5?** less **Do you increase the rounding place or leave it the same? Explain.** Leave it the same, because the digit to the right is less than 5.

So $6.34 rounded to the nearest ten cents is $6.30.

Repeat this activity several times with other examples. Have students round to the nearest ten cents and to the nearest dollar.

Name ______________________ Skill ______________________

Grade 4 **Skill 21**

Round Money Amounts

Use a number line to round money amounts to the nearest ten cents and to the nearest dollar.

Round to the Nearest Ten Cents

Round \$0.77 to the nearest ten cents.

- The amount \$0.77 is between \$0.70 and \$0.80. Look at the number line.

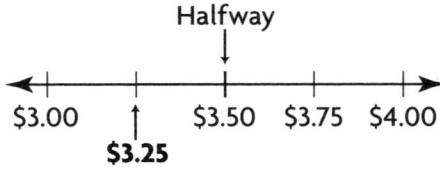

Since \$0.77 is closer to \$0.80 than to \$0.70, \$0.77 rounds to \$0.80.

Round to the Nearest Dollar

Round \$3.25 to the nearest dollar.

- The amount \$3.25 is between \$3.00 and \$4.00. Look at the number line.

Halfway

\$3.00 \$3.25 \$3.50 \$3.75 \$4.00

Since \$3.25 is closer to \$3.00 than to \$4.00, \$3.25 rounds to \$3.00.

Try These

Round each amount.

1. Round \$0.37 to the nearest ten cents.

   \$0.37

   \$0.30 \$0.35 \$0.40

   ____

2. Round \$0.82 to the nearest ten cents.

   \$0.82

   \$0.80 \$0.85 \$0.90

   ____

3. Round \$9.18 to the nearest dollar.

   \$9.18

   \$9.00 \$9.25 \$9.50 \$9.75 \$10.00

   ____

4. Round \$5.62 to the nearest dollar.

   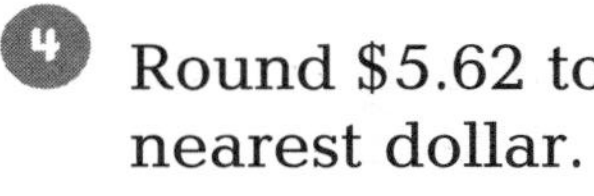

   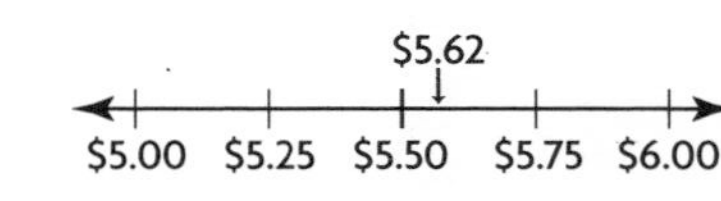

   ____

Go to the next side.

Name ______________________ Skill ______________________

Practice on Your Own

Skill 21

Round $7.50 to the nearest dollar.

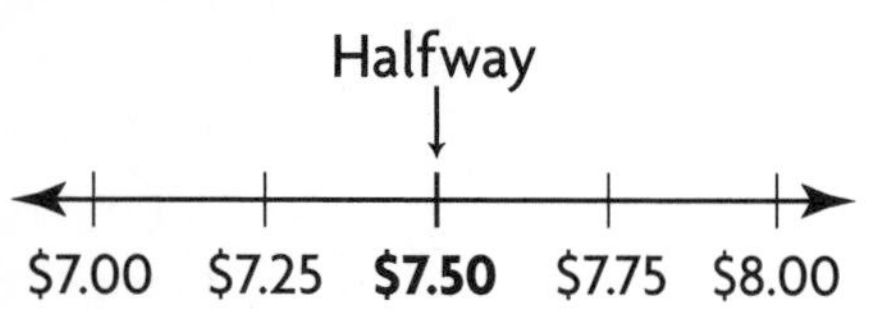

The amount $7.50 is halfway between $7.00 and $8.00.

So, $7.50 rounds to $8.00.

Round each amount.

1. Round $0.66 to the nearest ten cents.

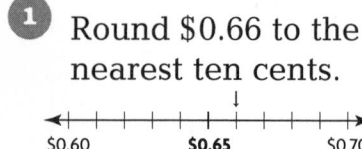

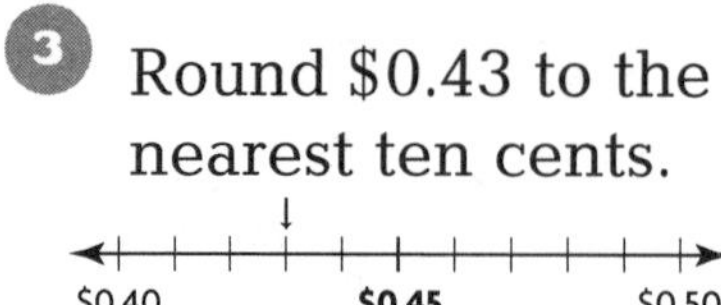

____

2. Round $0.34 to the nearest ten cents.

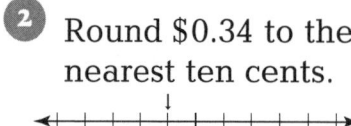

____

3. Round $0.43 to the nearest ten cents.

____

4. Round $3.56 to the nearest dollar.

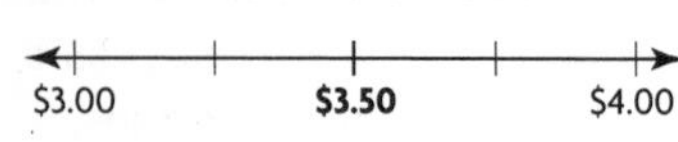

____

5. Round $1.28 to the nearest dollar.

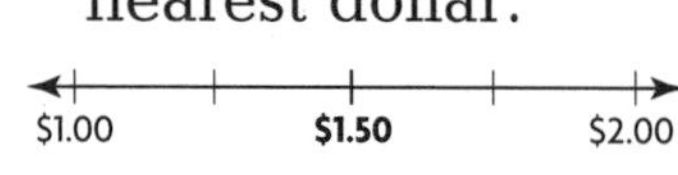

____

6. Round $1.85 to the nearest dollar.

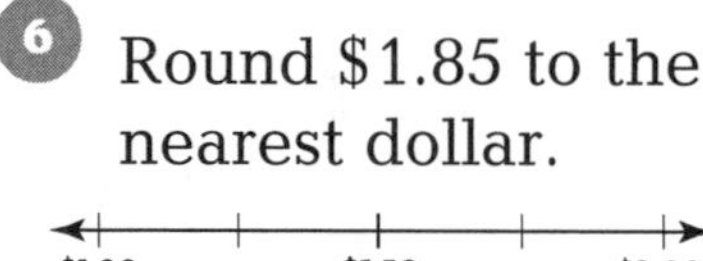

____

7. Round $0.14 to the nearest ten cents.

____

8. Round $1.55 to the nearest dollar.

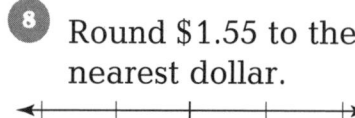

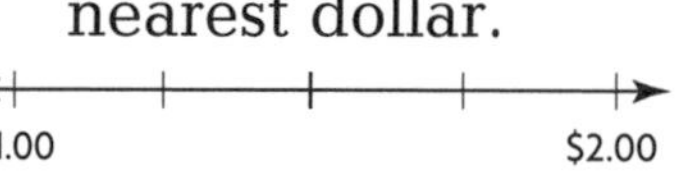

____

9. Round $0.22 to the nearest ten cents.

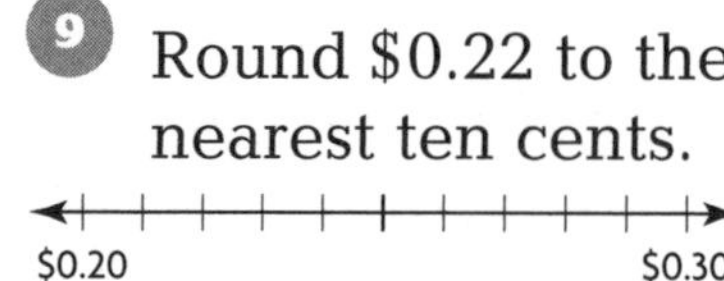

____

10. Round $0.37 to the nearest ten cents.

____

11. Round $2.82 to the nearest dollar.

____

12. Round $9.11 to the nearest dollar.

____

Check

Round each amount.

13. Round $0.89 to the nearest ten cents.

____

14. Round $7.02 to the nearest dollar.

____

15. Round $5.19 to the nearest dollar.

____

Skill 22 Grade 4

Add Money

OBJECTIVE Add money

Begin by reminding students that they add money the same way they add whole numbers. Explain that when adding money amounts, they place a decimal point to separate the dollars and cents, and a dollar sign is placed to the left of the dollars.

Then, have students note that the place-value labels are *dollars*, *dimes*, and *pennies*. Discuss how they can use these labels in the same way they use hundred, tens, and ones labels for whole numbers.

Have students look at Step 1 and recall that when adding money, always begin with the pennies.

Ask: **Do you have to regroup the pennies? Why or why not?** No, there are fewer than 10 pennies

In Step 2, students regroup the dimes.

Ask: **How do you know that the dimes can be regrouped?** 9 dimes + 2 dimes = 11 dimes. Since I have more than 9 dimes, I regroup. **What is the result when you regroup 11 dimes as dollars and dimes?** 1 dollar 1 dime **Where do you place the regrouped digit?** in the dollars column above the 3

Emphasize placing the digit in the correct column, and remembering to add the regrouped digit when adding dollars.

As students work through Step 3, point out that the sum is not complete until a dollar sign and decimal point is written.

TRY THESE Exercises 1–3 model the type of exercises students will find on the **Practice on Your Own** page.

- **Exercise 1** No regrouping.
- **Exercise 2** Regroup dimes.
- **Exercise 3** Regroup pennies and dimes.

PRACTICE ON YOUR OWN Review the example at the top of the page. Remind students that for some exercises they will regroup once or twice, while for others there may be no regrouping.

CHECK Determine if students regroup when adding money and place the decimal point and dollar sign in the sum. Success is indicated by 5 out of 6 correct responses.

Students who successfully complete the **Practice on Your Own** and **Check** are ready to move on to the next skill.

COMMON ERRORS

- Students may forget to add the regrouped digit.
- Students may forget to write the decimal point and dollar sign in the sum.
- Students may not know addition facts.

Students who made more than four errors in the **Practice on Your Own**, or who were not successful in the **Check** section, may benefit from the **Alternative Teaching Strategy** on the next page.

Optional

Alternative Teaching Strategy

15 Minutes

Model Adding Money

OBJECTIVE Use play money to model addition with money

MATERIALS play money

Distribute the play money to pairs of students. Discuss how adding and regrouping with pennies, dimes, and dollars is similar to regrouping with ones, tens, and hundreds. Recall with students that 10 pennies equal 1 dime and 10 dimes equal 1 dollar.

You may wish to begin by having students practice regrouping. Ask them to regroup 14 pennies. Explain that since there are more than 9 pennies, students can regroup 14 pennies as 1 dime 4 pennies.

Repeat the procedure with 19 dimes. When students are comfortable with regrouping pennies and dimes, present this example.

Dollars	dimes	pennies
$ 1 .	7	3
+ 0 .	3	5

Model reading the amounts of money.

Say: **We are adding one dollar and seventy-three cents plus thirty-five cents.**

Explain that when adding with money, the addends are aligned by the decimal points. Point out that zero is used in the dollars place when the amount is less than one dollar.

Suggest that one student model the addition using play money, as the partner records the addition on paper.

As students work through the addition, ask questions that guide the student through the regrouping process. Emphasize placing the regrouped digit carefully in the dollar column and remembering to add it as they add the dollars. Remind students to place the dollar sign and decimal point in the sum.

Dollars	dimes	pennies
$ 1 .	7	3
+ 0 .	3	5
$ 2 .	0	8

Have students change roles. Repeat the activity with dollar amounts allowing student to regroup pennies, and then dimes and pennies. When the students show understanding of adding money with regrouping, have them try exercises without using play money.

$1.73 + $0.35

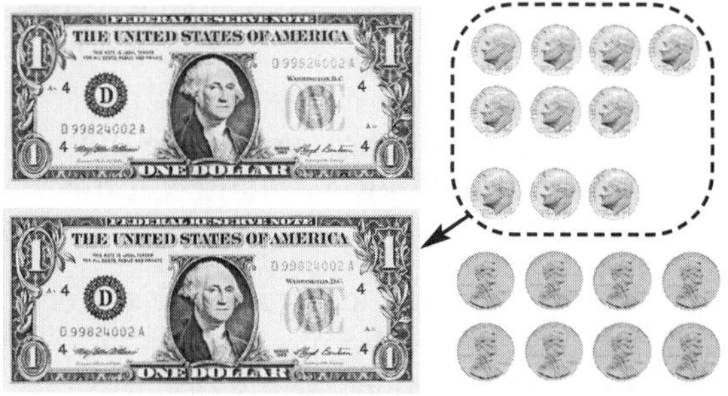

Regroup 10 Dimes as 1 Dollar

The sum is $2.08

Name ______________________ Skill ______________

Grade 4 Skill 22

Add Money

Add money the same way as you add whole numbers.

Find $3.95 + $1.24 = ■.

Step 1 Add the pennies.

dollars	.	dimes	pennies
$ 3	.	9	5
+$ 1	.	2	4
			9

Think:
5 pennies + 4 pennies = 9 pennies.

Step 2 Add the dimes. Regroup.

dollars	.	dimes	pennies
1			
$ 3	.	9	5
+$ 1	.	2	4
		1	9

Think:
9 dimes + 2 dimes = 11 dimes.
11 dimes = 1 dollar + 1 dime

Step 3 Add the dollars. Write the dollar sign and decimal point in the sum.

dollars	.	dimes	pennies
1			
$ 3	.	9	5
+$ 1	.	2	4
$ 5	.	1	9

Think:
1 dollar + 3 dollars + 1 dollar = 5 dollars

So, $3.95 + $1.24 = $5.19.

Try These

Find the sum.

1.

dollars	.	dimes	pennies
$ 3	.	0	0
+$ 2	.	9	2
$ ☐	.	☐	☐

2.

dollars	.	dimes	pennies
☐			
$ 0	.	9	0
+$ 0	.	9	9
$ ☐	.	☐	☐

3.

dollars	.	dimes	pennies
☐		☐	
$ 4	.	4	5
+$ 1	.	7	9
$ ☐	.	☐	☐

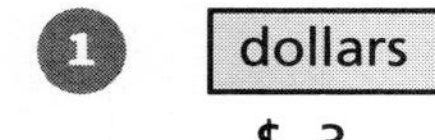

Name ______________________ Skill ______________________

Practice on Your Own

Find $8.07 + 0.95 = ■.
Add the pennies.
Regroup. 12 pennies = 1 dime + 2 pennies
Add the dimes.
Regroup. 10 dimes = 1 dollar
Add the dollars.

dollars	.	dimes	pennies
1		1	
$ 8	.	0	7
+ $ 0	.	9	5
$ 9	.	0	2

Find the sum.

dollars	.	dimes	pennies
$ 2	.	1	4
+$ 1	.	2	3
$ ☐	.	☐	☐

dollars	.	dimes	pennies
		☐	
$ 4	.	2	6
+$ 1	.	4	4
$ ☐	.	☐	☐

dollars	.	dimes	pennies
☐			
$ 3	.	9	0
+$ 2	.	3	5
$ ☐	.	☐	☐

4. $5.06 + $3.28 = $☐.☐☐

5. $2.87 + $2.42 = $☐.☐☐

6. $7.89 + $1.21 = $☐.☐☐

7. $4.56 + $4.07 = $☐.☐☐

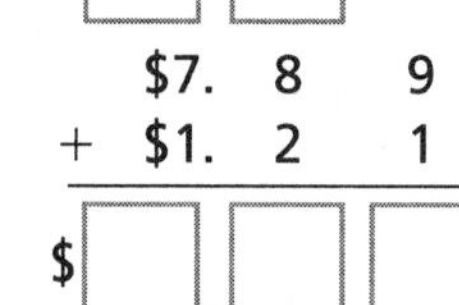

8. $2.98 + $4.01

9. $6.27 + $1.35

10. $4.81 + $3.79

11. $5.29 + $2.83

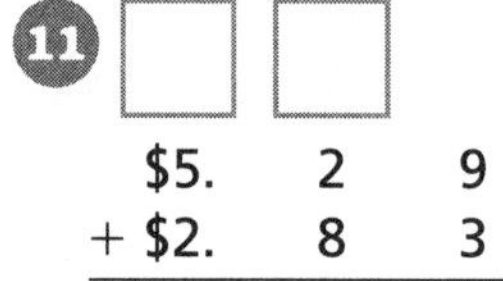

12. $1.79 + $0.11

13. $3.84 + $2.19

14. $3.46 + $4.74

15. $5.65 + $1.99

Check

Find the sum.

16. $1.37 + $2.04

17. $3.91 + $3.45

18. $6.15 + $1.95

19. $4.32 + $2.88

20. $5.09 + $3.23

21. $2.75 + $4.35

Skill 23 Grade 4

Subtract Money

OBJECTIVE Subtract money amounts

Begin by having students recall how many dimes are in 1 dollar and how many pennies are in 1 dime.

Direct students' attention to Step 1.

Ask: **Do you have enough pennies to subtract?** No, 5 pennies are less than 9 pennies. **What do you have to do before you can subtract?** Regroup 1 dime as 10 pennies. **How many pennies do you have now?** 15 **When you subtract, how many pennies will you have?** 6

Call attention to Step 2.

Ask: **Do you need to regroup the dimes?** Yes, 5 dimes are less than 6 dimes.

As you work through each step, note how dollars and dimes are regrouped, and the placement of the decimal point in the difference.

TRY THESE Exercises 1–4 model the type of exercises students will find on the **Practice on Your Own** page.

- **Exercise 1** No regrouping.
- **Exercise 2** Regrouping dimes.
- **Exercises 3–4** Regrouping dollars and dimes.

PRACTICE ON YOUR OWN Review the example at the top of the page. Ask students to explain why it was necessary to regroup dollars and dimes.

CHECK Determine if students are able to regroup dimes as pennies and dollars as dimes. Success is determined by 3 out of 4 correct responses.

Students who successfully complete the **Practice on Your Own** and **Check** are ready to move on to the next skill.

COMMON ERRORS

- Students may forget to subtract 1 from the dimes or dollars when they regroup them.
- Students may add 1 instead of 10 when they regroup in any place.

Students who made more than four errors in the **Practice on Your Own**, or who were not successful in the **Check** section, may benefit from the **Alternative Teaching Strategy** on the next page.

Optional

Alternative Teaching Strategy
Model Subtracting Money

20 Minutes

OBJECTIVE Use play money to model subtraction of money with regrouping

MATERIALS play money: dollars, dimes, and pennies only

You may wish to have students model a problem and work their way through it using the play money before trying to record the results with paper and pencil.

Distribute the play money. Write the subtraction in the place-value grid as shown on the board.

dollars	dimes	pennies
$ 3	.6	1
− $ 1	.4	3

Have students start by using the play money to model $3.61.

Ask: **Do you have enough pennies to subtract?** No, 1 penny is less than 3 pennies.

What can you do to have enough pennies to subtract? Regroup 1 dime as 10 pennies. **How many pennies do you have now?** 11 pennies

Have students subtract the pennies. Ask students to explain what step to do next. Have students complete the subtraction and tell you the difference. Then record each step in the place-value chart, as students summarize the modeling they completed.

Repeat this activity with similar examples. When the students show understanding of the regrouping process, remove the play money and have them try an exercise using only paper and pencil. Ask the students to explain each step in the subtraction process.

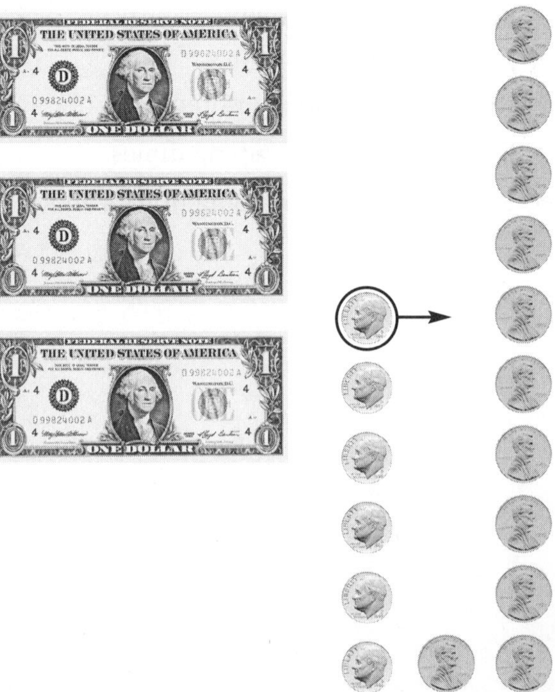

Name ______________________ Skill ______________________

Grade 4 Skill 23

Subtract Money

You subtract money the same way you subtract whole numbers.

Find $5.65 – $1.69 = ■.

Step 1 There are not enough pennies to subtract. Regroup. Subtract the pennies.

Dollars	Dimes	Pennies
	5	15
$5	~~6~~	~~5~~
– $1	.6	9
		6

Think:
Regroup 6 dimes as 5 dimes 10 pennies.
5 pennies + 10 pennies = 15 pennies
15 pennies – 9 pennies = 6 pennies

Step 2 There are not enough dimes to subtract. Regroup. Subtract the dimes.

Dollars	Dimes	Pennies
4	15	
	~~5~~	15
~~$5~~	~~6~~	~~5~~
– $1	.6	9
	.9	6

Think:
Regroup 5 dollars as 4 dollars 10 dimes.
5 dimes + 10 dimes = 15 dimes
15 dimes – 6 dimes = 9 dimes

Step 3 Subtract the dollars. Write the dollar sign and decimal point in the difference.

Dollars	Dimes	Pennies
4	15	15
~~$5~~	~~6~~	~~5~~
– $1	.6	9
$3	.9	6

Think:
4 dollars – 1 dollar = 3 dollars

So, $5.65 – $1.69 = $3.96.

Try These

Find the difference.

1.

Dollars	Dimes	Pennies
$6	.4	5
– $1	.3	2
$ ☐	. ☐	☐

2.

Dollars	Dimes	Pennies
	☐	☐
$1	.3	5
– $0	.2	7
$ ☐	. ☐	☐

3.

Dollars	Dimes	Pennies
	☐	
☐	☐	☐
$5	.2	5
– $0	.7	8
$ ☐	. ☐	☐

4.

Dollars	Dimes	Pennies
	☐	
☐	☐	☐
$2	.4	8
– $1	.8	9
$ ☐	. ☐	☐

Go to the next side.

Name ____________________ Skill ____________________

Practice on Your Own

Skill 23

Find $5.07 – $3.59.

There are not enough pennies or dimes to subtract. Regroup dollars and dimes.

Subtract.

Remember: Write the dollar sign and the decimal point in the difference.

Dollars	Dimes	Pennies
	9	
4	10	17
$5	.0	7
– $3	.5	9
$1	**.4**	**8**

Find the difference.

1

Dollars	Dimes	Pennies
$4	.3	7
– $2	.0	6
$☐	.☐	☐

2

Dollars	Dimes	Pennies
	☐	☐
$3	.8	1
– $0	.4	5
$☐	.☐	☐

3

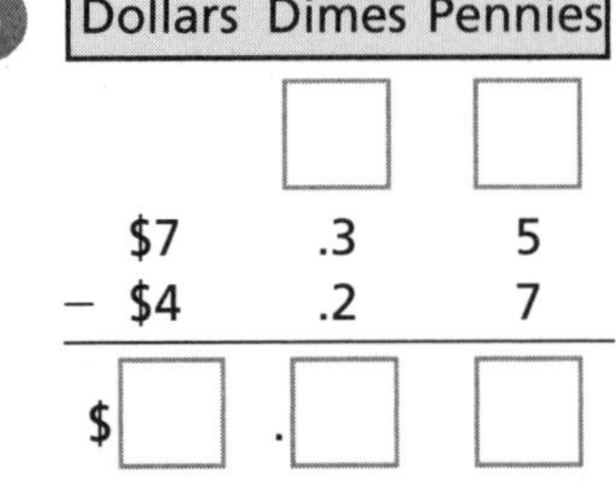

Dollars	Dimes	Pennies
	☐	☐
$7	.3	5
– $4	.2	7
$☐	.☐	☐

4

Dollars	Dimes	Pennies
	☐	☐
$8	.5	4
– $4	.3	8
$☐	.☐	☐

5

Dollars	Dimes	Pennies
☐	☐	
$5	.4	4
– $0	.8	4
$☐	.☐	☐

6

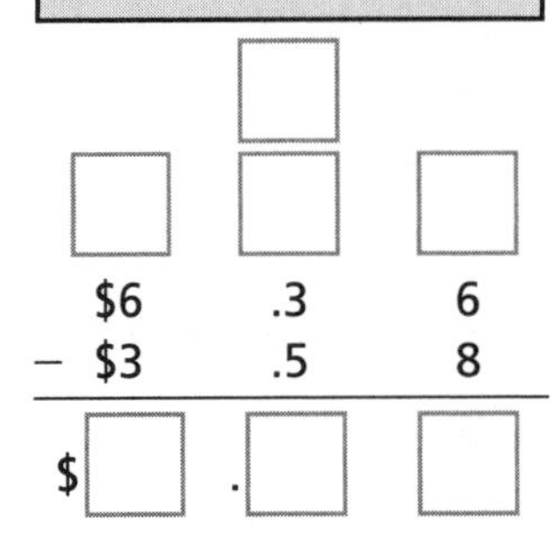

Dollars	Dimes	Pennies
	☐	
☐	☐	☐
$6	.3	6
– $3	.5	8
$☐	.☐	☐

7

☐	☐	
$2	.9	5
– $1	.3	7

8

☐	☐	
$5	.1	9
– $2	.9	0

9

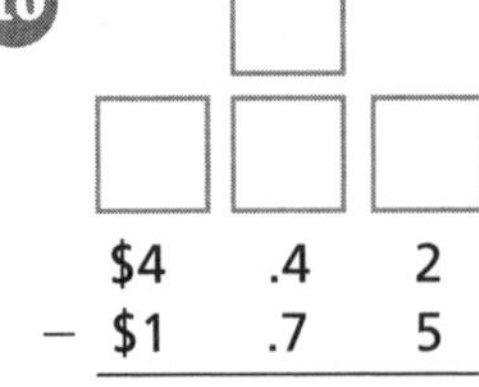

	☐	
☐	☐	☐
$7	.6	1
– $4	.7	8

10

	☐	
☐	☐	☐
$4	.4	2
– $1	.7	5

Check

Find the difference.

11 $3.71 – $1.40

12 $7.62 – $4.78

13 $9.23 – $3.64

14 $4.19 – $2.30

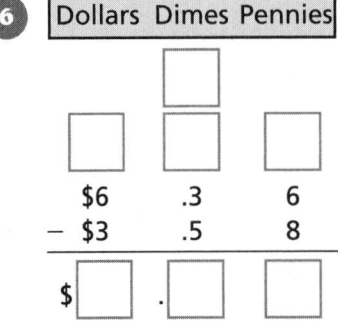

Answer Card
Money
Grade 4

Skill 21

TRY THESE

1. $0.40
2. $0.80
3. $9.00
4. $6.00

PRACTICE

1. $0.70
2. $0.30
3. $0.40
4. $4.00
5. $1.00
6. $2.00
7. $0.10
8. $2.00
9. $0.20
10. $0.40
11. $3.00
12. $9.00

CHECK

13. $0.90
14. $7.00
15. $5.00

Skill 22

TRY THESE

1. $5.92
2. 1, $1.89
3. 1, 1, $6.24

PRACTICE

1. $3.37
2. 1, $5.70
3. 1, $6.25
4. 1, $8.34
5. 1, $5.29
6. 1, 1, $9.10
7. 1, $8.63
8. $6.99
9. 1, $7.62
10. 1, 1, $8.60
11. 1, 1, $8.12
12. $1.90
13. $6.03
14. $8.20
15. $7.64

CHECK

16. $3.41
17. $7.36
18. $8.10
19. $7.20
20. $8.32
21. $7.10

Skill 23

TRY THESE

1. $5.13
2. 2, 15, $1.08
3. 4, 11, 1, 15, $4.47
4. 1, 13, 3, 18, $0.59

PRACTICE

1. $2.31
2. 7, 11, $3.36
3. 2, 15, $3.08
4. 4, 14, $4.16
5. 4, 14, $4.60
6. 5, 12, 2, 16, $2.78
7. 8, 15, $1.58
8. 4, 11, $2.29
9. 6, 15, 5, 11, $2.83
10. 3, 13, 3, 12, $2.67

CHECK

11. $2.31
12. $2.84
13. $5.59
14. $1.89

Number Sense

Whole Number Multiplication

Multiplication Facts to 5

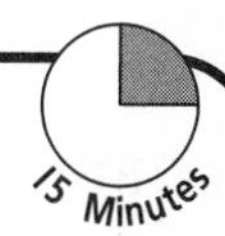

OBJECTIVE Recall multiplication facts through 5

MATERIALS tiles

You may wish to have students use tiles to make arrays that represent the number sentences in Steps 1–3.

Begin by recalling that the numbers that are multiplied are called *factors*. The result or answer is called the *product*. Remind students that multiplication facts can be shown horizontally or vertically.

$$2 \times 4 = 8 \qquad \begin{array}{r} 4 \\ \times\ 2 \\ \hline 8 \end{array}$$

Direct the students' attention to Step 1. Explain how an array can be used to show a multiplication sentence.

Ask: **How many rows do you see in the array of tiles?** 5 rows **How many tiles are in each row?** 3 tiles

Explain that the number of rows represents one factor. The number of tiles in a row represents the second factor.

Ask: **What multiplication sentence can you write to find how many tiles are shown?** $5 \times 3 = 15$ or $3 \times 5 = 15$

Have a student write the multiplication fact vertically.

TRY THESE Exercises 1–3 model other multiplication facts using multiplication number sentences.

- **Exercise 1** Factors 4 and 3.
- **Exercise 2** Factors 2 and 1.
- **Exercise 3** Factors 3 and 7.

PRACTICE ON YOUR OWN Work through the example at the top of the page. Have students identify the factors and product.

CHECK Determine if students understand the meaning of multiplication and can recall the basic multiplication facts for factors 1 through 5.

Success is indicated by 4 out of 5 correct responses.

Students who successfully complete the **Practice on Your Own** and **Check** are ready to move on to the next skill.

COMMON ERRORS

- Students may add instead of multiplying.
- Students may not recall their multiplication facts correctly.
- Some students may have difficulty using the horizontal or vertical format.

Students who made more than four errors in the **Practice on Your Own**, or who were not successful in the **Check** section, may benefit from the **Alternative Teaching Strategy** on the next page.

Optional

Alternative Teaching Strategy

Model Multiplication Facts to 5

OBJECTIVE Use concrete objects to recall multiplication facts through 5

MATERIALS tiles

Provide students with a group of 16 tiles. Have them organize the tiles into a 4 × 4 array. Review with students that an array is an arrangement of objects in rows and columns.

Ask: **How many rows of tiles did you make?** 4 **How many tiles are in each row?** 4 **How many tiles did you use in all?** 16

Remind students that the number of rows represents one factor, the number of tiles in a row represents the second factor. The total number of tiles is the product.

Ask: **What number sentence can you write to show the multiplication?** 4 × 4 = 16

4

4 □□□□
□□□□
□□□□
□□□□

4 × 4 = 16

Repeat the activity several times with other arrays of tiles. Have students record the facts they find.

When you feel confident that students can work without the tiles, provide three exercises for students to complete independently. When they have found the products, ask students to explain how they arrived at each result. Some students might offer that they pictured an array mentally to find the product. Others might say that they were able to recall the products from memory.

Grade 4 Skill 24

Multiplication Facts to 5

Use an array to multiply 5 × 3.

Step 1 Write the number of rows.

There are 5 rows.

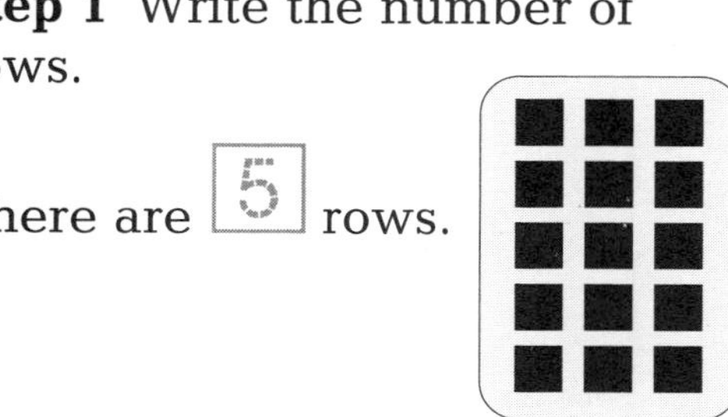

Step 2 Write the number of tiles in each row.

There are 3 tiles in each row.

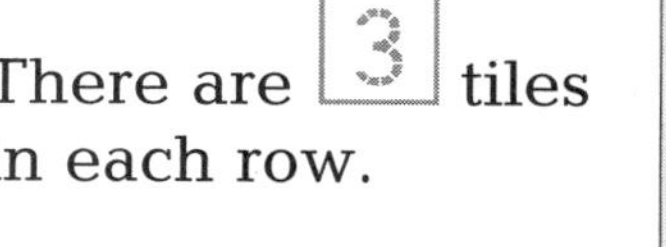

Step 3 Multiply to find how many tiles in all.

5	×	3	=	15
↓		↓		↓
number of rows		number in each row		total number of tiles

So, 5 × 3 = 15.

Try These

Find the product.

1. 

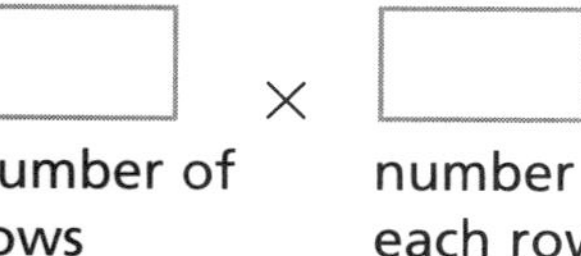

☐ × ☐ = ☐

number of rows × number in each row = total number of tiles

2.  

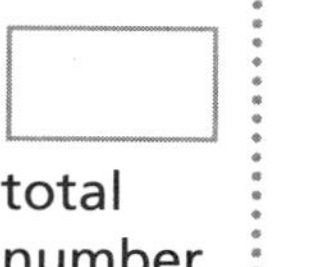

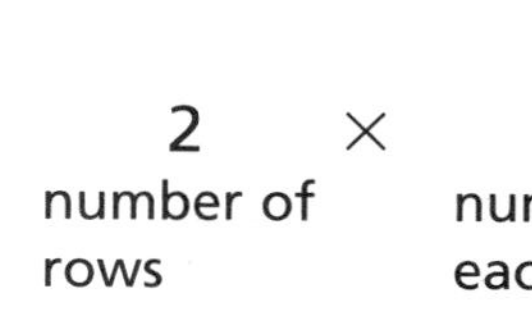

2 × 1 = ☐

number of rows × number in each row = total number of tiles

3. 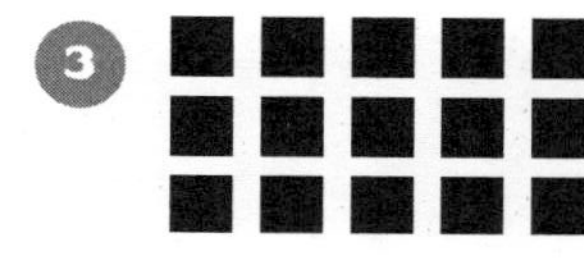

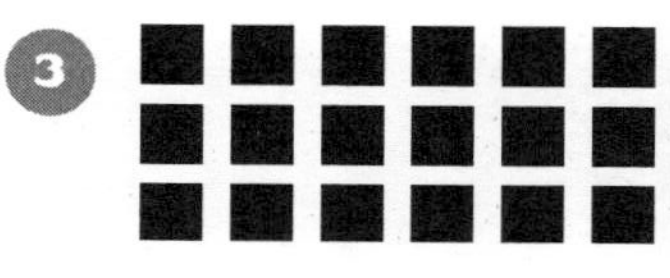

3 × 7 = ☐

number of rows × number in each row = total number of tiles

Name ____________________ Skill ____________________

Practice on Your Own

Skill 24

Find 5 × 5 = ■.

Think: Multiply the number of rows by the number in each row to find how many in all.

5	×	5	=	25
number of rows		number in each row		total number of tiles

Find the product.

1 

□ × □ = 10

2 

□ × □ = □

3 

□ × □ = □

4 2 × 3 = □

5 2 × 4 = □

6 2 × 6 = □

7 2 × 8 = □

8 3 × 1 = □

9 3 × 4 = □

10 3 × 8 = □

11 3 × 9 = □

12 7×3

13 4×4

14 5×4

15 6×4

16 9×2

17 5×3

18 8×4

19 9×5

 **Check**

Find the product.

20 1×2

21 6×3

22 2×4

23 7×5

24 8×5

Multiplication Facts to 10

OBJECTIVE Recall multiplication facts 6 through 10

MATERIALS tiles

Begin the lesson by discussing the example at the top of the Skill. How many rows are there in the array? **6** How many tiles are in each row? **5** How many tiles are there in all? **30** What multiplication fact does the array show? **6 × 5 = 30**

Show a 7 × 4 array. Ask questions similar to those above. Have a student write the multiplication fact 7 × 4 = 28. Continue with two or three other arrays if necessary.

You may wish to provide a context for multiplying, such as: **You have 4 boxes of 8 marbles. How many marbles do you have in all?** 32 **What multiplication fact can you use to answer the question?** 4 × 8 = 32

TRY THESE In Exercises 1–3, students use pictorial models of arrays to write facts.

- **Exercise 1** Array for 7 × 6.
- **Exercise 2** Array for 8 × 1.
- **Exercise 3** Array for 9 × 5.

PRACTICE ON YOUR OWN Go over the example at the top of the page. Review the meaning of *row* in multiplication. Discuss why multiplying is much more efficient than counting or using repeated addition. Only the first three exercises show arrays. If students are having trouble finding the products, allow them to use tiles for some of the remaining exercises.

CHECK Success is determined by 4 out of 5 correct responses.

Students who successfully complete the **Practice on Your Own**, are ready to move on to the next skill.

COMMON ERRORS

- Students may add instead of multiplying.
- Students may recall facts incorrectly.

Students who made more than three errors in the **Practice on Your Own,** or who were not successful in the **Check** section, may benefit from the **Alternative Teaching Strategy** on the next page.

Optional

Alternative Teaching Strategy

Model Multiplication Facts to 10

20 Minutes

OBJECTIVE Use manipulatives to understand and practice multiplication facts

MATERIALS tiles, triangle flash cards

Provide a supply of tiles. Ask each student to model 3 × 7.

How many rows? 3

How many in each row? 7

How many in all? 21

What multiplication fact does the array show? 3 × 7 = 21

Repeat two or three times with different facts.

Remind students that the numbers they multiply are called *factors* and the answer is called the *product*.

If you think that the students understand the concept of multiplication, but are not recalling the facts correctly, prepare a set of triangle flash cards. The factors are in two of the corners and the product is in the third corner. Let students work in pairs with the flash cards for a few minutes every day. Have them keep a list of the facts that give them trouble, and concentrate on those, using counters if necessary.

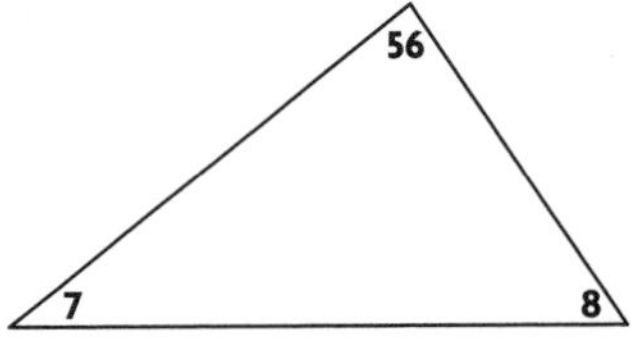

Students who do not understand what multiplication means may need further work with arrays, the number line, or equal groups.

Grade 4 Skill 25

Multiplication Facts to 10

Use an array to multiply 6 × 5.

Step 1 Write the number of rows.

There are 6 rows.

Step 2 Write the number of tiles in each row.

There are 5 tiles in each row.

Step 3 Multiply to find how many tiles in all.

6	×	5	=	30
number of rows		number in each row		total number of tiles

So, 6 × 5 = 30.

Try These

Find the product.

1.

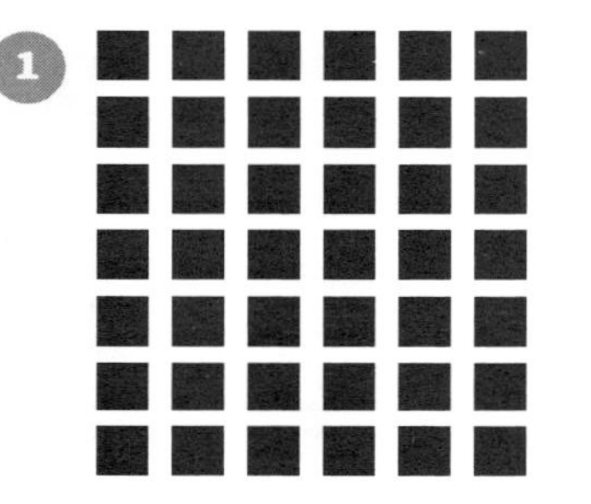

7	×	6	=	☐
number of rows		number in each row		total number of tiles

2.

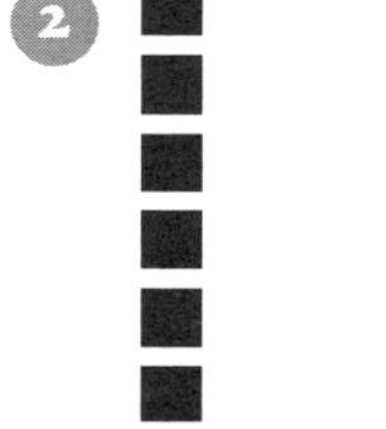

8	×	1	=	☐
number of rows		number in each row		total number of tiles

3.

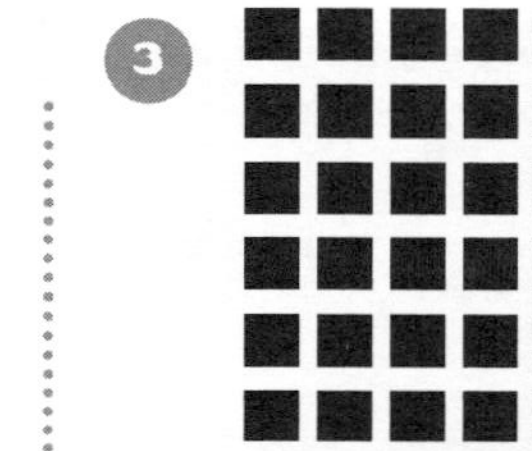
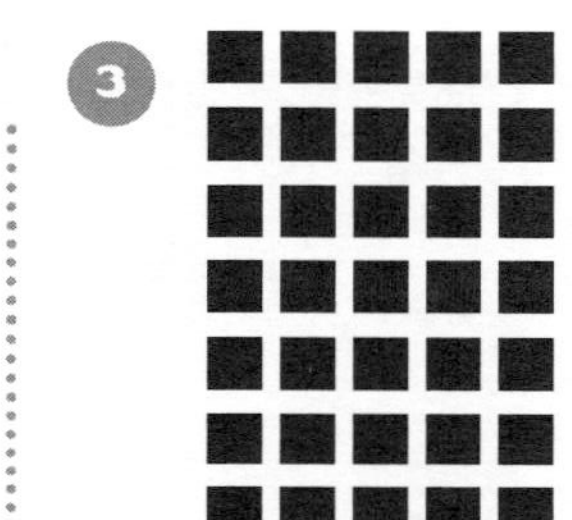
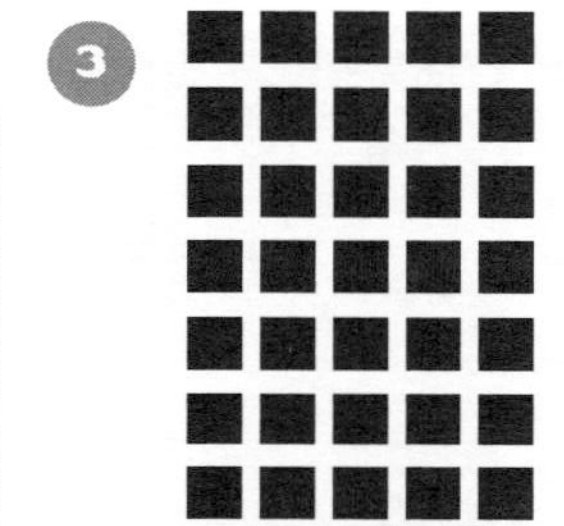

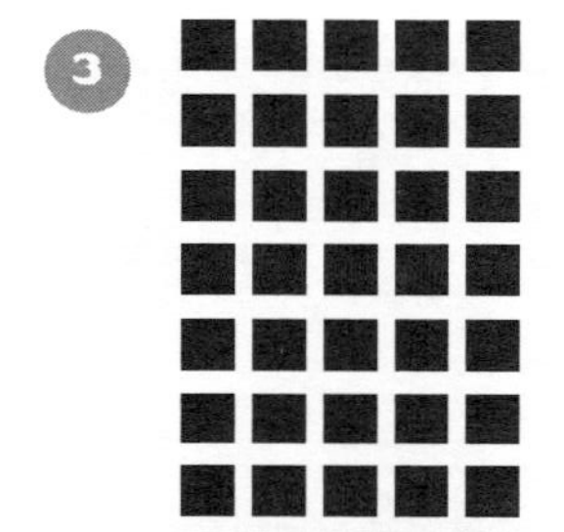

9	×	5	=	☐
number of rows		number in each row		total number of tiles

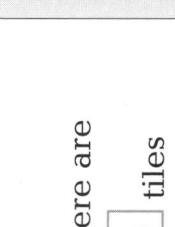

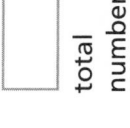

Name ______________________ Skill ______________________

Practice on Your Own

Find 6 × 6 = ■.

Think: Multiply the number of rows by the number in each row to find how many in all.

6	×	6	=	36
number of rows		number in each row		total number of tiles

Find the product.

1. 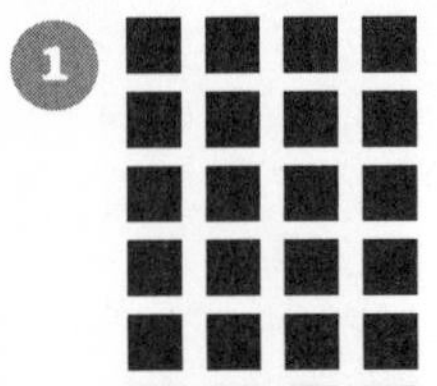

   6 × 4 = ☐

2. 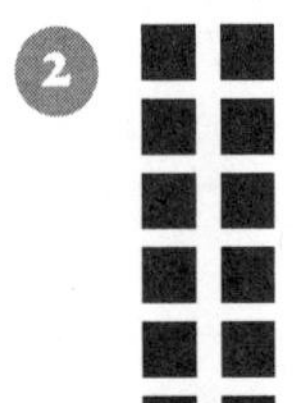

   ☐ × ☐ = ☐

3. 

   ☐ × ☐ = ☐

4. 7 × 1 = ☐
5. 7 × 4 = ☐
6. 7 × 7 = ☐
7. 7 × 8 = ☐
8. 8 × 2 = ☐
9. 8 × 5 = ☐
10. 8 × 8 = ☐
11. 8 × 9 = ☐
12. 9 × 3 = ☐
13. 4 × 9 = ☐
14. 7 × 9 = ☐
15. 9 × 9 = ☐
16. 7 × 6 = ☐
17. 6 × 3 = ☐
18. 6 × 8 = ☐
19. 7 × 9 = ☐

Check

20. 7 × 6 = ☐
21. 5 × 7 = ☐
22. 6 × 9 = ☐
23. 9 × 8 = ☐
24. 6 × 6 = ☐

Skill 26 Grade 4 — Meaning of Multiplication

OBJECTIVE Understand the meaning of multiplication

MATERIALS counters, tiles

Recall with students that multiplication is the operation that tells the total when there are groups with an equal number of items. The groups can be jumps on the number line, groups of objects, or an array. The numbers multiplied are called *factors*, and the total is called the *product*.

Point out that a product can be found by using repeated addition, but that multiplication is faster.

Direct the students' attention to Model A.

Ask: **How many jumps are there?** 2 **How many spaces are in each jump?** 4 **What multiplication sentence is shown?** $2 \times 4 = 8$

Be sure students understand that the first jump starts at zero.

Discuss Models B and C similarly.

Ask: **How many groups [or rows] are there?** 2 **How many are in each group [or row]?** 4 **How many are there in all?** 8 **What is the multiplication sentence?** $2 \times 4 = 8$

Ask a student how to show $3 \times 2 = 6$ on the number line. **3 jumps of 2 spaces each, 6 in all**

Give each student a handful of counters or tiles and have them illustrate $4 \times 3 = 12$ using equal groups or arrays.

TRY THESE In Exercises 1–3, students use different models for multiplication.

- **Exercise 1** Use the number line.
- **Exercise 2** Use equal groups.
- **Exercise 3** Use an array.

PRACTICE ON YOUR OWN Go over the example at the top of the Skill. Be sure students understand that 21 is the product of 3 and 7. Check that students understand the models.

CHECK You may want to allow students to draw a number line, an array, or equal sets of objects if they need help. Success is determined by 3 out of 4 correct responses.

COMMON ERRORS

- Students may think that 2 tick marks on the number line represent 2 jumps instead of 1 jump.
- Students may create "arrays" that do not have equal numbers of items in the rows.

Students who made more than three errors in the **Practice on Your Own** or were not successful in the **Check** section may benefit from the **Alternative Teaching Strategy** on the next page.

Optional

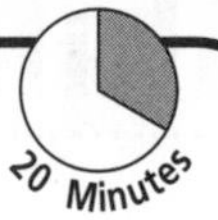

Alternative Teaching Strategy
Model Multiplication

OBJECTIVE Use models to understand the meaning of multiplication

MATERIALS counters, yarn, paper plates, construction paper, paper and pencil, triangle flash cards

Distribute a handful of counters to each student. Say a multiplication fact, for example 5×2. Have students illustrate the fact by putting out 5 groups of 2. They may find the total by repeated addition $2 + 2 + 2 + 2 + 2$ or by counting, if necessary. Have the students record the sentence: $5 \times 2 = 10$. Repeat two or three times.

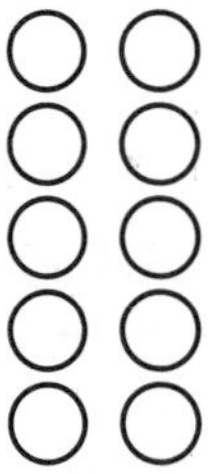

Provide students with yarn or paper plates to use for forming equal groups. Alternatively, provide large pieces of construction paper and have students make an equal-group work mat by folding the paper into thirds vertically and then into thirds horizontally.

Distribute counters. Call out different equal groups. For example, say: **Show me three groups of two.**

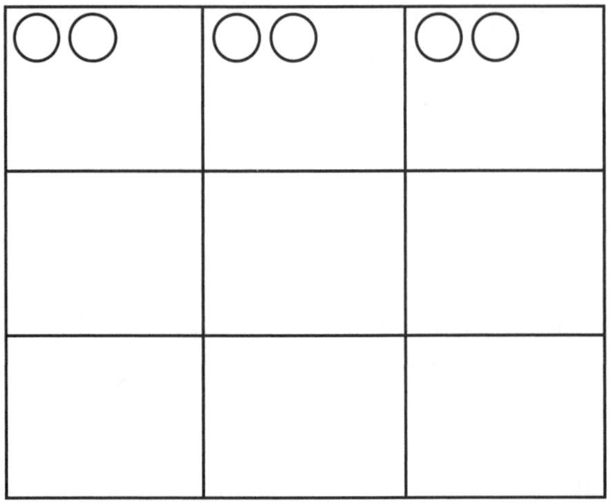

Have students verbalize the result. Students should be able to say that they made 3 groups of 2 counters and the total number is 6.

Continue in a similar way with other numbers of counters. When you think students are ready, have them record the multiplication sentence for each model they complete.

To check for understanding, display an expression such as 4×3. Ask students to describe the steps they would use for finding the product. Students might say that they would use an array, or equal groups of counters. Or they might suggest multiplying with pencil and paper.

Name ______________________ Skill ______________________

Grade 4
Skill 26

Meaning of Multiplication

You can use different models to show multiplication.

Model A Skip-count to find how many in all.

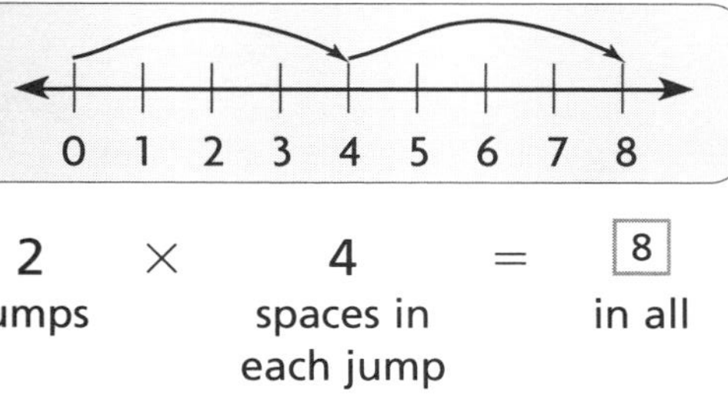

2 jumps × 4 spaces in each jump = 8 in all

Model B Use equal groups to find how many in all.

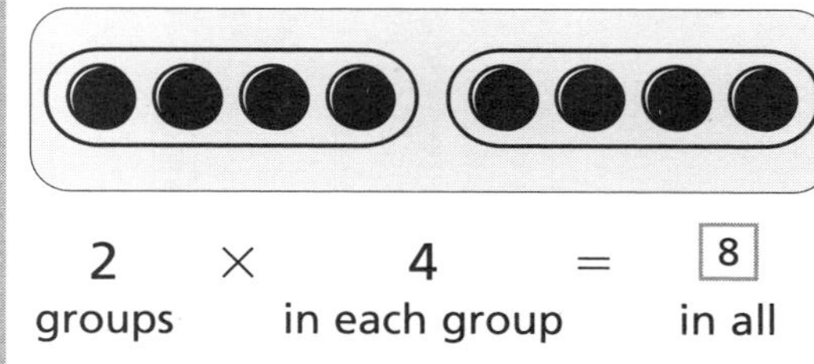

2 groups × 4 in each group = 8 in all

Model C Use an array to find how many in all.

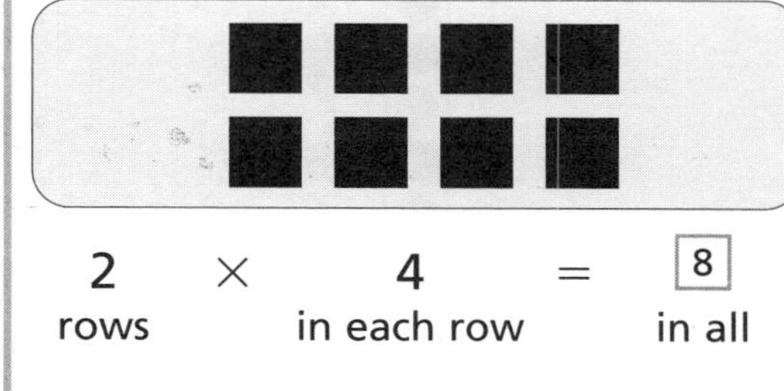

2 rows × 4 in each row = 8 in all

Try These

Complete the multiplication sentences.

1. 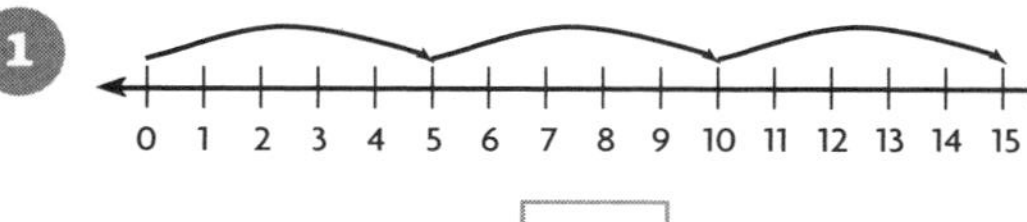

   3 jumps of 5 are ☐ in all.

   $3 \times 5 =$ ☐

2. 

   4 groups of 3 are ☐ in all.

   $4 \times 3 =$ ☐

3. 

   5 rows of 6 are ☐ in all.

   $5 \times 6 =$ ☐

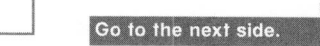

Go to the next side.

Name ______________________ Skill ______________________

Practice on Your Own

Skill 26

Find 3 × 7 = ■.

Use counters to find how many in all.

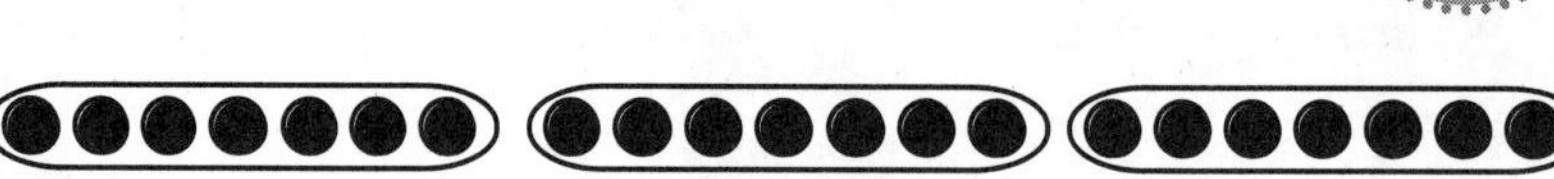

3 groups of 7 are 21 in all.

3 × 7 = 21

Complete the multiplication sentences.

1.
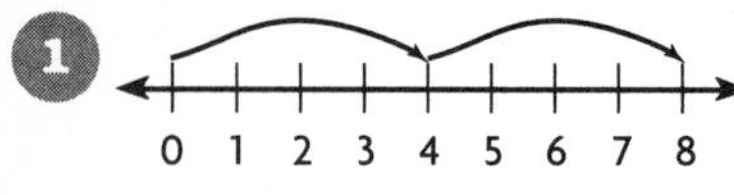

2 jumps of 4 are ☐ in all.

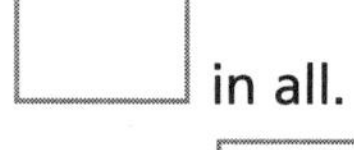

2 × 4 = ☐

2.

3 groups of 3 are ☐ in all.

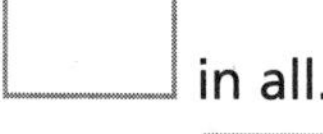

3 × 3 = ☐

3. 4 rows of 5 are ☐ in all.

4 × 5 = ☐

4.
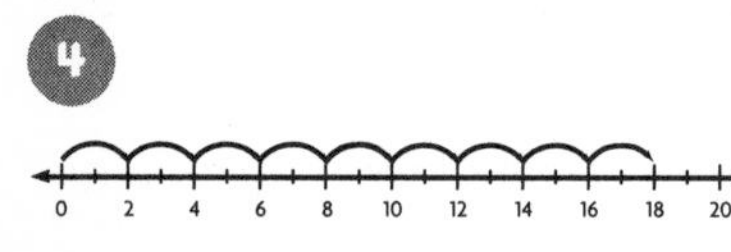

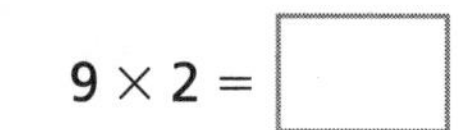

9 × 2 = ☐

5.

3 × 5 = ☐

6.

2 × 9 = ☐

Find the product.

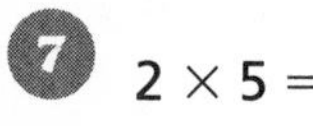

7. 2 × 5 = ☐

8. 3 × 4 = ☐

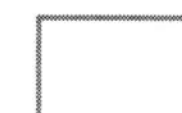

9. 4 × 6 = ☐

10. 6 × 3 = ☐

Check

Find the product.

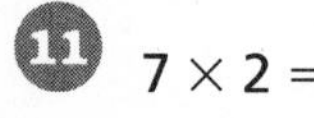

11. 7 × 2 = ☐

12. 6 × 5 = ☐

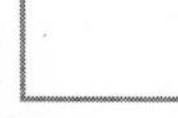

13. 5 × 3 = ☐

14. 8 × 4 = ☐

Skill 27 Grade 4

Model Multiplication (2-Digit by 1-Digit Numbers)

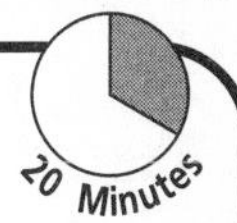

OBJECTIVE Model multiplication of a 2-digit number by a 1-digit number

MATERIALS place-value blocks

You may wish to begin by reviewing basic multiplication facts such as, 2 × 6 = 12, 3 × 5 = 15, 3 × 7 = 21, and so forth. Help students recall that there are different ways to model multiplication: equal groups of counters, arrays of tiles, and jumps on the number line.

Have students use place-value blocks to model the example.

Direct students' attention to Step 1. Note that 2 × 16 is pictured as 2 groups of 16 blocks.

Ask: **How many ones are in each group?** 6 **How many tens are in each group?** 1

Recall that to multiply, always begin with the ones digits. Have students look at Step 2.

Ask: **Do you need to regroup? Explain.** Yes, 2 x 6 ones is 12 ones; regroup 12 ones as 1 ten 2 ones. **What do you do with the regrouped ten?** Put it with the other tens blocks.

Continue with similar questioning for Step 3. Help students understand the connection between the pictorial models and the symbols that represent the multiplication.

TRY THESE Exercises 1–3 model regrouping ones as tens for multiplication. You may wish to work through each exercise with the students to make sure they understand regrouping tens as hundreds.

- **Exercise 1** Regroup 15 ones as 1 ten 5 ones.
- **Exercise 2** Regroup 21 ones as 2 tens 1 one, and 11 tens as 1 hundred 1 ten.
- **Exercise 3** Regroup 16 tens as 1 hundred 6 tens.

PRACTICE ON YOUR OWN Review the examples with students. Ask them to tell the steps to find the product.

CHECK Determine if students understand the meaning of multiplying a 2-digit number by a 1-digit number.

Success is indicated by 3 out of 3 correct responses.

Students who successfully complete the **Practice on Your Own** and **Check** are ready to move on to the next skill.

COMMON ERRORS

- Some students may write products without regrouping.
- Some students may forget to add the regrouped number.

Students who made more than four errors in the **Practice on Your Own**, or who were not successful in the **Check** section, may also benefit from the **Alternative Teaching Strategy** on the next page.

Optional

Alternative Teaching Strategy

Model Multiplication of 2-Digit Numbers

10 Minutes

OBJECTIVE Use place-value blocks to model multiplication of a 2-digit number by a 1-digit number

MATERIALS place-value blocks

Provide several exercises, such as:

2 × 14

3 × 15

4 × 18

Ask students to model the first exercise, which does not involve regrouping. Using the place-value blocks, keep the ones aligned with ones, and tens with tens.

Record the multiplication in a place-value grid on a large piece of paper.

Tens	Ones
1	4
x	2

Have students combine the ones, then the tens, to find the product.

Then ask the students to summarize the steps for multiplying.

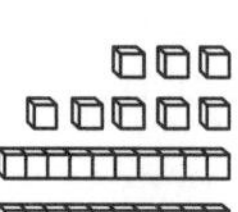

Next ask students to model 3 × 15.

As students combine the ones, say:

Now you have 15 ones. How can 15 ones be regrouped? 1 ten 5 ones **Where do you place the regrouped ten?** with the other tens

Have students complete the modeling and record the multiplication in a place-value grid.

Tens	Ones
1	
1	5
x	3
4	5

The third example involves regrouping 32 ones as 3 tens 2 ones. Use similar questioning as students model the multiplication.

When you feel confident that students understand the modeling and the recording steps, have them try 3 × 17 on their own. Monitor the way they regroup with the blocks and how they record the multiplication.

Model Multiplication (2-Digit by 1-Digit Numbers)

Grade 4 **Skill 27**

Name ______________________ Skill ______________________

Find $2 \times 16 = ■$.

Step 1 Make 2 groups of 16.

$$\begin{array}{r} 16 \\ \times\ 2 \\ \hline \end{array}$$

Step 2 Combine the ones. Regroup.

$$\begin{array}{r} \boxed{1} \\ 1\ 6 \\ \times\ \ 2 \\ \hline 2 \end{array}$$

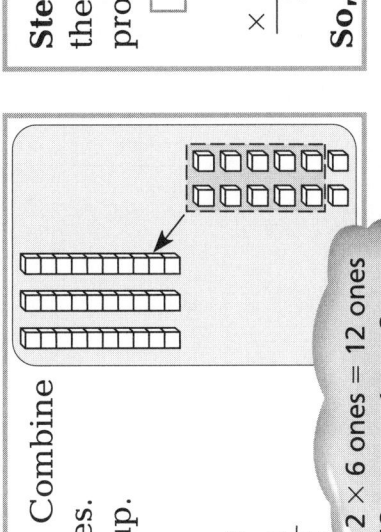

2 × 6 ones = 12 ones
12 ones = 1 ten 2 ones

Step 3 Multiply the tens. Add the regrouped ten. Record the product.

$$\begin{array}{r} \boxed{1} \\ 1\ 6 \\ \times\ \ 2 \\ \hline 3\ 2 \end{array}$$

2 × 1 ten = 2 tens
2 tens + 1 ten = 3 tens

So, $16 \times 2 = 32$

Try These

Find the product.

1. 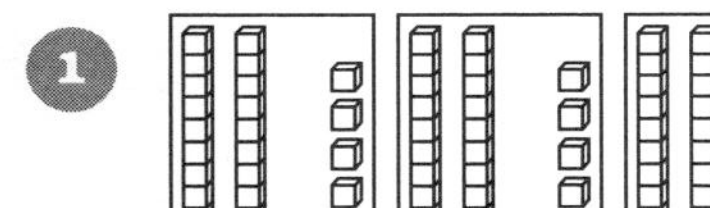

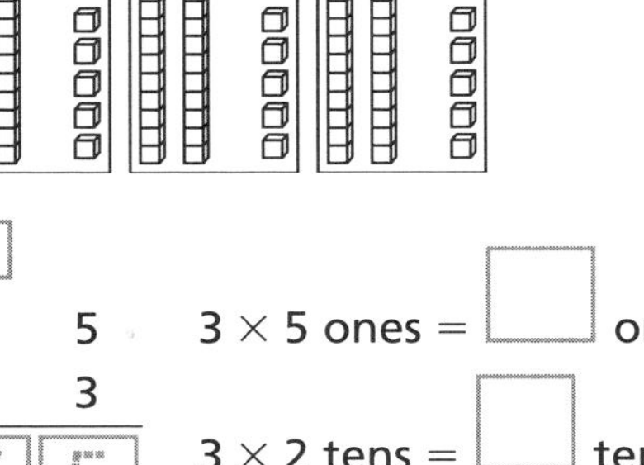

$$\begin{array}{r} \boxed{1} \\ 2\ 5 \\ \times\ \ 3 \\ \hline 7\ 5 \end{array}$$

3×5 ones = ☐ ones

3×2 tens = ☐ tens

2. $$\begin{array}{r} \square \\ 3\ 7 \\ \times\ \ 3 \\ \hline \square\square\square \end{array}$$

3×7 ones = ☐ ones

3×3 tens = ☐ tens

3. $$\begin{array}{r} 4\ 0 \\ \times\ \ 4 \\ \hline \square\square\square \end{array}$$

4×0 ones = ☐ ones

4×4 tens = ☐ tens

Go to the next side.

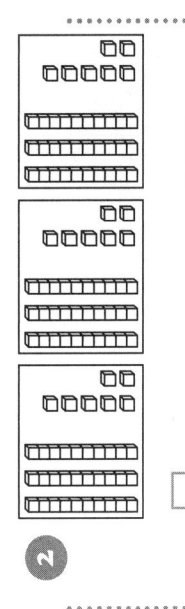

Name ______________________ Skill ______________________

Practice on Your Own

Skill 27

Find 5 × 35 = ■.

Think: Do I need to regroup?

5 × 5 ones = 25 ones
25 ones = 2 tens 5 ones
5 × 3 tens = 15 tens
15 tens + 2 tens = 17 tens

$$\begin{array}{r} \boxed{2} \\ 35 \\ \times\ 5 \\ \hline 175 \end{array}$$

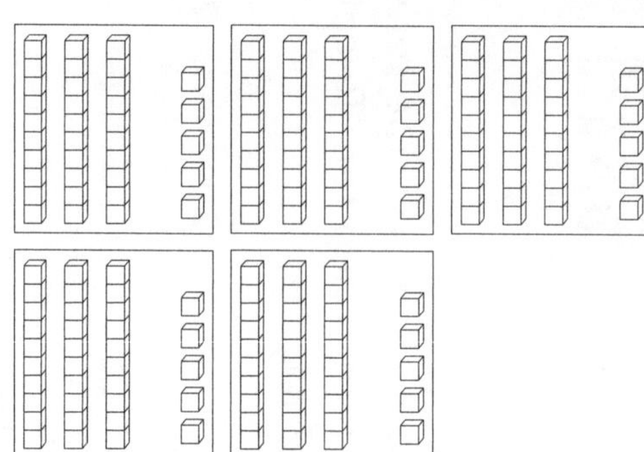

Find the product.

1.

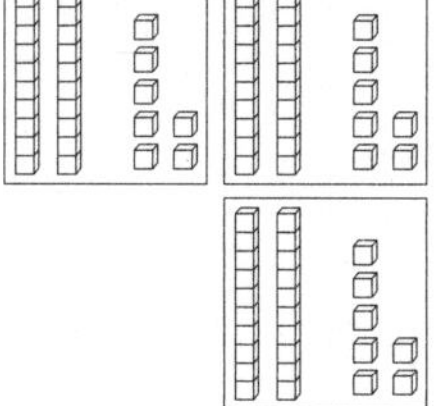

$$\begin{array}{r} \square \\ 2\ 7 \\ \times\ \ 3 \\ \hline \square\square \end{array}$$

2.

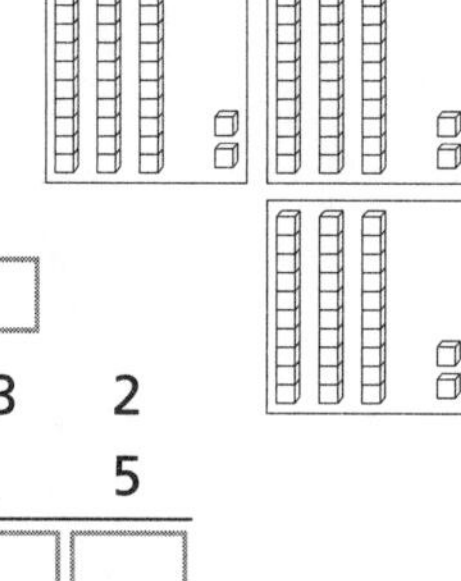

$$\begin{array}{r} \square \\ 3\ 2 \\ \times\ \ 5 \\ \hline \square\square\square \end{array}$$

3.

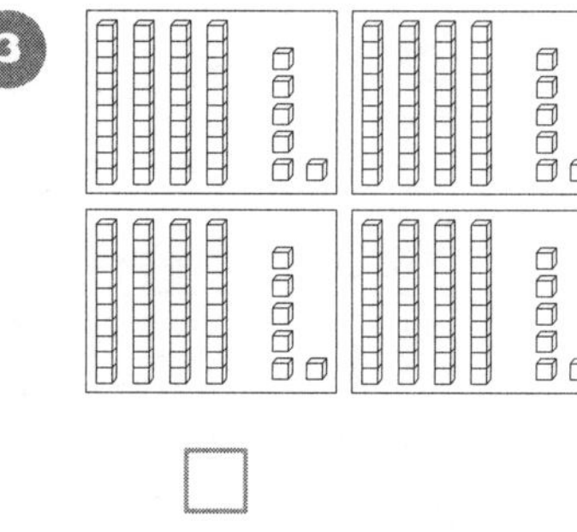

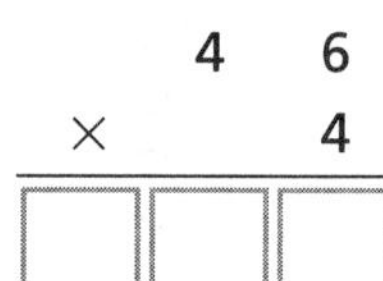

$$\begin{array}{r} \square \\ 4\ 6 \\ \times\ \ 4 \\ \hline \square\square\square \end{array}$$

4.

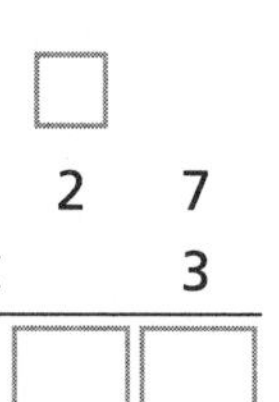

$$\begin{array}{r} \square \\ 1\ 5 \\ \times\ \ 3 \\ \hline \square\square \end{array}$$

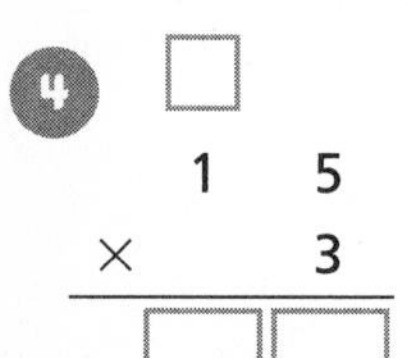

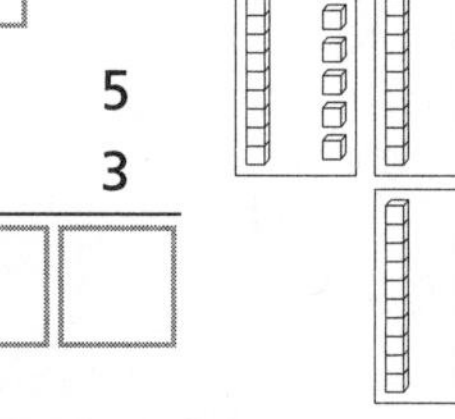

5.

$$\begin{array}{r} \square \\ 2\ 3 \\ \times\ \ 4 \\ \hline \square\square \end{array}$$

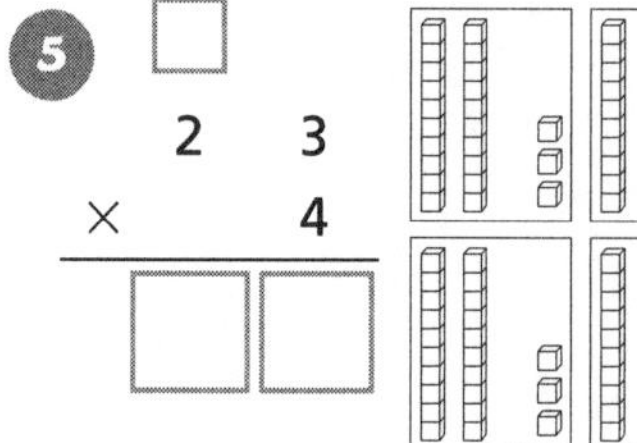

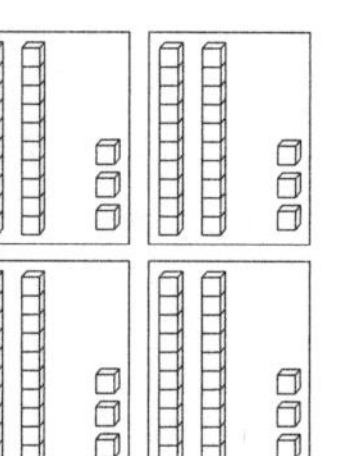

6.

$$\begin{array}{r} \square \\ 3\ 7 \\ \times\ \ 2 \\ \hline \square\square \end{array}$$

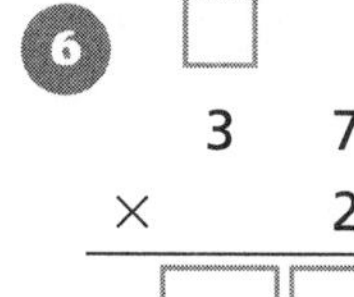

Check

7.

$$\begin{array}{r} \square \\ 3\ 9 \\ \times\ \ 2 \\ \hline \square\square \end{array}$$

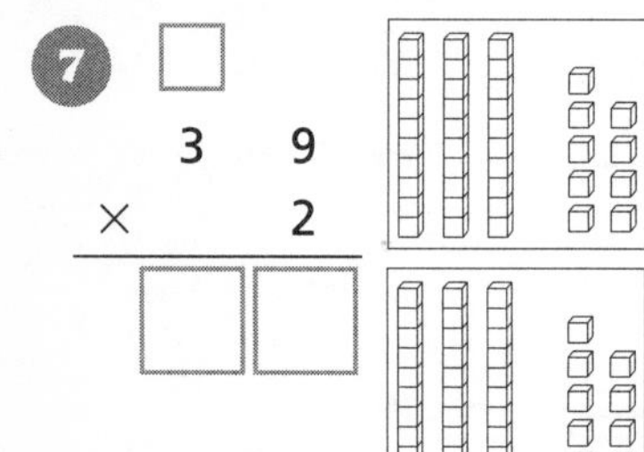

8.

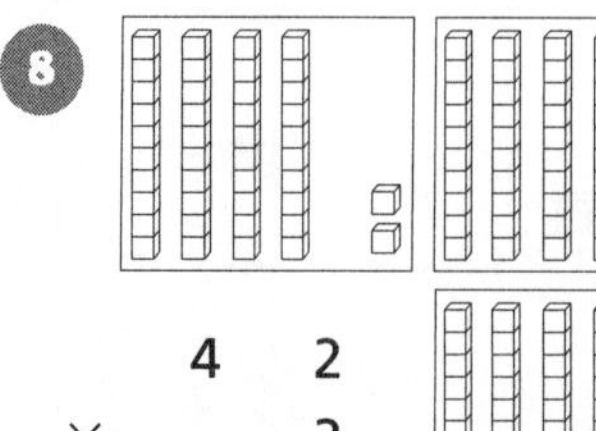

$$\begin{array}{r} 4\ 2 \\ \times\ \ 3 \\ \hline \square\square\square \end{array}$$

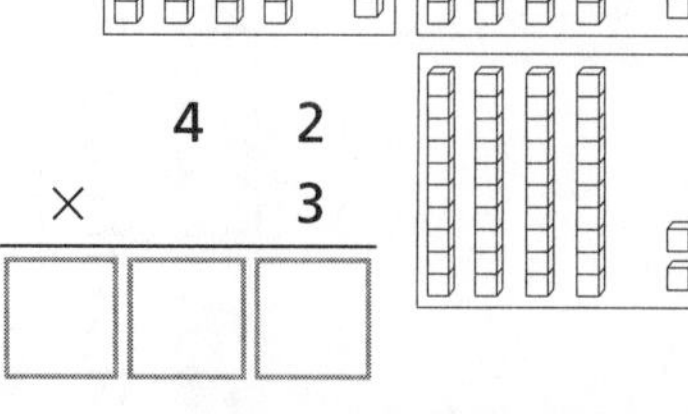

9.

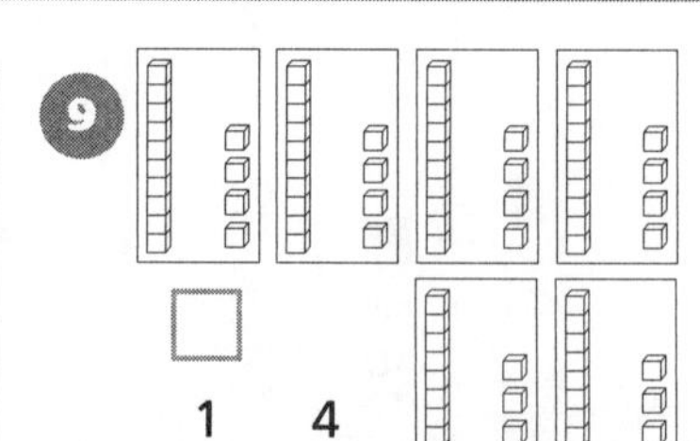

$$\begin{array}{r} \square \\ 1\ 4 \\ \times\ \ 6 \\ \hline \square\square \end{array}$$

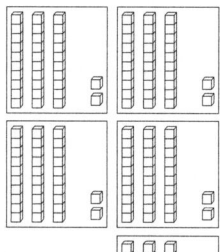

Skill 28 Grade 4

Record Multiplication

OBJECTIVE Multiply a 2-digit number by a 1-digit number and record the result

Begin by pointing to the place-value labels on the grids. You may wish to have the students use labels on their written work to help them remember the value of the digits when they regroup.

Direct students' attention to Step 1.

Ask: **What digit is in the ones place in the number 56?** 6 **What digit is in the ones place in the number you multiply by?** 2 **When you multiply 2 ones by 6 ones, what do you get?** 12 ones **Do you have to regroup the ones?** Yes, 12 ones is 1 ten 2 ones **What do you multiply in the next step?** The 5 tens by the 2 ones

Continue to ask similar questions as you work through Steps 2 and 3. Draw attention to the method of recording the partial products in Steps 2 and 3.

TRY THESE Exercises 1–3 model the type of exercises students will find on the **Practice on Your Own** page.

- **Exercise 1** Regroup tens.
- **Exercise 2** Regroup ones.
- **Exercise 3** Regroup ones and tens.

PRACTICE ON YOUR OWN Review the example at the top of the page. Ask students to explain the multiplication and regrouping process using the terms *ones*, *tens*, and *hundreds*.

CHECK Determine if the students know how to regroup ones as tens and tens as hundreds and can record the partial products in the correct positions.

Success is determined by 3 out of 4 correct responses.

Students who successfully complete the **Practice on Your Own** and **Check** are ready to move on to the next skill.

COMMON ERRORS

- Students may forget to multiply the tens in the first factor and think that the first partial product is the final product.
- Students may forget to regroup the tens.

Students who made more than three errors in the **Practice on Your Own**, or who were not successful in the **Check** section, may benefit from the **Alternative Teaching Strategy** on the next page.

Optional

Alternative Teaching Strategy
Model and Record Multiplication

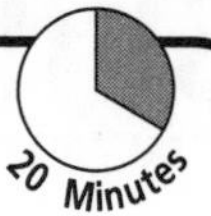

OBJECTIVE Use base-ten blocks to model multiplying a 2-digit number by a 1-digit number, recording the results

MATERIALS base-ten blocks, at least 10 tens and 50 ones, place-value tables

You may wish to have prepared and copied blank place-value tables for multiplication, as shown below.

100's	10's	1's
×		

Distribute the base-ten blocks. Have students model 4 × 23 and record the factors in the table.

Ask: **What is 4 times 3 ones?** 12 ones **Can you regroup the ones?** Yes, 12 ones = 1 ten 2 ones

100's	10's	1's
	2	3
×		4
	1	2
+	8	0
	9	2

Have the students regroup and record the first partial product in the table.

Draw attention to the tens rods.

Ask: **What is 4 times 2 tens?** 8 tens

Have students record the second partial product in the table. Then ask a student to add to find the final product. 92

Repeat the activity with other similar examples. When the students show understanding of the regrouping process, remove the base-ten blocks and have them try an exercise using only paper and pencil. Ask students to explain each step of the process as they do the multiplication.

Name ______________________ Skill ______________________

Grade 4 Skill 28

Record Multiplication

Find 2 × 56 = ■.

Step 1 Multiply the ones.

	Hundreds	Tens	Ones
		5	6
×			2
		1	2

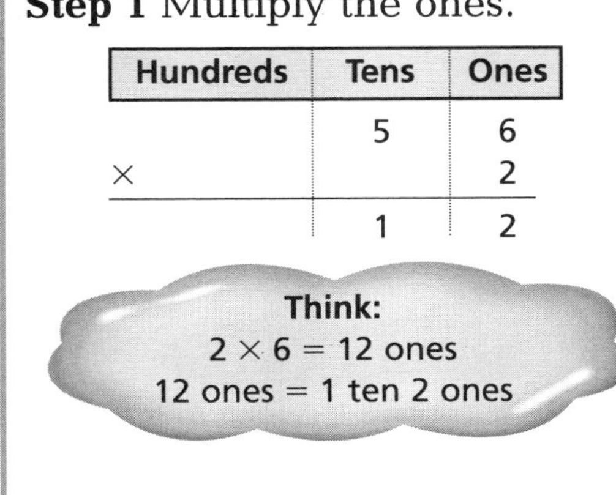

Think:
2 × 6 = 12 ones
12 ones = 1 ten 2 ones

Step 2 Multiply the tens.

	Hundreds	Tens	Ones
		5	6
×			2
		1	2
+	1	0	0

Think:
2 × 5 tens = 10 tens
10 tens = 1 hundred

Step 3 Add to find the product.

	Hundreds	Tens	Ones
		5	6
×			2
		1	2
+	1	0	0
	1	1	2

So, 2 × 56 = 112.

Try These

Find the product.

1

	Hundreds	Tens	Ones	
		6	4	
×			2	
			8	(2 × 4 ones)
+	1	2	0	(2 × 6 tens)
	☐	☐	☐	

2

	Hundreds	Tens	Ones	
		3	5	
×			2	
		☐	☐	(2 × 5 ones)
+		☐	☐	(2 × 3 tens)
		☐	☐	

3

	Hundreds	Tens	Ones	
		4	3	
×			7	
		☐	☐	(7 × 3 ones)
+	☐	☐	☐	(7 × 4 tens)
	☐	☐	☐	

Go to the next side.

Name ______________________ Skill ______________________

Practice on Your Own

Find 2 × 28 = ■.

a. Multiply the ones.

Tens	Ones	
2	8	
×	2	
1	6	(2 × 8 ones)

b. Multiply the tens.

Tens	Ones	
2	8	
×	2	
1	6	
4	0	(2 × 2 tens)

c. Add to find the product.

Tens	Ones
2	8
×	2
1	6
+ 4	0
5	6

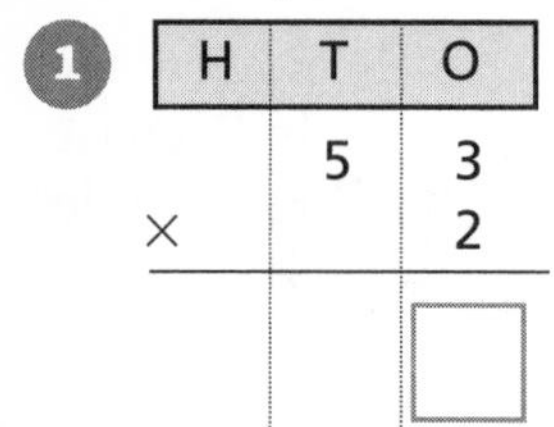

Find the product.

1.

H	T	O	
	5	3	
×		2	
		☐	(2 × 3 ones)
☐	☐	☐	(2 × 5 tens)
☐	☐	☐	

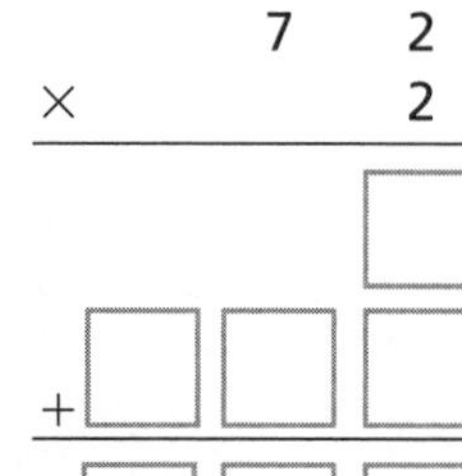

2.

H	T	O	
	3	2	
×		3	
		☐	(3 × 2 ones)
	☐	☐	(3 × 3 tens)
	☐	☐	

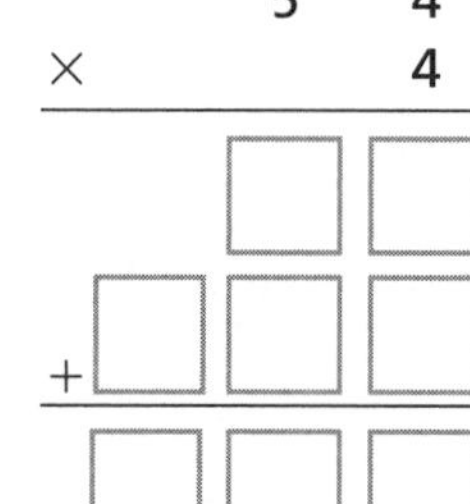

3.

H	T	O	
	3	5	
×		5	
	☐	☐	(5 × 5 ones)
☐	☐	☐	(5 × 3 tens)
☐	☐	☐	

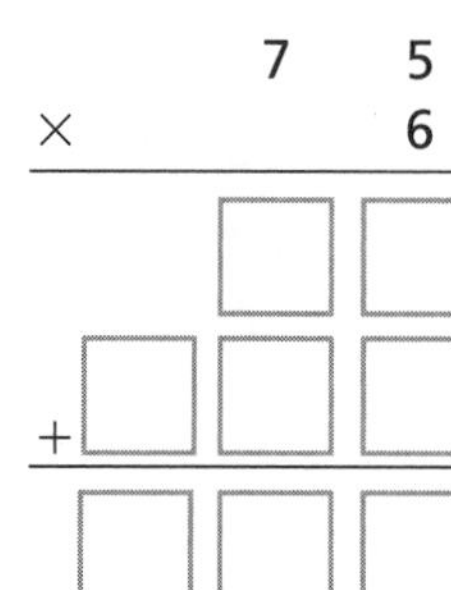

4. 72 × 2

5. 54 × 4

6. 75 × 6

7. 31 × 5

8. 42 × 7

9. 53 × 8

Check

Find the product.

10. 15 × 3

11. 57 × 5

12. 26 × 6

13. 64 × 7

Skill 29 Grade 4

Multiply by Multiples of 10

OBJECTIVE Use basic facts and patterns to multiply by multiples of 10

Explain to students that in this lesson they are asked to multiply by multiples of 10. Suggest that they can discover and use a pattern to multiply.

Have students look at the first number sentence in the first example, and note that it is a basic fact. $4 \times 1 = 4$ Direct their attention to the number of zeros in the next number sentence.

Say: **Each factor has a zero. How many zeros do the factors have altogether?** 2 **How many zeros are in the product?** 2

Repeat the questions for the next two number sentences.

Continue to ask similar questions as you work through the other examples. Have students note that each pattern begins with a basic multiplication fact, and that the pattern is the same in each example. Emphasize that they can use a basic fact and the pattern to multiply numbers by multiples of 10.

TRY THESE Exercises 1–3 model the types of exercises students will encounter on the **Practice on Your Own** page.

- **Exercise 1** Use the basic fact 2×1 and multiples of 10 to find a pattern.
- **Exercise 2** Use the basic fact 9×1 and multiples of 10 to find a pattern.
- **Exercise 3** Use the basic fact 3×1 and multiples of 10 to find a pattern.

PRACTICE ON YOUR OWN Review the example at the top of the page. In Exercises 1–6, students use a basic fact and multiples of 10 to find a pattern. In Exercises 7–14, they find the products of multiples of 10.

CHECK Determine if students can multiply by multiples of 10.

Success is indicated by 3 out of 4 correct responses.

Students who successfully complete the **Practice on Your Own** and **Check** are ready to move on to the next skill.

COMMON ERRORS

- Students may write the incorrect number of zeros in the product when multiplying by multiples of 10.
- Students may not know multiplication facts.

Students who made more than three errors in the **Practice on Your Own**, or who were not successful in the **Check** section, may benefit from the **Alternative Teaching Strategy** on the next page.

Optional

Alternative Teaching Strategy

Model Multiplying by Multiples of 10

20 Minutes

OBJECTIVE Use number cards to multiply by multiples of 10

MATERIALS index cards

Prepare sets of cards ahead of time or have students prepare the cards. On separate cards write the numerals 1 through 9, 5 zeros (one 0 per card), and the symbols +, =, and,.

Distribute two sets of cards to pairs of students. Begin by displaying these multiplication sentences and recalling the meaning of multiples of 10.

$6 \times 1 = 6$

$60 \times 10 = 600$

$60 \times 100 = 6{,}000$

$60 \times 1{,}000 = 60{,}000$

Point out the basic multiplication fact 6×1. Explain that in this pattern 60 is multiplied by 10, 100, 1,000, which are multiples of 10. Ask students to identify any patterns they see.

Then display 30×100 and have one partner show the expression with the cards.

Ask: **What basic multiplication fact will help you find the product?** $3 \times 1 = 3$ **How many zeros are in the factors?** 3 **How many zeros will the product have?** 3

Suggest that the second partner remove the zeros from the expression and use the three zeros from the factors to show the product of 30×100. **What is the product?** 3,000

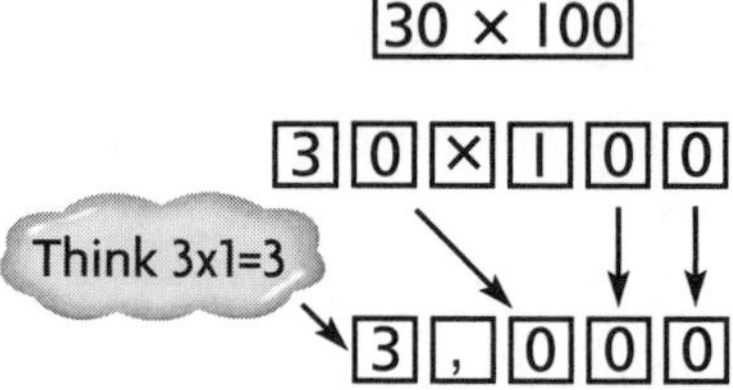

Repeat the activity for 300×100 and $3{,}000 \times 100$. Continue with other examples until students have internalized the pattern. Then have them multiply by multiples of 10 without the cards.

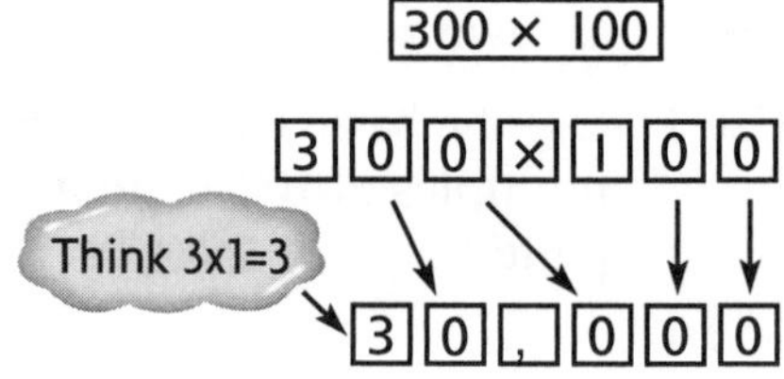

Name ____________________ Skill ____________________

Grade 4 Skill 29

Multiply by 10, 100, and 1,000

Use basic facts and patterns to help multiply by multiples of 10.

Find a pattern

Multiply by tens.

$4 \times 1 = 4$ ← basic fact
$40 \times 10 = 400$ ← 2 zeros
$40 \times 100 = 4,000$ ← 3 zeros
$40 \times 1,000 = 40,000$ ← 4 zeros

↑ ↑ factors ↑ product

The number of zeros in the products increases as the number of zeros in the factors increases.

Multiply by Hundreds

$6 \times 1 = 6$ ← basic fact
$600 \times 10 = 6,000$ ← 3 zeros
$600 \times 100 = 60,000$ ← 4 zeros
$600 \times 1,000 = 600,000$ ← 5 zeros

Multiply by Thousands

$7 \times 1 = 7$ ← basic fact
$7,000 \times 10 = 70,000$ ← 4 zeros
$7,000 \times 100 = 700,000$ ← 5 zeros
$7,000 \times 1,000 = 7,000,000$ ← 6 zeros

Try These

Find the products.

1. $2 \times 1 =$ ☐
   $20 \times 10 =$ ☐00
   $20 \times 100 =$ ☐000
   $20 \times 1,000 =$ ☐0,000

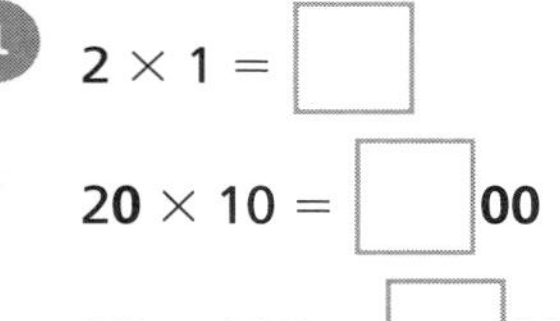

2. $9 \times 1 =$ ☐
   $90 \times 10 =$ ☐
   $90 \times 100 =$ ☐
   $90 \times 1,000 =$ ☐

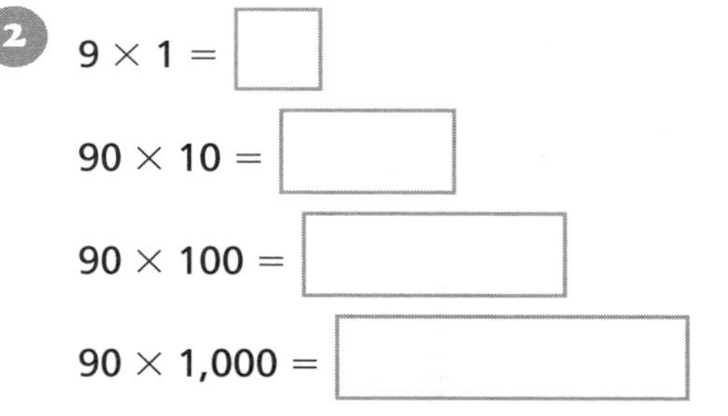

3. $3 \times 1 =$ ☐
   $300 \times 10 =$ ☐
   $300 \times 100 =$ ☐
   $300 \times 1,000 =$ ☐

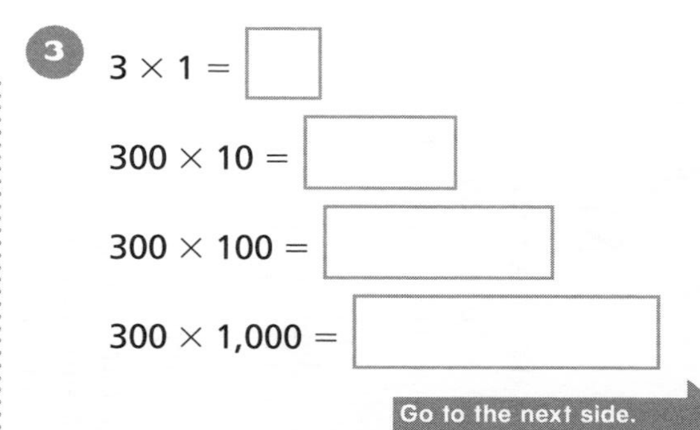

Go to the next side.

Name ____________________ Skill ____________________

Practice on Your Own

Skill 29

$8 \times 1 = 8$ ← basic fact
$800 \times 10 = 8{,}000$
$800 \times 100 = 80{,}000$
$800 \times 1{,}000 = 800{,}000$

Think:
You can use basic facts and patterns to multiply by multiples of 10.

Find the product.

1. $3 \times 1 =$ ☐
   $30 \times 10 =$ ☐
   $30 \times 100 =$ ☐
   $30 \times 1{,}000 =$ ☐

2. $2 \times 1 =$ ☐
   $200 \times 10 =$ ☐
   $200 \times 100 =$ ☐
   $200 \times 1{,}000 =$ ☐

3. $6 \times 1 =$ ☐
   $6{,}000 \times 10 =$ ☐
   $6{,}000 \times 100 =$ ☐
   $6{,}000 \times 1{,}000 =$ ☐

4. $5 \times 1 =$ ☐

   $50 \times 10 =$ ☐
   $50 \times 100 =$ ☐
   $50 \times 1{,}000 =$ ☐

5. $1 \times 1 =$ ☐
   $100 \times 10 =$ ☐
   $100 \times 100 =$ ☐
   $100 \times 1{,}000 =$ ☐

6. $2 \times 1 =$ ☐
   $2{,}000 \times 10 =$ ☐
   $2{,}000 \times 100 =$ ☐
   $2{,}000 \times 1{,}000 =$ ☐

7. 10×30

8. 400×100

9. $1{,}000 \times 20$

10. $2{,}000 \times 100$

11. $10 \times 4 =$ ☐

12. $500 \times 10 =$ ☐

13. $100 \times 900 =$ ☐

14. $1{,}000 \times 200 =$ ☐

Check

Find the product.

15. $9 \times 10 =$ ☐

16. $100 \times 60 =$ ☐

17. $1{,}000 \times 400$

18. $6{,}000 \times 100$

Skill 30 Grade 4 Multiply 2- and 3-Digit Numbers by 1-Digit Numbers Regrouping Once

OBJECTIVE Multiply 2-digit and 3-digit numbers by 1-digit numbers

MATERIALS base-ten blocks

You may wish to use the base-ten blocks to have the students model each step in the process.

Begin by directing students' attention to Step 1 and having them read the place-value labels.

Ask: **How many digits does the first number have?** 3 Point out that the top number has three digits.

Ask: **In what place do you begin multiplying?** ones place **When you multiply 2 × 6 ones, do you have to regroup?** Yes, 2 × 6 ones = 12 ones. **How many tens and ones is 12 ones?** 1 ten 2 ones **Where do you write the 2 ones in the product?** in the ones place **Where do you write the 1 ten that you regrouped?** above the 2 in the tens place

Continue to ask similar questions as you work through Step 2 and Step 3 with the students.

TRY THESE Exercises 1–4 model the type of exercises students will find on the **Practice on Your Own** page.

- **Exercises 1–2** Multiply a 2-digit number by a 1-digit number.
- **Exercises 3–4** Multiply a 3-digit number by a 1-digit number.

PRACTICE ON YOUR OWN Review the example at the top of the page. Ask students to explain how to regroup the 24 ones. 2 tens 4 ones Draw attention to the position of the 4 in the product and the regrouped digit in the tens place.

CHECK Determine if students know where to write the digits after they regroup and how to regroup the product. Success is determined by 3 out of 4 correct responses.

Students who successfully complete the **Practice on Your Own** and **Check** are ready to move on to the next skill.

COMMON ERRORS

- Students may add the digits they regrouped *before* they multiply, or they may forget to add the digits they regrouped *after* they multiply.
- Students may record a product such as 6 × 24 = 144 as 1224.

Students who made more than four errors in the **Practice on Your Own**, or who were not successful in the **Check** section, may benefit from the **Alternative Teaching Strategy** on the next page.

Optional

Alternative Teaching Strategy

15 Minutes

Model Multiplying 2- and 3-Digit Numbers by 1-Digit Numbers

OBJECTIVE Use base-ten blocks to model multiplying 2-digit and 3-digit numbers by 1-digit numbers

MATERIALS base-ten blocks

You may wish to have the students work in pairs. One student models the multiplication with base-ten blocks, while the other student records each step with paper and pencil.

Have the students write the multiplication in a place-value grid as shown.

Tens	Ones
2	5
×	3

Distribute the base-ten blocks.

Ask: **What does 3 × 25 mean?** 3 groups of 25 **How can you model this with the base-ten blocks?** Make 3 groups of 25 each.

Have the students use the base-ten blocks to model 3 × 25.

Say: **Begin with the ones. What is 3 × 5 ones?** 15 ones

Ask: **Do you have more than 10 ones?** Yes **What does having more than 10 ones mean you have to do?** Regroup the ones. **What do you get when you regroup the ones?** 1 ten 5 ones **How can you show that in the multiplication place-value grid?** Put a 5 in the ones place in the product and a 1 in the tens place above the 2.

Now look at the tens. What is 3 × 2 tens? 6 tens **What do you do with the regrouped ten?** Add it to 6 tens; 6 tens + 1 ten = 7 tens **Do you need to regroup the tens?** No, there are only 7 tens.

Have the students record the final product. 75

Repeat this activity with similar examples. When students show proficiency with the multiplication process, have them try an exercise using paper and pencil only.

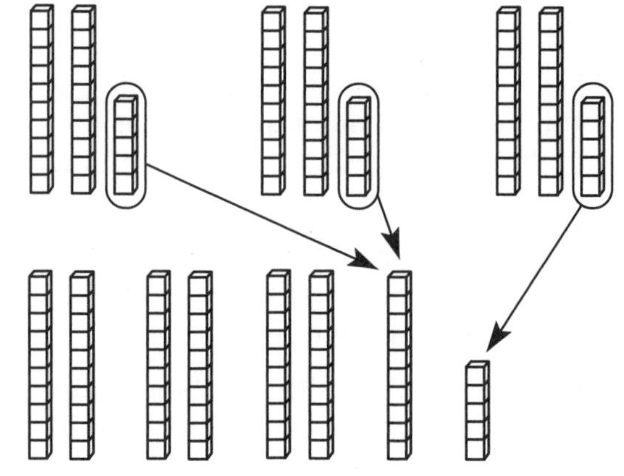

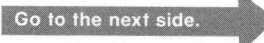

Name ______________________ Skill ______________________

Multiply 2- and 3-Digit Numbers by 1-Digit Numbers Regrouping Once

Grade 4 **Skill 30**

Find $2 \times 126 = \blacksquare$.

Step 1
Multiply the ones. $2 \times 6 = 12$ ones.
Regroup 12 ones as 1 ten 2 ones.

	Hundreds	Tens	Ones
		1	
	1	2	6
×			2
			2

Step 2
Multiply the tens. 2×2 tens = 4 tens. Add the regrouped ten.
4 tens + 1 ten = 5 tens.

	Hundreds	Tens	Ones
		1	
	1	2	6
×			2
		5	2

Sometimes when you multiply, you may need to regroup.

Step 3
Multiply the hundreds.
2×1 hundred = 2 hundreds.

	Hundreds	Tens	Ones
		1	
	1	2	6
×			2
	2	5	2

So, $2 \times 126 = 252$.

Try These

Find the product.

1.

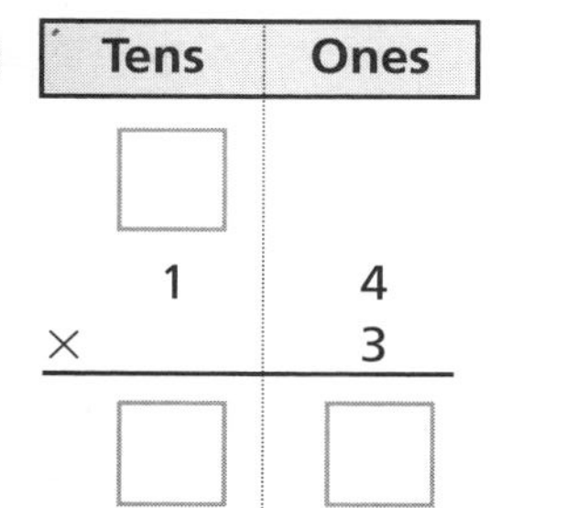

	Tens	Ones
	☐	
	1	4
×		3
	☐	☐

2.

	Tens	Ones
	☐	
	3	8
×		2
	☐	☐

3.

	Hundreds	Tens	Ones
		☐	
	1	1	7
×			4
	☐	☐	☐

4.

	Hundreds	Tens	Ones
		☐	
	1	2	4
×			3
	☐	☐	☐

Go to the next side.

Name ______________________ Skill ______________________

Practice on Your Own

Skill 30

Multiply the ones. Regroup.

Tens	Ones
[2]	
2	4
×	6
	4

Think: Do I need to regroup?

6×4 ones = 24 ones

24 ones = 2 tens 4 ones

Multiply the tens. Add the regrouped tens.

Hundreds	Tens	Ones
	[2]	
	2	4
×		6
1	4	4

6×2 tens = 12 tens

12 tens + 2 tens = 14 tens

14 tens = 1 hundred 4 tens

So, $6 \times 24 = 144$.

Find the product.

1. 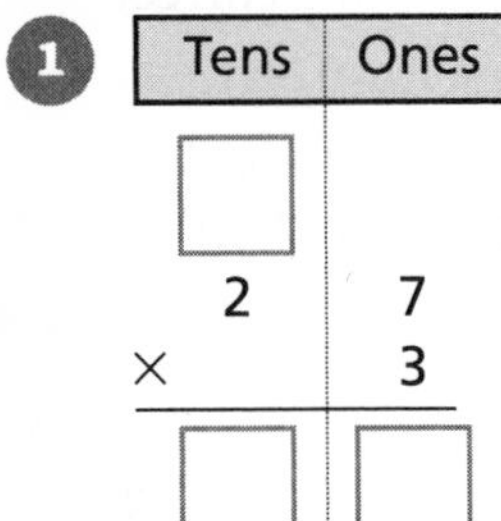

2. 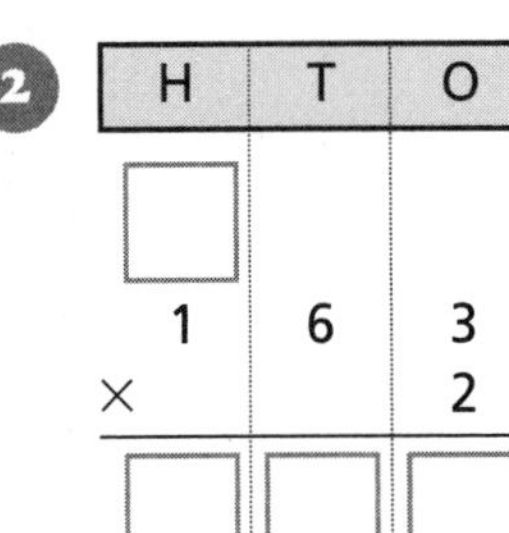

3. 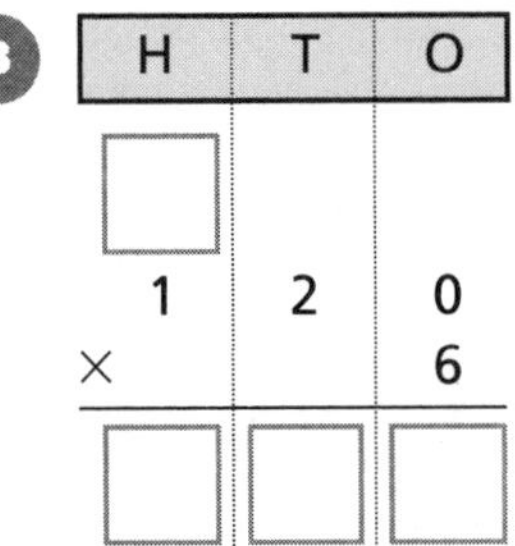

4.

H	T	O
	[ ]	
2	1	6
×		4
[ ]	[ ]	[ ]

5. 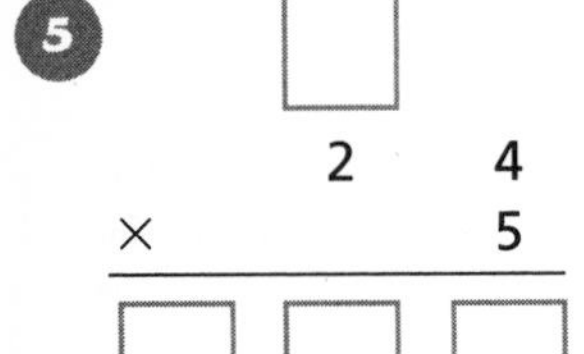

6. 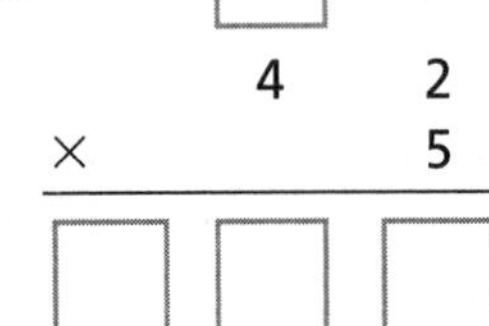

7. 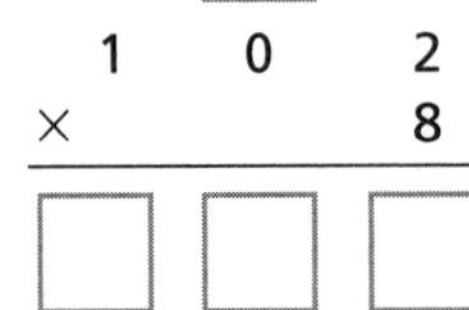

8. 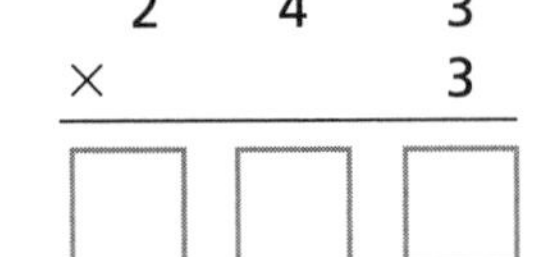

9. 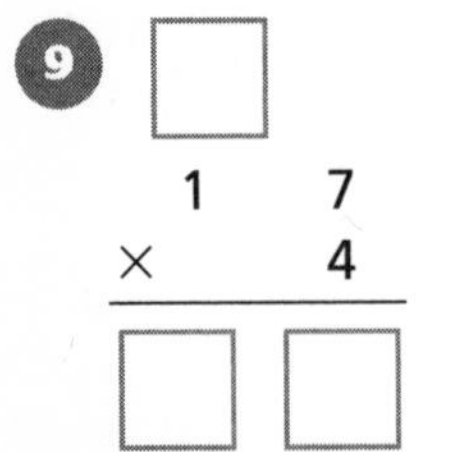

10. 75×3

11. 162×2

12. 353×2

Check

Find the product.

13. 23×4

14. 131×5

15. 345×2

16. 225×3

Grade 4

Multiply 3-Digit Numbers by 1-Digit Numbers Regrouping Twice

20 Minutes

OBJECTIVE Multiply three-digit numbers by one-digit numbers, regrouping twice

Direct the student's attention to Step 1. Discuss the place value of each digit, and point out that the labels may help them remember the value of the digits when they regroup.

As you work though the regrouping procedure, emphasize the need to place the regrouped digit in the appropriate column, and remind students to add the digit as they multiply.

Say: **Begin with the ones. What is 2 × 9 ones?** 18 ones **Do you need to regroup?** Yes, 18 ones is regrouped as 1 ten 8 ones. **Where do you place the regrouped ten?** in the tens column.

Then have students look at Step 2.

Continue: **Multiply the tens. What is 2 × 6 tens?** 12 tens **What do you do with the ten you regrouped?** Add it to the 12 tens. **How many tens are there now?** 13 tens **Do you need to regroup?** Yes, 13 tens is regrouped as 1 hundred 3 tens. **Where do you put the regrouped hundred?** in the hundreds column

Continue to ask similar questions as you work through Step 3.

TRY THESE Exercises 1–4 provide practice multiplying 3-digit numbers by one digit numbers.

- **Exercises 1–4** Provide place-value labels and regrouping boxes to help students find the 3-digit product, regrouping twice.

PRACTICE ON YOUR OWN Review the example at the top of the page. As students multiply, ask them to explain when to regroup a digit, and in what place to put the regrouped digit.

CHECK Determine if students know how to multiply a three-digit number by a one-digit number and can regroup twice. Success is indicated by 3 out of 4 correct responses.

Students who successfully complete the **Practice on Your Own** and **Check** are ready to move on to the next skill.

COMMON ERRORS

- Students may place the regrouped digit in the incorrect place.
- Students may forget to add the regrouped digit.
- Students may multiply the digits in the hundreds or tens place before multiplying the digits in the ones place.

Students who made more than three errors in the **Practice on Your Own**, or were not successful in the **Check** section, may benefit from the **Alternative Teaching Strategy** on the next page.

Optional

20 Minutes

Alternative Teaching Strategy

Model Multiplying 3-Digit Numbers

OBJECTIVE Multiply 3-digit numbers regrouping both ones and tens

MATERIALS base-ten blocks, pencil, paper

Distribute the base-ten blocks. You may wish to demonstrate the regrouping procedure.

Have students use the base-ten blocks to show how to regroup 14 ones as 1 ten 4 ones. Explain that the blocks were regrouped but the value of the number that the blocks represent is the same, since 14 ones is equal to 1 ten 4 ones.

Then have students use the base-ten blocks to regroup 21 tens as 2 hundreds 1 ten.

Have students model 2 × 176 by showing 2 groups of 176.

Say: **Begin with the ones. How many ones do you have in 2 groups of 6 ones?** 12 ones **You have more than 9 ones. When you have more than 9 in any place, what can you do?** Regroup **How do you regroup the ones?** Regroup 12 ones as 1 ten 2 ones.

How many tens do you have in 2 groups of 7 tens? 14 tens **What happens to the regrouped ten?** It is added to the 14 tens; 14 + 1 = 15 tens **Can you regroup tens?** Yes, there are more than 9 tens; regroup 15 tens as 1 hundred 5 tens. **How many hundreds are in 2 groups of 1 hundred?** 2 hundreds **What do you do next?** Add the regrouped hundred; 2 + 1 = 3 hundreds. **What is the product?** 352

Have students repeat the modeling as they record the multiplication steps on paper. Stress the placement of the regrouped digit in the appropriate column.

Repeat the modeling and recording with different numbers. When students understand the regrouping process, have them multiply without using the base-ten blocks.

Model: 2 × 176

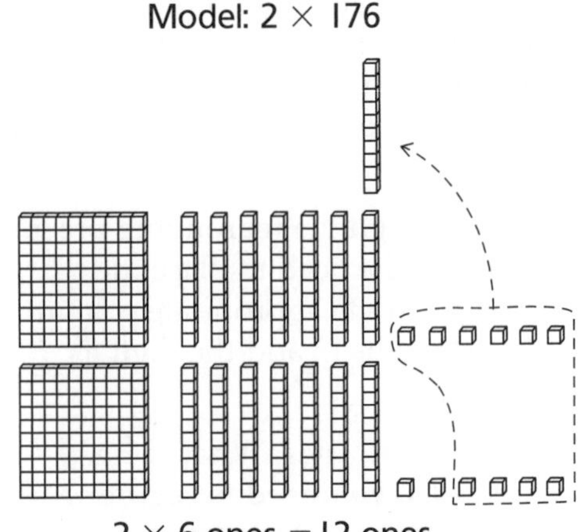

2 × 6 ones = 12 ones
12 ones = 1 ten 2 ones

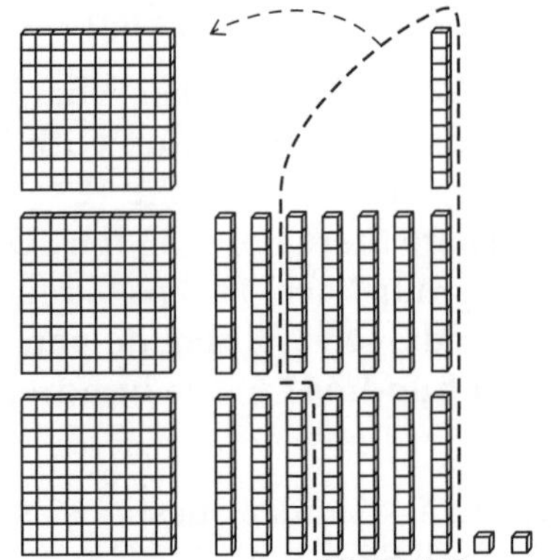

2 × 7 tens = 14 tens
14 tens + 1 ten = 15 tens
15 tens = 1 hundred 5 tens

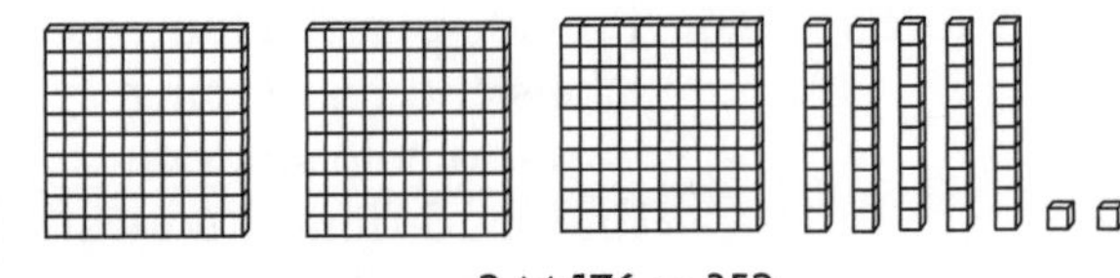

2 × 176 = 352

Name ____________________ Skill ____________________

Multiply 3-Digit Numbers by 1-Digit Numbers Regrouping Twice

Grade 4 Skill 31

Find 2 × 169 = ■.

Step 1 Multiply the ones.

Hundreds	Tens	Ones
	[1]	
1	6	9
×		2
		8

2 × 9 ones = 18 ones
18 ones = 1 ten 8 ones

Step 2 Multiply the tens. Add the regrouped ten. Regroup the tens as hundreds.

Hundreds	Tens	Ones
[1]	[1]	
1	6	9
×		2
	3	8

2 × 6 tens = 12 tens
12 tens + 1 ten = 13 tens
13 tens = 1 hundred 3 tens

Step 3 Multiply the hundreds. Add the regrouped hundred.

Hundreds	Tens	Ones
[1]	[1]	
1	6	9
×		2
3	3	8

2 × 1 hundred = 2 hundreds
2 hundreds + 1 hundred = 3 hundreds

When you multiply and regroup, check that you have added the regrouped digit.

Try These

Find the product.

1

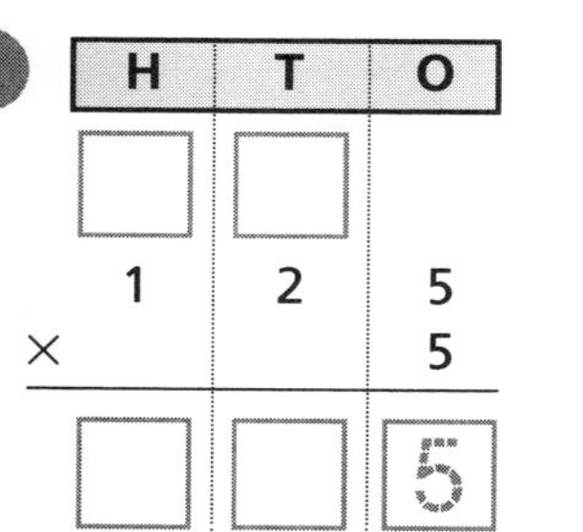

2

H	T	O
☐	☐	
2	7	9
×		2
☐	☐	☐

3

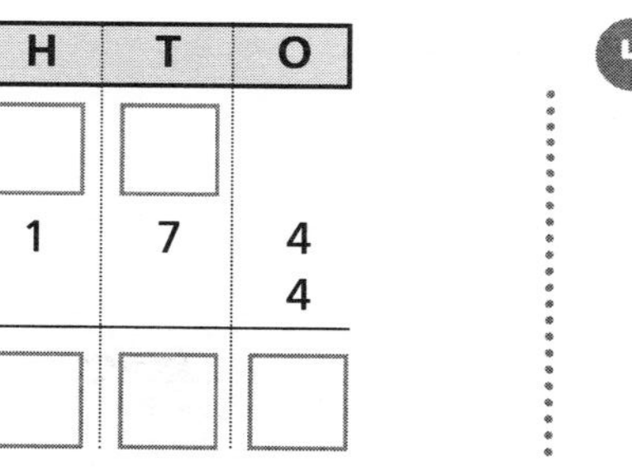

4

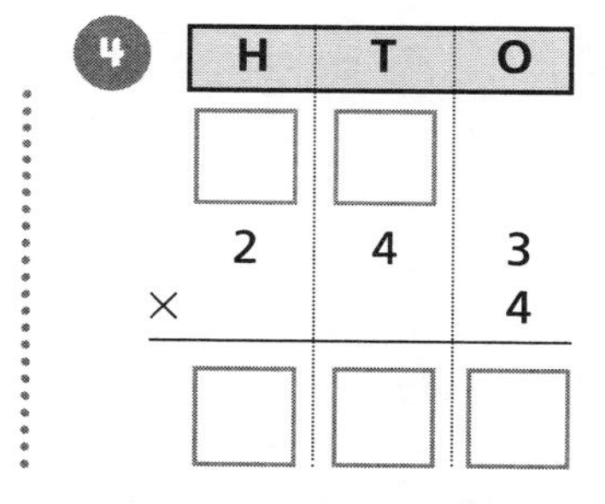

Go to the next side.

Name ______________________ Skill ______________________

Skill 31

Practice on Your Own

Think:
Multiply the ones. Regroup.
Multiply the tens. Regroup.
Multiply the hundreds.

Remember to regroup ones as tens. Then regroup tens as hundreds.

Hundreds	Tens	Ones
[1]	[3]	
1	3	6
×		5
6	8	0

Find the product.

1.
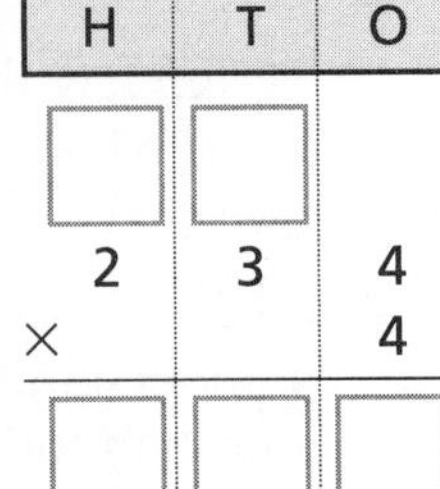

H	T	O
☐	☐	
2	3	4
×		4
☐	☐	☐

2.
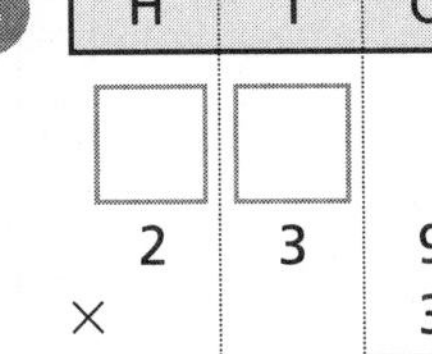

H	T	O
☐	☐	
2	3	9
×		3
☐	☐	☐

3.
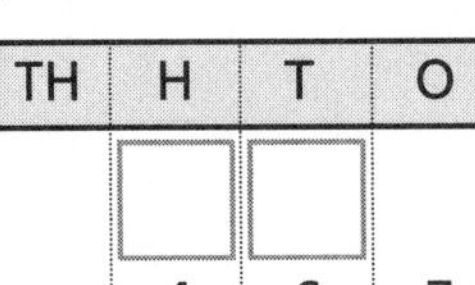

TH	H	T	O
	☐	☐	
	4	6	7
×			3
☐	☐	☐	☐

4.

TH	H	T	O
	☐	☐	
	3	4	6
×			5
☐	☐	☐	☐

5.

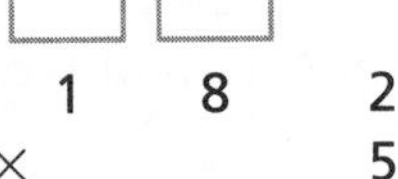

$$\begin{array}{r} 182 \\ \times \quad 5 \\ \hline \end{array}$$

6.
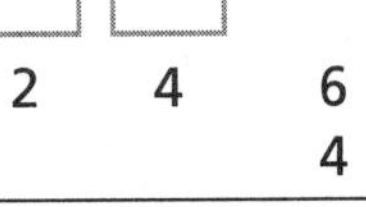

$$\begin{array}{r} 246 \\ \times \quad 4 \\ \hline \end{array}$$

7.
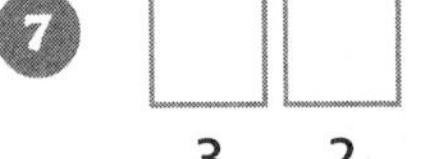
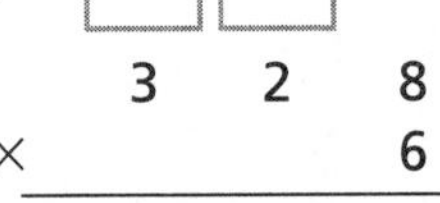

$$\begin{array}{r} 328 \\ \times \quad 6 \\ \hline \end{array}$$

8.
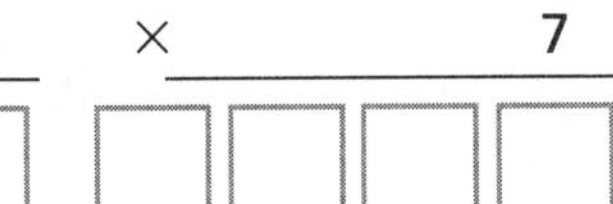

$$\begin{array}{r} 369 \\ \times \quad 7 \\ \hline \end{array}$$

9. $\begin{array}{r} 174 \\ \times \ 5 \\ \hline \end{array}$

10. $\begin{array}{r} 285 \\ \times \ 3 \\ \hline \end{array}$

11. $\begin{array}{r} 213 \\ \times \ 9 \\ \hline \end{array}$

12. $\begin{array}{r} 329 \\ \times \ 4 \\ \hline \end{array}$

Check

Find the product.

13. $\begin{array}{r} 182 \\ \times \ 6 \\ \hline \end{array}$

14. $\begin{array}{r} 159 \\ \times \ 4 \\ \hline \end{array}$

15. $\begin{array}{r} 254 \\ \times \ 3 \\ \hline \end{array}$

16. $\begin{array}{r} 287 \\ \times \ 5 \\ \hline \end{array}$

Skill 32 Grade 4

Estimate Products

OBJECTIVE Estimate products

You may wish to begin the lesson by discussing with students why estimating a product is useful. Use a real-life situation: Suppose there are 17 trays with 5 cans of juice on each tray in the cafeteria. About how many cans of juice are there?

Do you need an exact amount? No, it says "about how many."

Explain to students that when they estimate a product they find out "about how many."

Review the rounding rules with the students. Then direct their attention to the first sentence.

Ask: **What does "round to the greatest place" mean?** Round to the left-most place in the number.

Work through the rounding process for factors 17 and 5.

Ask: **Do you need to round both factors?** No, because you can multiply 20 by 5 using mental math **Do you think the estimated product will be greater or less than the actual product?** greater than **Why?** 17 is rounded up to 20.

Continue modeling the skill in the same way as you review rounding hundreds and rounding thousands.

TRY THESE Exercises 1–3 give the students the opportunity to estimate products for all three places.

- **Exercise 1** Round to the nearest ten.
- **Exercise 2** Round to the nearest hundred.
- **Exercise 3** Round to the nearest thousand.

PRACTICE ON YOUR OWN Review the example at the top of the page. Have students name the digit in the place to be rounded and the digit to its right.

CHECK Make sure students understand that the estimate will be greater than the actual product if they round up, and less than the actual product if they round down. Success is determined by 3 out of 4 correct responses.

Students who successfully complete the **Practice on Your Own** and **Check** are ready to move on to the next skill.

COMMON ERRORS

- Students may use the wrong digit to round, for example, when rounding to the tens place, they may use the tens digit.
- Students may use an incorrect number of zeros in an estimated product.

Students who made more than four errors in the **Practice on Your Own**, or who were not successful in the **Check** section, may benefit from the **Alternative Teaching Strategy** on the next page.

Optional

Alternative Teaching Strategy

Estimate Products on the Number Line

OBJECTIVE Use a number line to estimate products

MATERIALS three number lines: one number line showing marks for ones from 10 to 20, another showing marks for tens from 200 to 300, and a third number line showing marks for hundreds from 2,000 to 3,000

Before estimating products, review the rounding rules. Distribute the first number line. Provide students with two numbers, one that rounds up, and one that rounds down. Use 18 and 13, for example.

Ask students to find 18 on the number line. Then have students find the halfway number, 15.

Say: **You can use the halfway number to make decisions about how to round.**

Ask: **Is 18 to the left or to the right of the halfway number?** to the right

Explain that since 18 is to the right of the halfway number and closer to 20 than to 10, it rounds to 20.

Use the same questioning strategy for rounding 13. Help students understand that because 13 is to the left of the halfway number, 13 rounds down to 10.

Then ask students to estimate 4×18 by rounding 18 to 20. Have students note that 4×20 can be multiplied using mental math.

Explain that since they rounded 18 up, the estimated product will be greater than the actual product.

Then have students estimate 3×276 and $2 \times 2{,}595$ using the number lines.

Repeat the activity several times. As students recognize that the halfway number has a 5 in the ones place, tens place, or hundreds place, link the activity with the number line to the rounding rules. Have them conclude the lesson by stating the rounding rules.

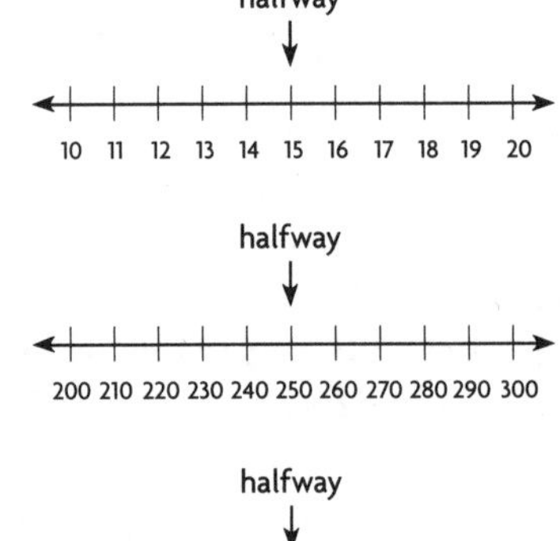

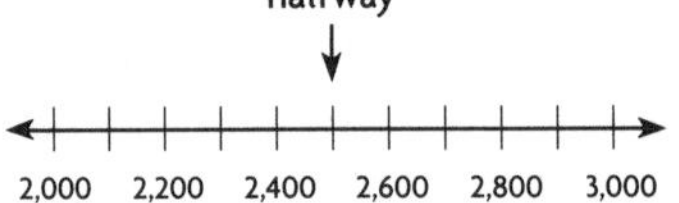

Name ______________________ Skill ______________________

Grade 4 **Skill 32**

Estimate Products

Round to the greatest place value to estimate products.

Round to the nearest ten.
Find $5 \times 17 = ■$.
Estimate. Round 17 to the nearest ten. Then multiply.

$$\begin{array}{r} 17 \\ \times\ 5 \\ \hline \end{array} \rightarrow \begin{array}{r} 20 \\ \times\ 5 \\ \hline 100 \end{array}$$

Look at the *ones* digit. It is greater than 5, so 17 rounds to 20.

So, 5×17 is about 100.

Round to the nearest hundred.
Find $4 \times 239 = ■$.
Estimate. Round 239 to the nearest hundred. Then multiply.

$$\begin{array}{r} 239 \\ \times\ 4 \\ \hline \end{array} \rightarrow \begin{array}{r} 200 \\ \times\ 4 \\ \hline 800 \end{array}$$

Look at the *tens* digit. It is less than 5, so 239 rounds to 200.

So, 4×239 is about 800.

Round to the nearest thousand.
Find $6 \times 4{,}531 = ■$.
Estimate. Round 4,531 to the nearest thousand. Then multiply.

$$\begin{array}{r} 4{,}531 \\ \times\ 6 \\ \hline \end{array} \rightarrow \begin{array}{r} 5{,}000 \\ \times\ 6 \\ \hline 30{,}000 \end{array}$$

Look at the *hundreds* digit. It is equal to 5, so 4,531 rounds to 5,000.

So, $6 \times 4{,}531$ is about 30,000.

Try These

Estimate the product.

1. Round to the nearest ten.

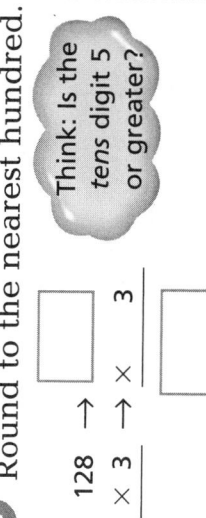

$$\begin{array}{r} 41 \\ \times\ 4 \\ \hline \end{array} \rightarrow \begin{array}{r} 40 \\ \times\ 4 \\ \hline \square \end{array}$$

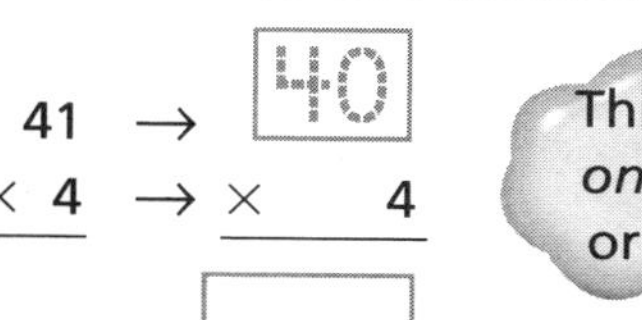

Think: Is the *ones* digit 5 or greater?

2. Round to the nearest hundred.

$$\begin{array}{r} 128 \\ \times\ 3 \\ \hline \end{array} \rightarrow \begin{array}{r} \square \\ \times\ 3 \\ \hline \square \end{array}$$

Think: Is the *tens* digit 5 or greater?

3. Round to the nearest thousand.

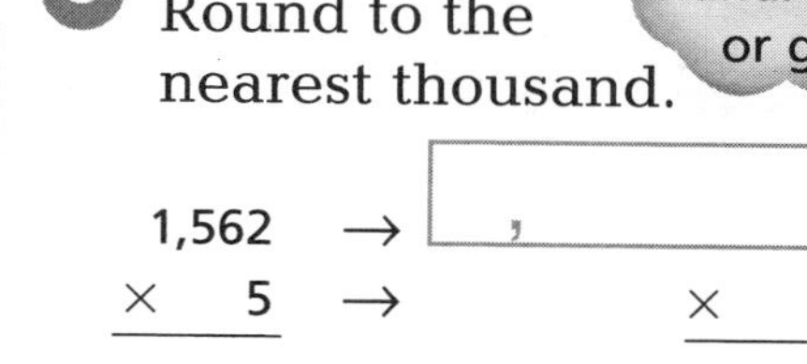

Think: Is the *hundreds* digit 5 or greater?

$$\begin{array}{r} 1{,}562 \\ \times\ 5 \\ \hline \end{array} \rightarrow \begin{array}{r} \square,\square \\ \times\ 5 \\ \hline \square,\square \end{array}$$

Go to the next side.

Name ______________________ Skill ______________________

Practice on Your Own

Skill 32

Think:
Look at the digit to the right of the greatest place value.

$$\begin{array}{r} 322 \\ \times\ 6 \\ \hline \end{array} \rightarrow \begin{array}{r} 300 \\ \times\ 6 \\ \hline 1{,}800 \end{array}$$

Look at the *tens* digit. It is less than 5, so 322 rounds to 300.

So, 6 × 322 is about 1,800.

Round to the greatest place value. Estimate the product.

1. 32 → ☐ ; × 5 → × 5 ; ☐
2. 49 → ☐ ; × 2 → × 2
3. 178 → ☐ ; × 4 → × 4
4. 4,621 → ☐ ; × 4 → × 4
5. 18 → ☐ ; × 3 → × 3
6. 123 → ☐ ; × 5 → × 5
7. 529 → ☐ ; × 4 → × 4
8. 7,193 → ☐ ; × 3 → × 3
9. 12 → ☐ ; × 6 → × 6
10. 327 → ☐ ; × 5 → × 5
11. 567 → ☐ ; × 3 → × 3
12. 6,515 → ☐ ; × 5 → × 5

Check

Estimate the product.

13. 22 × 3
14. 332 × 4
15. 1,134 × 2
16. 5,825 × 2

Skill 33 Grade 4 Multiplication Facts and Patterns

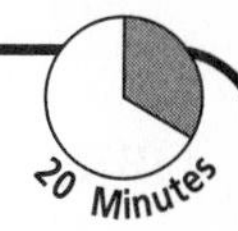

OBJECTIVE Multiply by 10 and multiples of 10

Begin by reviewing basic multiplication facts with the students. Review the meaning of the terms *factor* and *product*. Explain to students that they can use basic multiplication facts and multiples of 10 to build multiplication patterns.

Direct students' attention to the first example of a multiplication pattern.

Ask: **What basic fact do you see in the first number sentence at the beginning of the pattern?** 4 × 1

Continue: **Now look at the next number sentence in the pattern. How has the second factor changed?** **Possible responses: the 1 is now 10; there is a zero written after the 1.**

Reconfirm: **There is a zero in the ones place of the *factor*. How has the product changed?** **Possible responses: the 4 becomes 40; a zero was written after the 4.**

Reconfirm: **There is a zero in the ones place of the *product*. Continue this questioning through the remainder of the pattern.**

Reinforce that the product of 10 and any other factor has a zero in the ones place, the product of 100 and any other factor has zeros in both the ones and tens places, and so on.

Discuss the patten for facts that end in zero, for example:

$$5 \times 4 = 20$$
$$5 \times 40 = 200$$
$$5 \times 400 = 2{,}000$$

Point out that they can still use the pattern.

Have them note that the extra zero comes from 5 × 4, the multiplication fact. The pattern tacks zeros onto the product of the fact.

TRY THESE Exercises 1–3 model the type of exercises students will find on the **Practice on Your Own** page.

- **Exercises 1–2** The number of zeros are given for each term of the pattern. Students supply the basic fact.
- **Exercise 3** Find missing factors.

PRACTICE ON YOUR OWN In the example at the top of the page, have students focus on the multiplication pattern. Then ask them to explain how they can use the basic fact and multiples of 10 to find the missing factor.

CHECK Determine if students can multiply by multiples of 10.

Success is determined by 3 out of 4 correct responses.

Students who successfully complete the **Practice on Your Own** and **Check** are ready to move on to the next skill.

COMMON ERRORS

- Students may write the wrong number of zeros in the product.
- Students may multiply the factor by the product when looking for a missing factor.

Students who made more than four errors in the **Practice on Your Own**, or who were not successful in the **Check** section, may benefit from the **Alternative Teaching Strategy** on the next page.

Optional

Alternative Teaching Strategy
Model Multiplication Patterns

20 Minutes

OBJECTIVE Use base-ten blocks to model multiplying by tens and multiples of 10

MATERIALS base-ten blocks: ones, tens, hundreds; paper, pencil

Begin by having the students write the following:

5 × 1 = _____

5 × 10 = _____

5 × 100 = _____

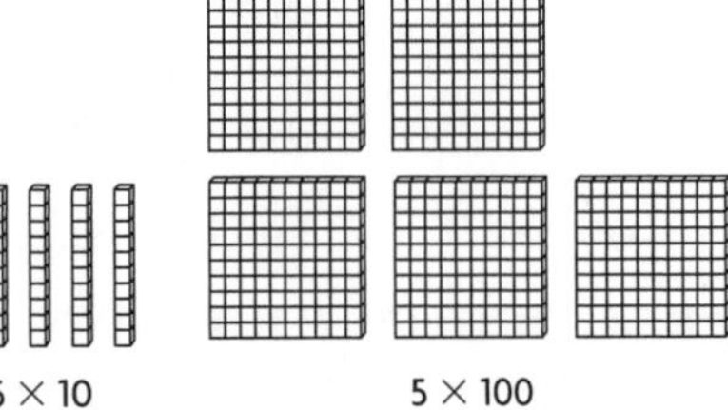

Distribute the base-ten blocks. Have the students use the base-ten blocks to model each problem.

Say: **What is the product of 5 × 1** 5

What is the product of 5 × 10 50

What is the product of 5 × 100 500

Point out that each product is ten times greater than the previous product. For example, 50 = 5 × 10; 500 = 50 × 10.

Now have students write the number sentence 5 × 1,000 = _______.

Ask: **How many zeros will be in the product?** 3

Have the students write the product.

Continue by having the students write and model 2 × 1, 2 × 10, 2 × 100; 3 × 1, 3 × 10, 3 × 100. Then ask them to write and find the product of 2 × 1,000 and 3 × 1,000.

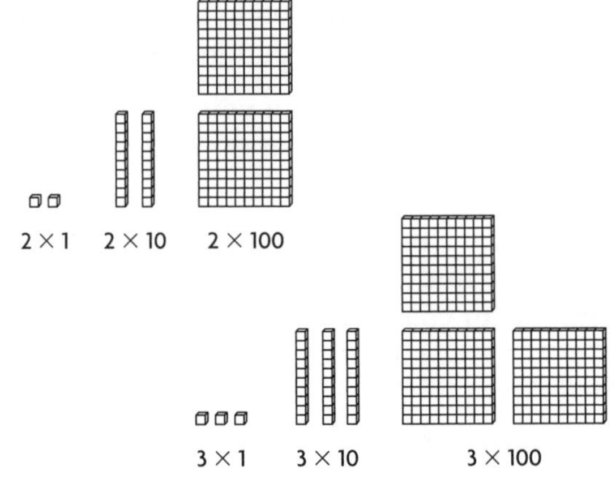

When the students show understanding of multiplying by factors of 10, have them try patterned exercises using only paper and pencil.

Have the students write and find the products for:

1 × 2	2 × 2	3 × 2
1 × 20	2 × 20	3 × 20
1 × 200	2 × 200	3 × 200
1 × 2,000	2 × 2,000	3 × 2,000

You may wish to continue with similar patterns.

Grade 4 Skill 33

Multiplication Facts and Patterns

Look for the basic fact.
Then find a pattern.

4 × 1 = 4 ← basic fact
4 × **10** = **40** ← 1 zero
4 × **100** = **400** ← 2 zeros
4 × **1,000** = **4,000** ← 3 zeros
4 × **10,000** = **40,000** ← 4 zeros

The number of zeros in the product increases as the number of zeros in the factor increases.

Find 5 × ■ = 20,000.
To find the missing factor, first think of a basic fact.
5 × ■ = 20

Use the pattern to find the missing factor.

5 × 4 = 20
5 × **40** = **200**
5 × **400** = **2,000**
5 × **4,000** = **20,000**

5 times what number is 20?
5 × 4 = 20

So, the missing factor is 4,000.

Then 5 × 4,000 = 20,000.

Try These

Find the products.

1.
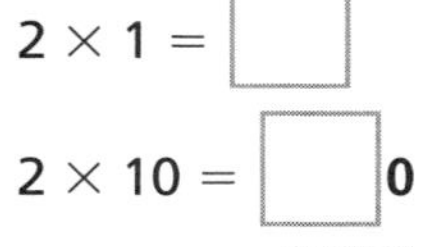
2 × 1 = ☐
2 × 10 = ☐0
2 × 100 = ☐00

2.
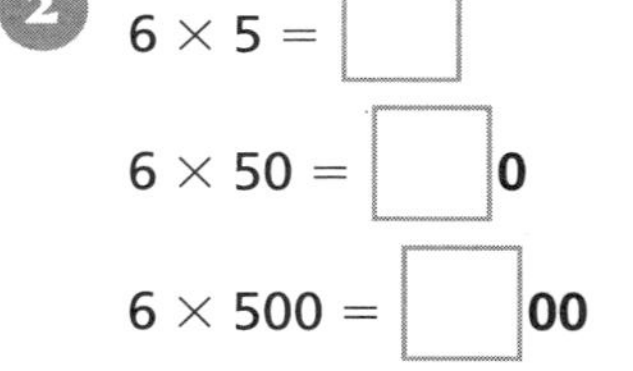
6 × 5 = ☐
6 × 50 = ☐0
6 × 500 = ☐00

Find the missing factor.

3.

7 × ■ = 1,400

Missing Factor: ☐

Basic Fact → 7 × ☐ = 14
7 × ☐ = 140
7 × ☐ = 1,400

Go to the next side.

Name ______________________ Skill ______________________

Practice on Your Own

Think:
To find the missing factor, look for a basic fact. Then use a pattern.

3 × ■ = 2,400
The missing factor is 800.

Basic fact → 3 × 8 = 24
3 × 80 = 240
3 × 800 = 2,400

Find the products.

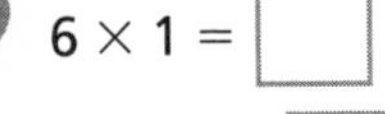

1	2	3	4
6 × 1 = ☐	3 × 7 = ☐	6 × 4 = ☐	5 × 7 = ☐
6 × 10 = ☐	3 × 70 = ☐	6 × 40 = ☐	5 × 70 = ☐
6 × 100 = ☐	3 × 700 = ☐	6 × 400 = ☐	5 × 700 = ☐
6 × 1,000 = ☐	3 × 7,000 = ☐	6 × 4,000 = ☐	5 × 7,000 = ☐

Find the missing factor.

5. 4 × ■ = 1,200

Basic Fact→ 4 × ☐ = 12
4 × ☐ = 120
4 × ☐ = 1,200

Missing Factor: ☐

6. ■ × 5 = 10,000

Basic Fact→ ☐ × 5 = 10
☐ × 5 = 100

☐ × 5 = 1,000
☐ × 5 = 10,000

Missing Factor: ☐

Find the product.

7. 5 × 10 = ☐
8. 6 × 30 = ☐
9. 3 × 400 = ☐
10. 6 × 500 = ☐

Check

Find the product or missing factor.

11. 6 × 70 = ☐
12. 3 × 600 = ☐
13. 4 × ☐ = 160
14. ☐ × 90 = 2,700

Skill 34 Grade 4 — Multiplication Properties

OBJECTIVE Use multiplication properties to recall basic facts

MATERIALS tiles

You might have students model the activity using tiles. Begin by reminding students that multiplication properties can help them recall multiplication facts.

Direct students' attention to the Order Property. Explain how an array can help them write two multiplication facts.

Ask: **Look at the first array. How many rows are there?** 2 **How many tiles in each row?** 6 **What multiplication fact can you write for the array?** $2 \times 6 = 12$

Repeat the questions for the second array. Ask students to look at the two multiplication sentences.

Ask: **How are they alike? How are they different?** The product remains the same. The order of the factors is different.

For the Grouping Property, explain that the parentheses group the factors two different ways. Point out to students how one way of grouping the factors may make it easier to find the product. For example, 6×5 is a basic fact that might be easier for some students to multiply than 2×15.

As you work through the Property of One and the Zero Property, ask: **How can these properties help you to remember basic facts for multiplication?** Possible response: By knowing the Property of One and the Zero Property, I can build all the facts for 1 and 0.

TRY THESE In Exercises 1–4, students complete a multiplication sentence using each of the properties.

- **Exercise 1** Order Property.
- **Exercise 2** Grouping Property.
- **Exercise 3** Property of One.
- **Exercise 4** Zero Property.

PRACTICE ON YOUR OWN Review the examples with students. Caution students to take their time and check their work as they complete the page.

CHECK Determine if students know the multiplication properties and can use them to recall basic multiplication facts. Success is indicated by 3 out of 3 correct responses.

Students who successfully complete the **Practice on Your Own** and **Check** are ready to move on to the next skill.

COMMON ERRORS

- Students may confuse the Order and Grouping Properties.
- Students may multiply incorrectly.
- Students may mistakenly use the Property of One and either add 1 to the factor or find a product of 1.

Students who made more than three errors in the **Practice on Your Own**, or who were not successful **Check** section, may benefit from the **Alternative Teaching Strategy** on the next page.

Optional

Alternative Teaching Strategy
Model Multiplication Properties

OBJECTIVE Use multiplication properties to help remember basic facts

MATERIALS 40 tiles for each pair of students

You may wish students to work in pairs. Distribute tiles.

Demonstrate to students how knowing the Multiplication Properties can make multiplying easier. Ask students to show 5 rows of 3. Then have them show 3 rows of 5.

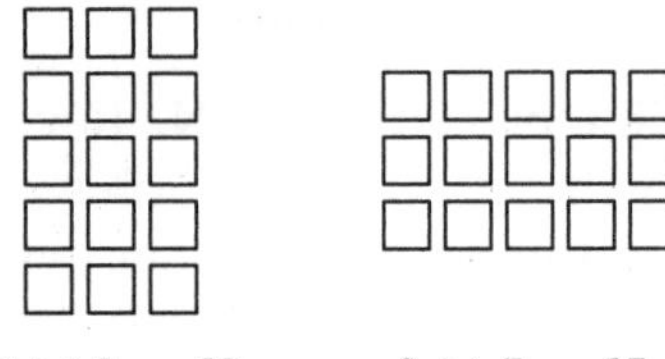

$5 \times 3 = 15$ $3 \times 5 = 15$

Point out to students that their arrays now model the Order Property.

Have students write a multiplication sentence for each array.

Ask: **If 3 × 5 = 15, then does 5 × 3 = 15? How do you know?** Possible response: Yes; the Order Property of Multiplication tells us that since the factors are the same, 3 and 5, then the product is the same, 15. **So, when the order of the factors changes, the product is always the same.**

If necessary, repeat the activity several times by having students suggest other factors.

To model the Grouping Property have students write the following :

$$(2 \times 4) \times 3 = 2 \times 4 \times 3$$

Have them show three 2 by 4 arrays to model the first expression.

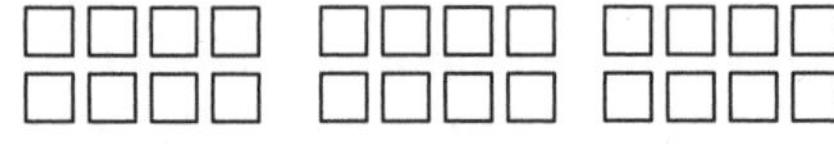

$(2 \times 4) \times 3 = 24$

Then have them show two 4 by 3 arrays to model the second.

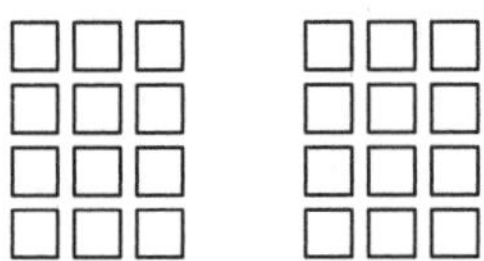

$2 \times (4 \times 3) = 24$

Repeat this activity several times until the students see that they can change the grouping of the factors without changing the product.

You can use arrays to demonstrate the Property of One. Then demonstrate the Zero Property by showing, for example, why 4 groups of 0 or 0 groups of 4 will result in 0 as the product.

As students show an understanding of the Multiplication Properties, have them show the properties using number sentences only.

Name ______________________ Skill ______________________

Grade 4 **Skill**

Multiplication Properties

Use multiplication properties to help you remember basic facts.

Order Property
Two factors can be multiplied in any order. The product is the same.

2 × 6 and 6 × 2 both equal 12.

So, if you know $2 \times 6 = 12$, then you also know that $6 \times 2 = 12$.

Grouping Property
When the grouping of factors is changed, the product remains the same.

$2 \times (3 \times 5) = (2 \times 3) \times 5$
$2 \times 15 = 6 \times 5$
same $\rightarrow 30 = 30$

When you multiply 3 addends, you can choose any 2 of the factors to multiply first. The product is the same.

So, $2 \times (3 \times 5)$ can be grouped as $(2 \times 3) \times 5$, and the product stays the same.

Property of One
When one of the factors is 1, the product is the other factor.

$1 \times 4 = 4$
$4 \times 1 = 4$

Zero Property
When one of the factors is 0, the product is 0.

$0 \times 3 = 0$
$3 \times 0 = 0$

Try These

Complete to show each multiplication property.

1. Order Property

   $5 \times 7 = \square \times 5$

2. Grouping Property

   $(4 \times 6) \times 3 = \square \times (6 \times 3)$

3. Property of One

   $9 \times \square = 9$

4. Zero Property

   $\square \times 7 = 0$

Go to the next side.

Name ________________________ Skill ________________________

Practice on Your Own

Skill 34

Order Property

$4 \times 5 = 20$
$5 \times 4 = 20$

4 × 5 and 5 × 4 have the same product, 20.

So, $4 \times 5 = 5 \times 4$.

Grouping Property

$(3 \times 2) \times 4 = 3 \times (2 \times 4)$
$6 \times 4 = 3 \times 8$
$24 = 24$

(3 × 2) × 4 and 3 × (2 × 4) have the same product, 24.

So, $(3 \times 2) \times 4 = 3 \times (2 \times 4)$.

Write the multiplication property.

1. $0 \times 9 = 0$

   ________ Property

2. $6 \times 3 = 18$
   $3 \times 6 = 18$

   ________ Property

3. $2 \times (3 \times 3) = (2 \times 3) \times 3$

   ________ Property

Complete to show the multiplication property.

4. Property of One

   5 ×  = 5

5. Grouping Property

   $4 \times (2 \times 5) =$

   (☐ × ☐) × ☐

6. Order Property

   3 × ☐ = 12

   ☐ × 3 = 12

7. Grouping Property

   $(3 \times 6) \times 4 =$

   ☐ × (☐ × ☐)

8. Zero Property

   6 × ☐ = 0

9. Order Property

   ☐ × 5 = 15

   5 × ☐ = 15

Check

Write the multiplication property.

10. $6 \times 1 = 6$

    ________ Property

11. $2 \times 7 = 14$
    $7 \times 2 = 14$

    ________ Property

12. $4 \times (5 \times 2) =$
    $(4 \times 5) \times 2$

    ________ Property

Skill 35 Grade 4

Multiplication Patterns

OBJECTIVE Find and extend a pattern involving multiplication

Begin by recalling the meaning of the term *multiple*: a number that is the product of a given number and another whole number. You may wish to have students count by 2s to 20. Explain the connection between counting by 2s and the meaning of multiples of 2.

Direct students' attention to Step 1.

Ask: **What is the first number in the pattern?** 3 **What times 3 equals 3?** 1

Explain that the first term in the pattern is the result of finding 3 × 1. Continue with: **What times 3 equals 6?** 2 **What times 3 equals 9?** 3 **What times 3 equals 12?** 4

Explain that since the pattern increases by 3 each time, the numbers are multiples of 3.

As the students look at Step 2, ask: **What is the last number in the pattern?** 12 **What is the next multiple of 3?** 15

You may wish to have students find the next three numbers in the pattern. 18, 21, 24

TRY THESE Exercises 1–3 model the type of exercises students will find on the **Practice on Your Own** page.

- **Exercise 1** Multiples of 4.
- **Exercise 2** Multiples of 3.
- **Exercise 3** Multiples of 12.

PRACTICE ON YOUR OWN Review the example at the top of the page. Monitor students' work as they look for the factor of which each number is a multiple. Some students may need help finding the pattern.

CHECK Determine if students can identify each pattern.

Success is determined by 2 out of 2 correct responses.

Students who successfully complete the **Practice on Your Own** and **Check** are ready to move on to the next skill.

COMMON ERRORS

- Students may use the wrong factor as the multiple.
- Students may add the factors instead of multiplying.

Students who made more than two errors in the **Practice on Your Own**, or who were not successful in the **Check** section, may benefit from the **Alternative Teaching Strategy** on the next page.

Optional

Alternative Teaching Strategy

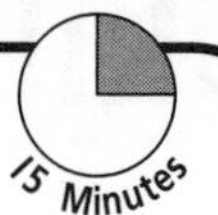

Modeling Multiplication Patterns

OBJECTIVE Use tiles to model multiplication patterns to find the multiples

MATERIALS tiles

Explain that patterns can be illustrated using tiles. Have students write the pattern 4, 8, 12, 16 on their papers.

Distribute tiles to the students.

Say: **Start by forming a group for the first number in the pattern.**

Ask: **How many tiles did you use to make the group?** 4 **If you have 1 group of 4 tiles, what multiplication expression can you write to show this?** 1 × 4

Have the students display tiles to represent 8. Guide students to display 2 rows of 4 tiles.

Ask: **How many groups of 4 tiles did you show?** 2

If you have 2 groups of 4 tiles, what multiplication expression can you write to show this? 2 × 4

Continue guiding students to build arrays to show 12 and to show 16. Have students study the patterns of tiles to see that each succeeding model increases by one row of 4 each time.

Have students predict the number of rows of 4 that will be in the next array. 5 Ask students to verify their prediction by writing the multiplication expression and telling the number it represents in the pattern.

You may wish to continue the activity with other multiples.

4

□□□□

1 × 4

8

□□□□
□□□□

2 × 4

12

□□□□
□□□□
□□□□

3 × 4

16

□□□□
□□□□
□□□□
□□□□

4 × 4

Name ______________________ Skill ______________________

Grade 4 Skill 35

Multiplication Patterns

Step 1 Find the pattern.

Each number is 3 more than the number before it.

3×1	3×2	3×3	3×4	?
3,	6,	9,	12,	☐

The pattern shows multiples of 3.

Step 2 Use the pattern to find the next number. Begin with the last number in the pattern. Find the next multiple of 3.

last number $3 \times 4 = 12$

next multiple of 3 $3 \times 5 = 15$

3, 6, 9, 12, 15

So, 15 is the next number.

Try These

Describe the pattern. Write the next number.

1 4, 8, 12, 16, ☐

$1 \times$ ☐ $= 4$

$2 \times$ ☐ $= 8$

$3 \times$ ☐ $= 12$

$4 \times$ ☐ $= 16$

$5 \times$ ☐ $=$ ☐

The pattern shows multiples of 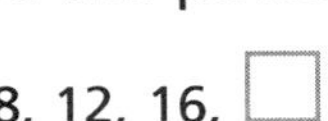.

The next number is ☐.

2 15, 18, 21, 24, ☐

$5 \times$ ☐ $= 15$

$6 \times$ ☐ $= 18$

$7 \times$ ☐ $= 21$

$8 \times$ ☐ $= 24$

$9 \times$ ☐ $=$ ☐

The pattern shows multiples of ☐.

The next number is ☐.

3 12, 24, 36, 48, ☐

$1 \times$ ☐ $= 12$

$2 \times$ ☐ $= 24$

$3 \times$ ☐ $= 36$

$4 \times$ ☐ $= 48$

$5 \times$ ☐ $=$ ☐

The pattern shows multiples of ☐.

The next number is ☐.

Go to the next side.

Name ______________________ Skill ______________________

Practice on Your Own

Write the next number: 20, 24, 28, 32, ■. Find the pattern.

4 × 5	4 × 6	4 × 7	4 × 8	?
20	24	28	32	☐

The pattern shows multiples of 4. The next multiple of 4 is 4 × 9 or 36.

Describe the pattern. Write the next number.

1. 24, 27, 30, 33, ☐

   8 × ☐ = 24   11 × ☐ = 33
   9 × ☐ = 27   12 × ☐ = ☐
   10 × ☐ = 30

   Multiples of: ☐ Next number: ☐

2. 48, 60, 72, 84, ☐

   4 × ☐ = 48   7 × ☐ = 84
   5 × ☐ = 60   8 × ☐ = ☐
   6 × ☐ = 72

   Multiples of: ☐ Next number: ☐

3. 16, 24, 32, 40, ☐

   2 × ☐ = 16   5 × ☐ = 40
   3 × ☐ = 24   6 × ☐ = ☐
   4 × ☐ = 32

   Multiples of: ☐ Next number: ☐

4. 32, 48, 64, 80, ☐

   2 × ☐ = 32   5 × ☐ = 80
   3 × ☐ = 48   6 × ☐ = ☐
   4 × ☐ = 64

   Multiples of: ☐ Next number: ☐

5. 36, 39, 42, 45, ☐
   Multiples of: ☐ Next number: ☐

6. 32, 36, 40, 44, ☐
   Multiples of: ☐ Next number: ☐

Describe the pattern. Write the next number.

7. 18, 24, 30, 36, ☐
   Multiples of: ☐ Next number: ☐

8. 28, 35, 42, 49, ☐
   Multiples of: ☐ Next number: ☐

Skill 36 Grade 4

Multiply Three Factors

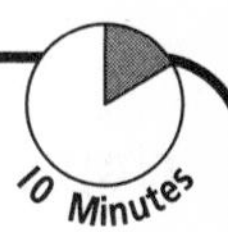

OBJECTIVE Use the Grouping Property of Multiplication to multiply with three factors

Provide students with a simple example of multiplying three factors orally. Begin with just two factors (7 × 2). Then have students multiply the product 14, by 5. Explain that the multiplication can be expressed as:

(7 × 2) × 5 = 70 or 7 × (2 × 5) = 70.

Recall that the Grouping (Associative) Property of Multiplication states that factors can be grouped differently and the product is always the same. Have students verify this in the above example. Remind them that when they see parentheses it means to do that operation within the parentheses first.

Work through the examples. Have students discuss the two ways that the factors are grouped.

TRY THESE In Exercises 1–4, students choose one of the two ways to group three factors and then find the product. You may wish to have students explain the reason for each choice. After two factors are multiplied in each exercise, the results for multiplying by the third factor are as follows:

- **Exercise 1** Multiply 20 × 5 or 4 × 25.
- **Exercise 2** Multiply 60 × 2 or 12 × 10.
- **Exercise 3** Multiply 32 × 5 or 16 × 10.
- **Exercise 4** Multiply 18 × 10 or 9 × 20.

Ask students to explain why they chose the grouping that they did.

PRACTICE ON YOUR OWN Review the example. Remind students that they have learned to multiply by 10 and multiples of 10 (20, 30, . . .).

In Exercises 1–3, students are given the two ways of grouping. In Exercises 4–6, they fill in the blanks for the grouping methods. In Exercises 7–12, they are given the factors and asked to do the grouping on their own.

CHECK Students are asked to insert parentheses to establish which factors they will multiply first. Success is determined by 3 out of 3 correct responses.

Students who successfully complete **Practice on Your Own** and **Check** are ready to move on to the next skill.

COMMON ERRORS

- Students may forget to multiply by the third factor or may add the third factor to the product of the first two.
- Students may not recall multiplication facts correctly or may be unable to multiply by multiples of 10.

Students who made more than four errors in the **Practice on Your Own**, or who were not successful in the **Check** section, may benefit from the **Alternative Teaching Strategy** on the next page.

Optional

Alternative Teaching Strategy
Model Multiplying Three Factors

OBJECTIVE Use counters to model multiplying three factors

MATERIALS counters

Present a multiplication expression such as $2 \times 5 \times 3$. Have the students first represent 2×5 with the counters. Have students verbalize that the product is 10.

Return to the expression.

Ask: **What do you multiply by next?** 3
What are you multiplying? 10

Display 3×10. Ask students how they can show this with the counters. **Display 3 groups of 10 counters.** Have students use the counters to show the product.

Have the students summarize the steps they used to find the product and to record it in symbols.

$$2 \times 5 \times 3$$

$$(2 \times 5) \times 3$$

$$\downarrow$$

$$10 \times 3$$

$$30$$

Then, working symbolically, have the students explain how to group the factors differently to find the product.

$$2 \times (5 \times 3)$$

$$\downarrow$$

$$2 \times 15$$

$$30$$

Point out that the product is the same. If needed, work through another example in a similar way.

Check to make sure that students are multiplying all three factors, and not adding one of them.

If students are having trouble multiplying by multiples of 10, review patterns for multiplying.

If students are having trouble recalling the basic multiplication facts, give them a short daily review or let them work with flash cards with a partner. They should keep a list of the facts that give them trouble and concentrate on memorizing those.

Name ____________ Skill ____________

Grade 4 Skill 36

Multiply Three Factors

Use the Grouping Property to help you multiply 3 factors. Find $7 \times 2 \times 5$.

Group the factors.

$(7 \times 2) \times 5 = ■$

↓

$14 \times 5 = ■$

Remember: Multiply inside the parentheses first.

Can you use mental math to multiply 14×5?

If not, group the factors another way.

$7 \times (2 \times 5) = ■$

↓

$7 \times 10 = ■$

Think: It is easier to multiply 7 × 10 than 14 × 5.

Find the product.

$7 \times (2 \times 5) = ■$

↓

$7 \times 10 = 70$

So, $7 \times 2 \times 5 = 70$.

Try These

Circle the grouping of factors that you use. Find the product.

1. $4 \times 5 \times 5$

   $(4 \times 5) \times 5$
   or
   $4 \times (5 \times 5)$

   The product is ☐.

2. $12 \times 5 \times 2$

   $(12 \times 5) \times 2$
   or
   $12 \times (5 \times 2)$

   The product is ☐.

3. $16 \times 2 \times 5$

   $(16 \times 2) \times 5$
   or
   $16 \times (2 \times 5)$

   The product is ☐.

4. $9 \times 2 \times 10$

   $(9 \times 2) \times 10$
   or
   $9 \times (2 \times 10)$

   The product is ☐.

Go to the next side.

Name ______________________ Skill ______________________

Practice on Your Own

Skill 36

Think:
One way of grouping 3 factors may be easier to multiply than another way.

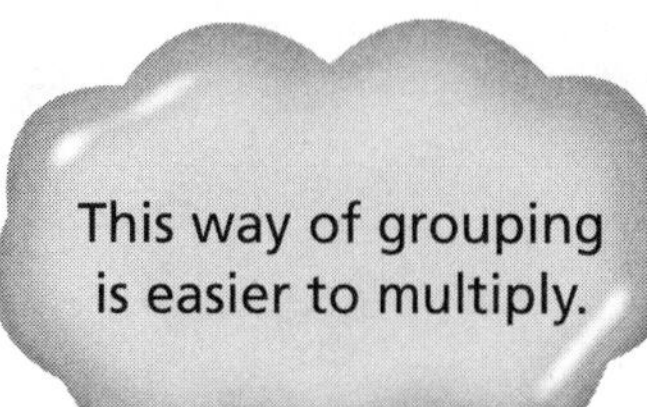

This way of grouping is easier to multiply.

Find $15 \times 2 \times 8$.

$(15 \times 2) \times 8 = \blacksquare$	$15 \times (2 \times 8) = \blacksquare$
↓	↓
➡ $30 \times 8 = \square$	$15 \times 16 = \square$
$30 \times 8 = 240$	$15 \times 16 = 240$

So, the product is 240.

Circle the grouping of factors you use. Find the product.

1. $3 \times 5 \times 4$
   $(3 \times 5) \times 4 = \square$
   $3 \times (5 \times 4) = \square$
   The product is $\square$.

2. $7 \times 2 \times 2$
   $(7 \times 2) \times 2 = \square$
   $7 \times (2 \times 2) = \square$
   The product is $\square$.

3. $4 \times 5 \times 6$
   $(4 \times 5) \times 6 = \square$
   $4 \times (5 \times 6) = \square$
   The product is $\square$.

Group the factors two ways. Find the product.

4. $6 \times 5 \times 9$
   $\square \times \square \times \square$
   $\square \times \square \times \square$
   The product is $\square$.

5. $5 \times 4 \times 12$
   $\square \times \square \times \square$
   $\square \times \square \times \square$
   The product is $\square$.

6. $10 \times 8 \times 9$
   $\square \times \square \times \square$
   $\square \times \square \times \square$
   The product is $\square$.

Find the product. Show how you grouped the factors, using parentheses.

7. $3 \times 6 \times 2 = \square$
8. $2 \times 6 \times 5 = \square$
9. $4 \times 2 \times 5 = \square$
10. $12 \times 2 \times 5 = \square$
11. $17 \times 2 \times 10 = \square$
12. $9 \times 2 \times 15 = \square$

Find the product. Show how you grouped the factors, using the parentheses.

13. $12 \times 5 \times 9 = \square$
14. $8 \times 10 \times 9 = \square$
15. $11 \times 3 \times 5 = \square$

Answer Card

Multiplication

Grade 4

Skill 24

TRY THESE

1. 3, 4, 12
2. 2
3. 21

PRACTICE

1. 2, 5
2. 4, 5, 20
3. 5, 6, 30
4. 6
5. 8
6. 12
7. 16
8. 3
9. 12
10. 24
11. 27
12. 21
13. 16
14. 20
15. 24
16. 18
17. 15
18. 32
19. 45

CHECK

20. 2
21. 18
22. 8
23. 35
24. 40

Skill 25

TRY THESE

1. 42
2. 8
3. 45

PRACTICE

1. 24
2. 6, 2, 12
3. 6, 8, 48
4. 7
5. 28
6. 49
7. 56
8. 16
9. 40
10. 64
11. 72
12. 27
13. 36
14. 63
15. 81
16. 42
17. 18
18. 48
19. 63

CHECK

20. 42
21. 35
22. 54
23. 72
24. 36

Skill 26

TRY THESE

1. 15, 15
2. 12, 12
3. 30, 30

PRACTICE

1. 8, 8
2. 9, 9
3. 20, 20
4. 18
5. 15
6. 18
7. 10
8. 12
9. 24
10. 18

CHECK

11. 14
12. 30
13. 15
14. 32

Skill 27

TRY THESE

1. 15 ones, 6 tens, 75
2. 2, 21 ones, 9 tens, 111
3. 0 ones, 16 tens, 160

PRACTICE

1. 2, 81
2. 1, 160
3. 2, 184
4. 1, 45
5. 1, 92
6. 1, 74

CHECK

7. 1, 78
8. 126
9. 2, 84

Answer Card

Multiplication

Grade 4

SKILL 28

TRY THESE

1. 128
2. 10, 60, 70
3. 21, 280, 301

PRACTICE

1. 6, 100, 106
2. 6, 90, 96
3. 25, 150, 175
4. 4, 140, 144
5. 16, 200, 216
6. 30, 420, 450
7. 155
8. 294
9. 424

CHECK

10. 45
11. 285
12. 156
13. 448

SKILL 29

TRY THESE

1. 2, 2, 2, 2
2. 9; 900; 9,000; 90,000
3. 3; 3,000; 30,000; 300,000

PRACTICE

1. 3; 300; 3,000; 30,000
2. 2; 2,000; 20,000; 200,000
3. 6; 60,000; 600,000; 6,000,000
4. 5; 500; 5,000; 50,000
5. 1; 1,000; 10,000; 100,000
6. 2; 20,000; 200,000; 2,000,000
7. 300
8. 40,000
9. 20,000
10. 200,000
11. 40
12. 5,000
13. 90,000
14. 200,000

CHECK

15. 90
16. 6,000
17. 400,000
18. 600,000

SKILL 30

TRY THESE

1. 1, 42
2. 1, 76
3. 2, 468
4. 1, 372

PRACTICE

1. 2, 81
2. 1, 326
3. 1, 720
4. 2, 864
5. 2, 120
6. 1, 210
7. 1, 816
8. 1, 729
9. 2, 68
10. 1, 225
11. 1, 324
12. 1, 706

CHECK

13. 92
14. 655
15. 690
16. 675

Answer Card
Multiplication
Grade 4

SKILL 31

TRY THESE

1. 1; 2; 625
2. 1; 1; 558
3. 2; 1; 696
4. 1; 1; 972

PRACTICE

1. 1; 1; 936
2. 1; 2; 717
3. 1; 2; 2; 1,401
4. 1; 2; 3; 1,730
5. 4; 1; 910
6. 1; 2; 984
7. 1; 4; 1,968
8. 4; 6; 2,583
9. 870
10. 855
11. 1,917
12. 1,316

CHECK

13. 1,092
14. 636
15. 762
16. 1,435

SKILL 32

TRY THESE

1. 40, 160
2. 100, 300
3. 2,000; 10,000

PRACTICE

1. 30, 150
2. 50, 100
3. 200, 800
4. 5,000; 20,000
5. 20, 60
6. 100, 500
7. 500; 2,000
8. 7,000; 21,000
9. 10, 60
10. 300; 1,500
11. 600; 1,800
12. 7,000; 35,000

CHECK

13. 60
14. 1,200
15. 2,000
16. 12,000

SKILL 33

TRY THESE

1. 2, 2, 2
2. 30; 300; 3,000
3. 200, 2, 20, 200

PRACTICE

1. 6; 60; 600; 6,000
2. 21; 210; 2,100; 21,000
3. 24; 240; 2,400; 24,000
4. 35; 350; 3,500; 35,000
5. 3; 30; 300; 300
6. 2; 20; 200; 2,000; 2,000
7. 50
8. 180
9. 1,200
10. 3,000

CHECK

11. 420
12. 1,800
13. 40
14. 30

Answer Card
Multiplication
Grade 4

SKILL 34

TRY THESE

1. 7
2. 4
3. 1
4. 0

PRACTICE

1. Zero Property
2. Order Property
3. Grouping Property
4. 1
5. (4 × 2) × 5
6. 4, 4
7. 3 × (6 × 4)
8. 0
9. 3, 3

CHECK

10. Property of One
11. Order Property
12. Grouping Property

SKILL 35

TRY THESE

1. 20, 4, 4, 4, 4, 4, 20, 4, 20
2. 27, 3, 3, 3, 3, 3, 27, 3, 27
3. 60, 12, 12, 12, 12, 12, 60, 12, 60

PRACTICE

1. 36, 3, 3, 3, 3, 3, 36, 3, 36
2. 96, 12, 12, 12, 12, 12, 96, 12, 96
3. 48, 8, 8, 8, 8, 8, 48, 8, 48
4. 96, 16, 16, 16, 16, 16, 96 16, 96
5. 48, 3, 48
6. 48, 4, 48

CHECK

7. 42, 6, 42
8. 56, 7, 56

SKILL 36

TRY THESE

Student may circle either expression in Ex. 1–4.

1. 100
2. 120
3. 160
4. 180

PRACTICE

Student may circle either expression in Ex. 1–3.

1. 60, 60, 60
2. 28, 28, 28
3. 120, 120, 120
4. (6 × 5) × 9; 6 × (5 × 9); 270
5. (5 × 4) × 12; 5 × (4 × 12); 240
6. (10 × 8) × 9; 10 × (8 × 9); 720

Check students' work.

7. 36
8. 60
9. 40
10. 120
11. 340
12. 270

CHECK

Check students' work.

13. 540
14. 720
15. 165

Number Sense

Whole Number Division

Skill 37 Grade 4

Division Facts (Divide by 2 and 5)

OBJECTIVE Divide by 2 and 5

MATERIALS counters

You may wish to use counters to model Steps 1–3 with the students. As you begin the lesson, remind the students that division is the inverse, or opposite, of multiplication. So, they can use multiplication facts to help them with division facts.

Then ask: **Two times what number is 6?** $2 \times 3 = 6$

Emphasize the idea that $6 \div 2 = 3$ because $2 \times 3 = 6$.

Ask: **How can you show the division with counters?** Separate 6 counters into 2 equal groups.

Look at the model. How many counters are there in all? 6 **How many groups are there?** 2 **What does the 3 represent?** The number of counters in each group.

Explain that the division can be written two ways, as a number sentence and in vertical form.

Continue in a similar way for $15 \div 5$. Have students note that when they divide, they separate a total number into equal groups.

TRY THESE In Exercises 1–4, students use models in ways similar to the exercises they will encounter in the **Practice on Your Own** section.

- **Exercise 1** Division facts for 2.
- **Exercises 2–3** Division facts for 5.

PRACTICE ON YOUR OWN Review the examples with students. For Exercises 1–8 student circle groups to show facts for 2 and 5. Exercises 9–16 are done without visual clues.

CHECK Determine if students can divide without models. Success is indicated by 3 out of 4 correct responses.

Students who successfully complete the **Practice on Your Own** and **Check** are ready to move on to the next skill.

COMMON ERRORS

- Students may not know multiplication facts and may use an incorrect fact to find a quotient.
- Students may add or multiply instead of dividing.

Students who made more than three errors in the **Practice on Your Own**, or were not successful in the **Check** section, may benefit from the **Alternative Teaching Strategy** on the next page.

Optional

Alternative Teaching Strategy
Model Division Facts for 2 and 5

OBJECTIVE Use concrete objects to divide by 2 and 5

MATERIALS paper plates, counters

Have students form small groups. Before you begin the activity, discuss times when they have shared objects equally among friends or family members. Explain that the sharing can be shown as a division number sentence.

Present this example: $8 \div 2 =$ ___. Have students model the division by sharing 8 counters between 2 students. Have them note that the result is 2 groups of 4 counters each.

Help students see that 2 groups of 4 counters or 2×4 suggests the multiplication fact $2 \times 4 = 8$.

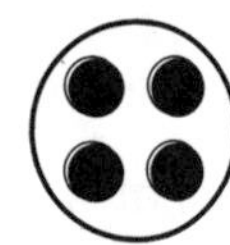

$8 \div 2 = 4$

Continue with other examples for dividing by two. Encourage students to think of a multiplication fact that can help them find the quotient.

Show students how the results of the division can be written in this form:

$$\text{divisor} \longrightarrow 2\overline{)\,8\,} \longleftarrow \text{dividend}, \quad 4 \longleftarrow \text{quotient}$$

Label the divisor, dividend, and quotient.

Repeat the activity for division by 5.
As the students show understanding, provide less guidance. Have students record the division two ways.

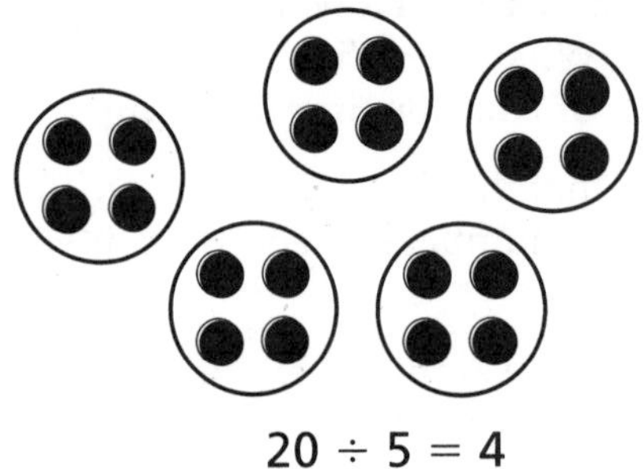

$20 \div 5 = 4$

Grade 4 **Skill 37**

Division Facts (Divide by 2 and 5)

Find 6 ÷ 2 = ■.

Step 1
Show 6 counters.

Think: 2 times what number equals 6?

Step 2

Put the counters into 2 equal groups.

Step 3
Record the division fact.

6 ÷ 2 = 3

So, 6 divided by 2 equals 3.

Find 15 ÷ 5 = ■.

Step 1
Show 15 counters.

Think: 5 times what number equals 15?

Step 2
Put the counters into 5 equal groups.

Step 3
Record the division fact.

15 ÷ 5 = 3

So, 15 divided by 5 equals 3.

Try These

Find the quotient.

1. 

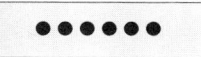

4 ÷ 2 = ☐

2.   

10 ÷ 5 = 

3.  

5 ÷ 5 = 

Go to the next side.

Name ______________________________ Skill ______________________________

Practice on Your Own

Skill 37

Example: Use multiplication facts to help you divide.

Find 10 ÷ 2 = ■.

2 × 5 = 10

10 ÷ 2 = 5 ← quotient

Find 25 ÷ 5 = ■.

5 × 5 = 25

25 ÷ 5 = 5 ← quotient

Show equal groups. Then record the quotient.

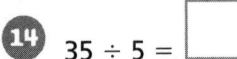

1. 2 ÷ 2 = ☐

2. 4 ÷ 2 = ☐

3. 6 ÷ 2 = ☐

4. 8 ÷ 2 = ☐

5. 5 ÷ 5 = ☐

6. 10 ÷ 5 = ☐

7. 15 ÷ 5 = ☐

8. 20 ÷ 5 = ☐

Find the quotient.

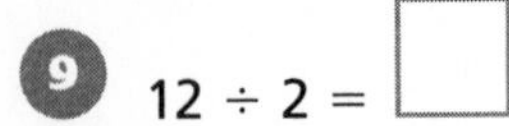

9. 12 ÷ 2 = ☐
10. 14 ÷ 2 = ☐
11. 16 ÷ 2 = ☐
12. 18 ÷ 2 = ☐
13. 30 ÷ 5 = ☐
14. 35 ÷ 5 = ☐
15. 40 ÷ 5 = ☐
16. 45 ÷ 5 = ☐

Check

17. 15 ÷ 5 = ☐

18. 2 ÷ 2 = ☐

19. 20 ÷ 5 = ☐

20. 16 ÷ 2 = ☐

Division Facts (Divide by 3 and 4)

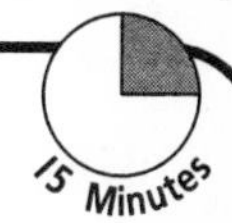

OBJECTIVE Divide by 3 and 4

MATERIALS counters

You may wish to use counters to model Steps 1–3 with the students.

Begin the lesson by writing:

$12 \div 3 = \_\_\_$

$3 \times \_\_\_ = 12$

Have students note that the numbers 12 and 3 are the same for each fact.

Ask: **What multiplication fact can help you find the quotient? Explain.** 3 × 4 = 12; if 3 × 4 = 12, then 12 ÷ 3 = 4 **What does the 3 represent? The 4?** the number of groups; the number of counters in a group

Continue in a similar way for 8 ÷ 4. Emphasize that division means separating the total number of counters into equal groups.

TRY THESE Exercises 1–3 provide models in this order:

- **Exercises 1 and 3** Division facts for 3.
- **Exercise 2** Division facts for 4.

PRACTICE ON YOUR OWN Review the examples with students. For Exercises 1–8 students circle groups to show facts for 3 and 4. Exercises 9–16 are done without visual cues.

CHECK Determine if students can divide models.

Success is indicated by 3 out of 4 correct responses.

Students who successfully complete the **Practice on Your Own** and **Check** are ready to move on to the next skill.

COMMON ERRORS

- Students may not be able to connect the activity of separating of counters into groups into a division sentence.
- Students may not know multiplication facts.

Students who made more than three errors in the **Practice on Your Own**, or who were not successful in the **Check** section, may benefit from the **Alternative Teaching Strategy** on the next page.

Optional

Alternative Teaching Strategy

Model Division Facts for 3 and 4

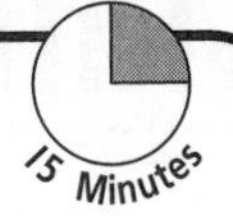

OBJECTIVE Use concrete objects to divide by 3 and 4

MATERIALS paper plates, counters

Have students form small groups. Present this example:

> You have 24 marbles. You want to give 4 marbles each to some of your friends. How many friends will get marbles?

Suggest students solve the problem by using models and repeated subtraction to find the answer. As the students work with models, show the subtraction on the chalkboard.

$$\begin{array}{r}24\\-4\\\hline 20\end{array}\quad\begin{array}{r}20\\-4\\\hline 16\end{array}\quad\begin{array}{r}16\\-4\\\hline 12\end{array}\quad\begin{array}{r}12\\-4\\\hline 8\end{array}\quad\begin{array}{r}8\\-4\\\hline 4\end{array}\quad\begin{array}{r}4\\-4\\\hline 0\end{array}$$

Have students note that they subtracted 6 times. So, 6 friends will each get 4 marbles.

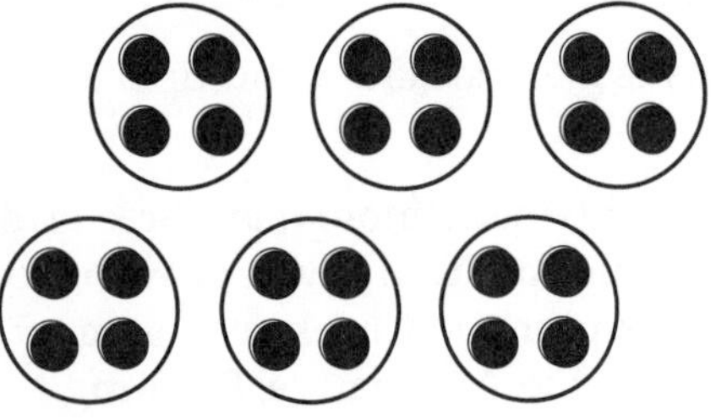

Explain that they can use division to find the answer. Have the students model the division fact. Then write the division fact two ways.

$24 \div 4 = 6$ $\qquad$ $4\overline{)24}$ with quotient 6

Students may conclude that division is easier than repeated subtraction. Remind them also that they can use multiplication facts to remember division facts they may have forgotten.

Repeat the activity several times by modeling other division facts for 3 and 4. Finally, have students divide by 3 and 4 symbolically on their own.

Name ______________________ Skill ______________________

Division Facts (Divide by 3 and 4)

Grade 4 Skill 38

Find 12 ÷ 3 = ■.

Step 1
Show 12 counters.

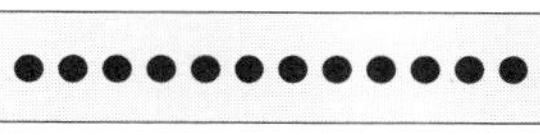

Think: 3 times what number equals 12?

Step 2

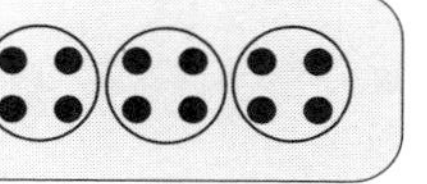

Put the counters into 3 equal groups.

Step 3
Record the division fact.

12 ÷ 3 = 4

So, 12 divided by 3 equals 4.

Find 8 ÷ 4 = ■.

Step 1
Show 8 counters.

Think: 4 times what number equals 8?

Step 2
Put the counters into 4 equal groups.

Step 3
Record the division fact.

8 ÷ 4 = 2

So, 8 divided by 4 equals 2.

Try These

Find the quotient.

1. 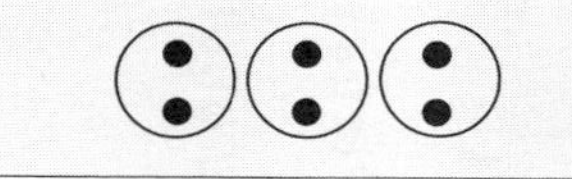

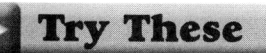

6 ÷ 3 = ☐

2. 

16 ÷ 4 = 

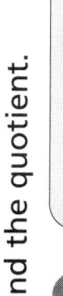

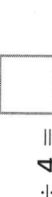

3. 15 ÷ 3 = ☐

Go to the next side.

Name ______________________ Skill ______________________

Practice on Your Own

Skill 38

Example: Use multiplication facts to help you divide.

Find 15 ÷ 3 = ■.

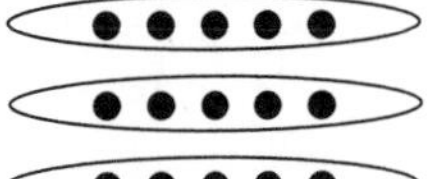

3 × 5 = 15

15 ÷ 3 = 5 ← quotient

Find 20 ÷ 4 = ■.

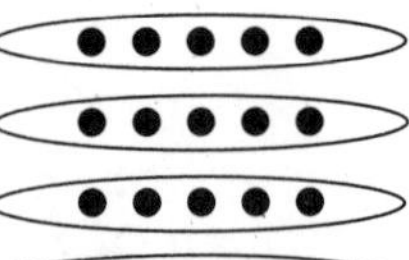

4 × 5 = 20

20 ÷ 4 = 5 ← quotient

Show equal groups. Then record the quotient.

1. 3 ÷ 3 = ☐

2. 6 ÷ 3 = ☐

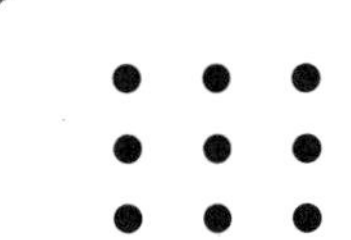

3. 9 ÷ 3 = ☐

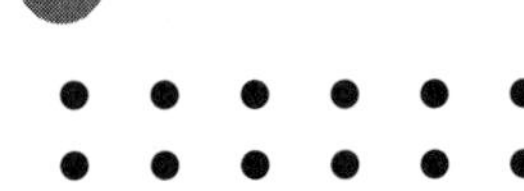

4. 12 ÷ 3 = ☐

5. 4 ÷ 4 = ☐

6. 8 ÷ 4 = ☐

7. 12 ÷ 4 = ☐

8. 16 ÷ 4 = ☐

9. 18 ÷ 3 = ☐
10. 21 ÷ 3 = ☐
11. 24 ÷ 3 = ☐
12. 27 ÷ 3 = ☐

13. 24 ÷ 4 = ☐
14. 28 ÷ 4 = ☐
15. 32 ÷ 4 = ☐
16. 36 ÷ 4 = ☐

Check

Find the quotient.

17. 28 ÷ 4 = ☐

18. 18 ÷ 3 = ☐

19. 27 ÷ 3 = ☐
20. 12 ÷ 4 = ☐

Skill 39 Grade 4

Division Facts (Divide by 6 and 7)

OBJECTIVE Divide by 6 and 7

MATERIALS counters

You may wish to model Steps 1–3 with the students using counters.

Begin by asking: **How are multiplication and division related? Multiplication is the inverse or opposite of division.**

Explain that because they are opposite operations, students can use a multiplication fact to help them remember a division fact.

Ask: **Six times what number equals 12?** $6 \times 2 = 12$

Present this example:

$6 \times 2 = 12$

$12 \div 6 =$ ___

Say: **So, when 12 counters are divided into 6 groups, how many counters are in each group?** 2

Continue in a similar way with $21 \div 7$.

Emphasize that division means separating the counters into equal groups.

TRY THESE Students use models in this order:

- **Exercises 1 and 3** Division facts for 6.
- **Exercise 2** Division facts for 7.

PRACTICE ON YOUR OWN Review the examples with students. Suggest that students refer to Exercises 1–8 when drawing their own models.

CHECK Determine if students can divide by 6 and 7 without models.

Success is indicated by 3 out of 4 correct responses.

Students who successfully complete the **Practice on Your Own** and **Check** are ready to move on to the next skill.

COMMON ERRORS

- Students may not be able to connect the activity of separating of counters into groups into a division sentence.
- Students may not know the multiplication facts.

Students who made more than three errors in the **Practice on Your Own**, or who were not successful in the **Check** section, may benefit from the **Alternative Teaching Strategy** on the next page.

Optional

Alternative Teaching Strategy

Model Division Facts for 6 and 7

20 Minutes

OBJECTIVE Model division facts for 6 and 7

MATERIALS flash cards for multiplication and division facts for 6 and 7, counters

Because multiplication and division are related, many students will use a multiplication fact to find a division fact. So, the learning of multiplication facts is crucial to mastering division facts. This activity reinforces the learning of both skills.

Have students work in pairs and take turns showing flash cards and modeling operations.

Begin with multiplication. One partner shows a multiplication fact, the other student shows the appropriate array. Repeat the activity several times.

Then have one student show a division flash card, while the partner divides counters into groups. Repeat several times.

Finally, using only flash cards, one student shows a division fact, the partner shows the related multiplication fact.

Continue until students show proficiency with the facts for 6 and 7 without using models.

Name ____________________ Skill ____________________

Grade 4

Division Facts (Divide by 6 and 7)

Find 12 ÷ 6 = ■.

Step 1
Show 12 counters.

Think: 6 times what number equals 12?

Step 2
Put the counters into 6 equal groups.

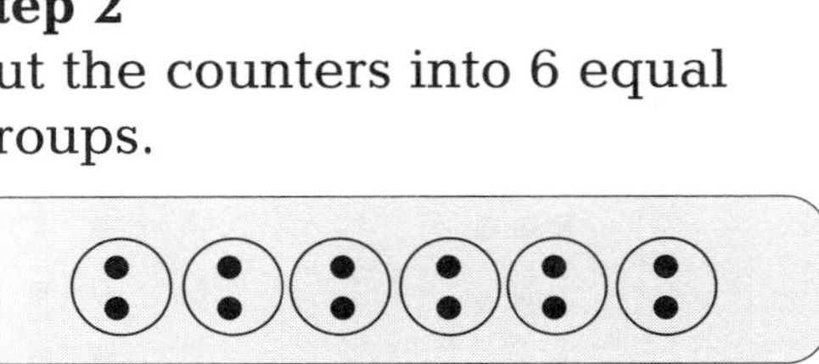

Step 3
Record the division fact.

12 ÷ 6 = 2

So, 12 divided by 6 equals 2.

Find 21 ÷ 7 = ■.

Step 1
Show 21 counters.

Think: 7 times what number equals 21?

Step 2
Put the counters into 7 equal groups.

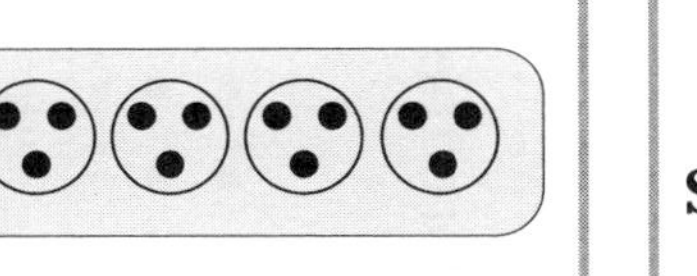

Step 3
Record the division fact.

21 ÷ 7 = 3

So, 21 divided by 7 equals 3.

Try These

Find the quotient.

1. 6 ÷ 6 = ☐

2. 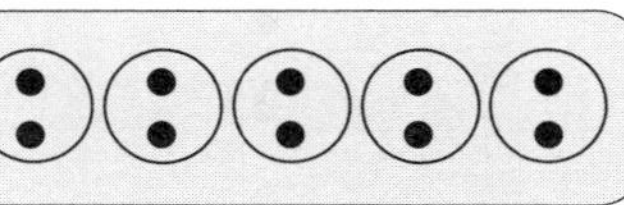

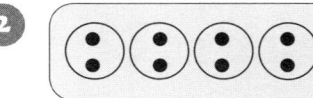

14 ÷ 7 = 

3. 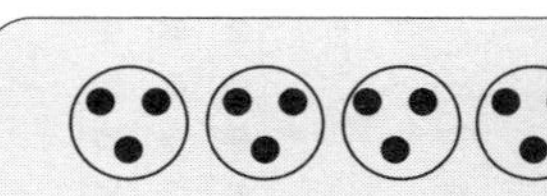

18 ÷ 6 = 

Go to the next side.

Name ______________________ Skill ______________________

Practice on Your Own

Skill 39

Example: Use multiplication facts to help you divide.

Find 30 ÷ 6 = ■.

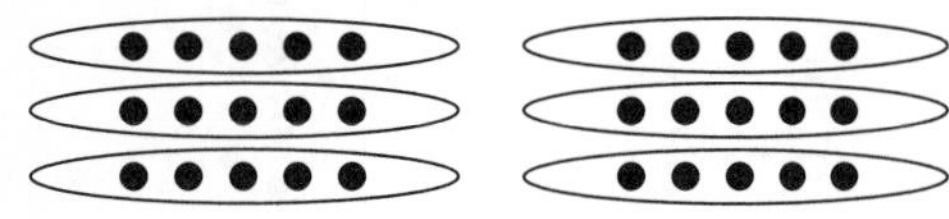

5 × 6 = 30

30 ÷ 6 = 5 ← quotient

Find 35 ÷ 7 = ■.

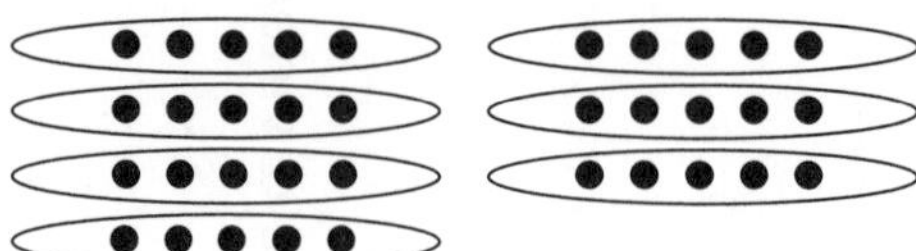

7 × 5 = 35

35 ÷ 7 = 5 ← quotient

Show equal groups. Then record the quotient.

1. 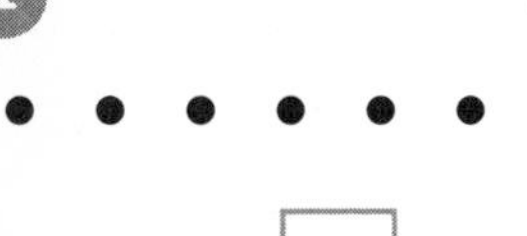 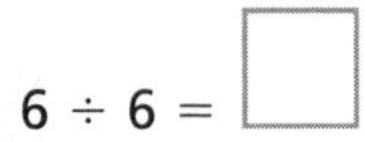 6 ÷ 6 = ☐
2.  12 ÷ 6 = ☐
3. 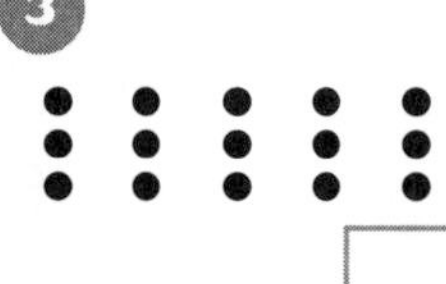  18 ÷ 6 = ☐
4. 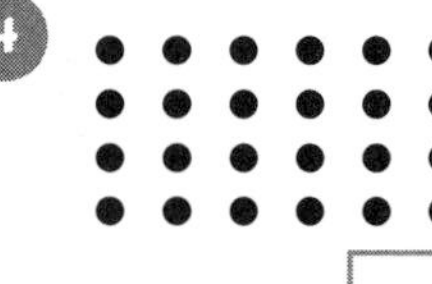 24 ÷ 6 = ☐

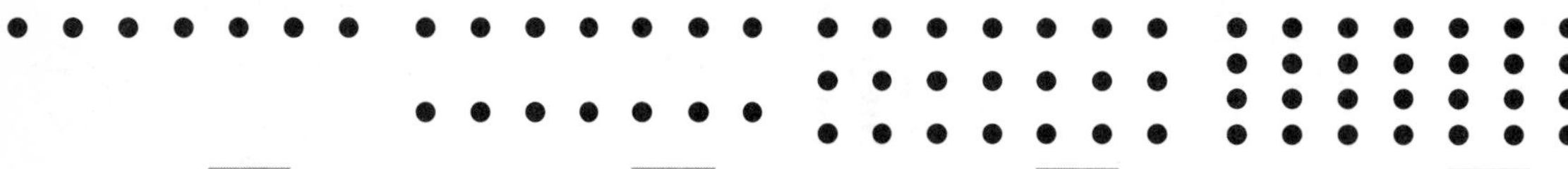

5. 7 ÷ 7 = ☐
6. 14 ÷ 7 = ☐
7. 21 ÷ 7 = ☐
8. 28 ÷ 7 = ☐

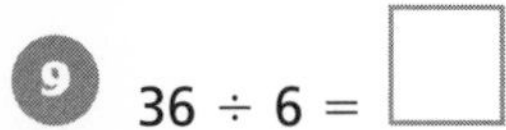    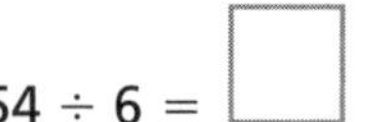

9. 36 ÷ 6 = ☐
10. 42 ÷ 6 = ☐
11. 48 ÷ 6 = ☐
12. 54 ÷ 6 = ☐

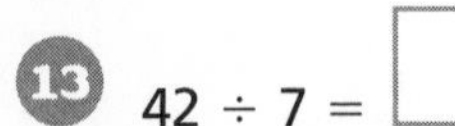 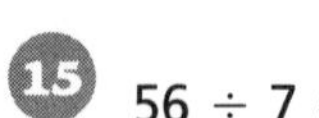 

13. 42 ÷ 7 = ☐
14. 49 ÷ 7 = ☐
15. 56 ÷ 7 = ☐
16. 63 ÷ 7 = ☐

Check

Find the quotient.

17. 56 ÷ 7 = ☐
18. 48 ÷ 6 = ☐
19. 42 ÷ 6 = ☐
20. 49 ÷ 7 = ☐

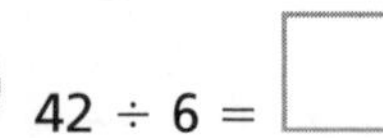 

Division Facts (Divide by 8 and 9)

OBJECTIVE Divide by 8 and 9

MATERIALS counters

You may wish to have students model the steps in the lesson with counters.

Ask a student to explain the meaning of division and how it relates to multiplication. Guide students to understand that since multiplication is the inverse, or opposite, of division, multiplication facts can be used to help divide.

Direct students' attention to the first example.

Ask: **What multiplication fact can help you find 16 ÷ 8?** 8 × 2 = 16

Explain. If 8 × 2 = 16, then 16 ÷ 8 = 2

After the counters are divided into equal groups, what number represents the numbers of groups? 8 **The number of counters in a group?** 2

Write the following on the board:

$$8 \times 2 = 16$$

$$16 \div 8 = 2$$

Repeat the activity to find 27 ÷ 9.

TRY THESE Exercises 1–3 use models to divide facts for 8 and 9.

- **Exercise 1** Division facts for 8.
- **Exercise 2** Division facts for 9.
- **Exercise 3** Division facts for 8.

PRACTICE ON YOUR OWN Review the examples at the top of the page. In Exercises 1–8, students show equal groups for division facts for 8 and 9, and record the quotient. In Exercises 9–16, students divide without models. The exercises are sequenced so students may use a pattern to find each succeeding quotient.

CHECK Determine if students can divide by 8 and 9. Success is indicated by 3 out of 4 correct responses.

Students who successfully complete the **Practice on Your Own** and **Check** are ready to move on to the next skill.

COMMON ERRORS

- Students may not know multiplication facts for 8 and 9, and thus use an incorrect fact to help them find a quotient.
- Students may not understand how multiplication and division are related.

Students who made more than three errors in the **Practice on Your Own**, or who were not successful in the **Check** section, may benefit from the **Alternative Teaching Strategy** on the next page.

Optional

Alternative Teaching Strategy

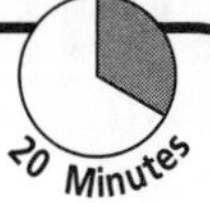

Model Division Facts for 8 and 9

OBJECTIVE Use models to divide by 8 and 9

MATERIALS paper plates, counters

Have students form small groups. Distribute paper plates and counters.

Have students put 3 counters on each of 8 plates.

Ask: **What multiplication fact represents the model?** 8 × 3 = 24

Suggest students gather the 24 counters together, then model 24 ÷ 8.

Explain that 8 × 3 = 24 and 24 ÷ 8 = 3 are *related multiplication and division facts.* If students know that 8 × 3 = 24, they also know that 24 ÷ 8 = 3.

Repeat the activity for other facts for 8 and facts for 9.

Point out that students can use what they have already learned to remember the facts for 8 and 9. Write the following on a chart.

Already Learned	Related Fact
8 ÷ 1 = 8	8 ÷ 8 = 1
16 ÷ 2 = 8	16 ÷ 8 = 2
24 ÷ 3 = 8	24 ÷ 8 = 3
32 ÷ 4 = 8	32 ÷ 8 = 4
9 ÷ 1 = 9	9 ÷ 9 = 1
18 ÷ 2 = 9	18 ÷ 9 = 2
27 ÷ 3 = 9	27 ÷ 9 = 3
36 ÷ 4 = 9	36 ÷ 9 = 4

Conclude by having students try some exercises without using models. The goal is to have students use related fact strategies to divide symbolically.

Name ______________________ Skill ______________________

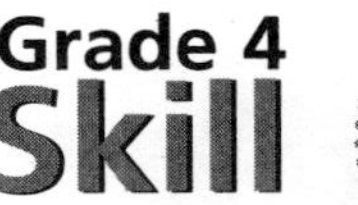

Grade 4 Skill 40

Division Facts (Divide by 8 and 9)

Find 16 ÷ 8 = ■

Step 1
Show 16 counters.

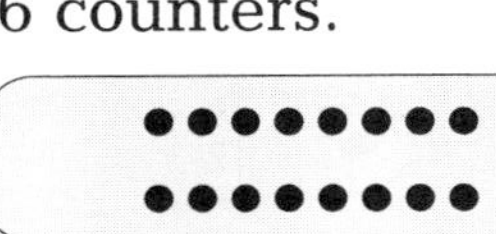

Think: 8 times what number equals 16?

Step 2
Put the counters into 8 equal groups.

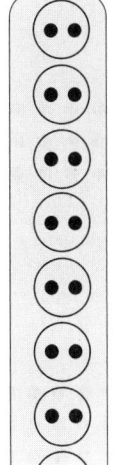

Step 3
Record the division fact.

16 ÷ 8 = 2

So, 16 divided by 8 equals 2.

Find 27 ÷ 9 = ■

Step 1
Show 27 counters.

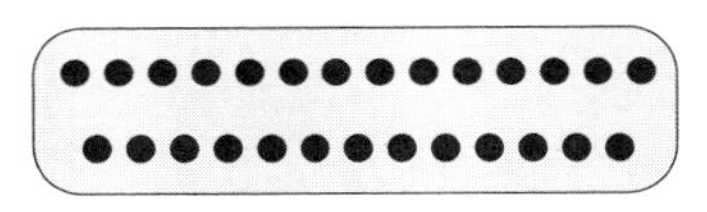

Think: 9 times what number equals 27?

Step 2
Put the counters into 9 equal groups.

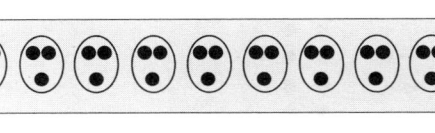

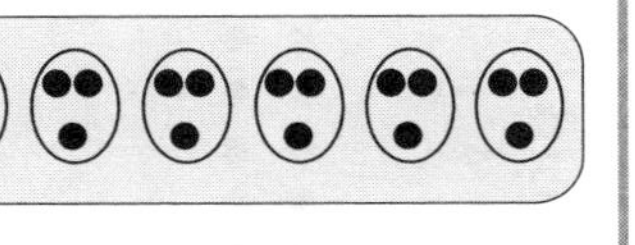

Step 3
Record the division fact.

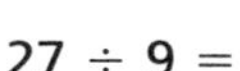

27 ÷ 9 = 3

So, 27 divided by 9 equals 3.

Try These

Find the quotient.

1. 8 ÷ 8 = ☐

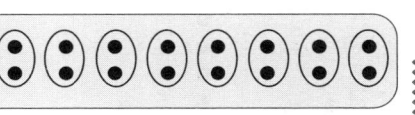

2. 

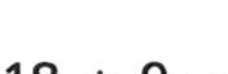
18 ÷ 9 = ☐

3. 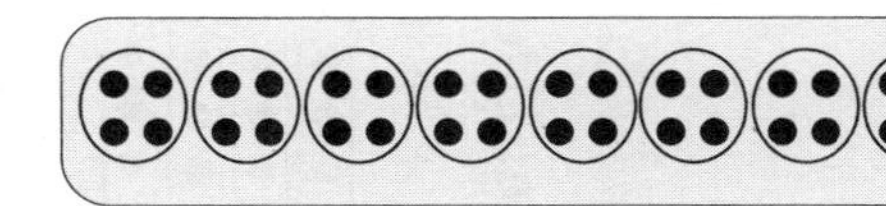
32 ÷ 8 = ☐

Go to the next side.

Name ____________________ Skill ____________________

Practice on Your Own

Skill 40

Example: Use multiplication facts to help you divide.

Find $48 \div 8 = \blacksquare$.

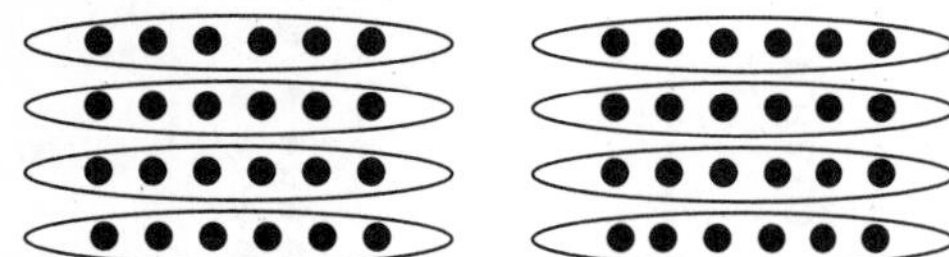

$6 \times 8 = 48$

48 ÷ 8 = 6 ← quotient

Find $36 \div 9 = \blacksquare$.

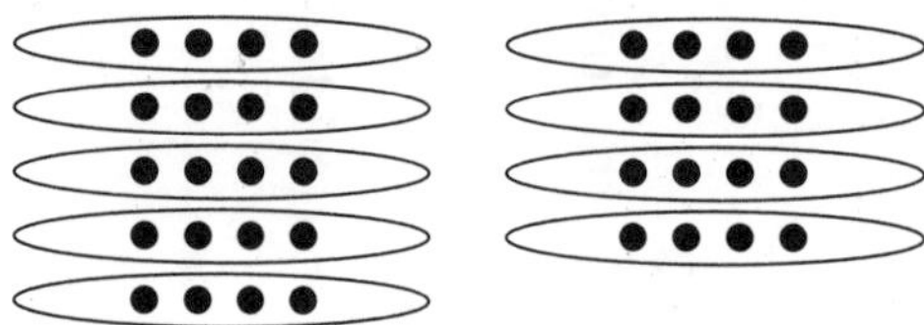

$4 \times 9 = 36$

36 ÷ 9 = 4 ← quotient

Show equal groups. Then record the quotient.

1. 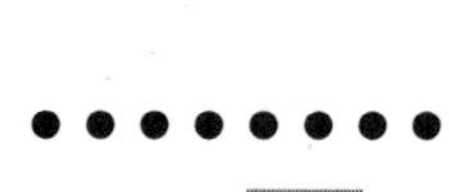 $8 \div 8 =$ ☐
2. 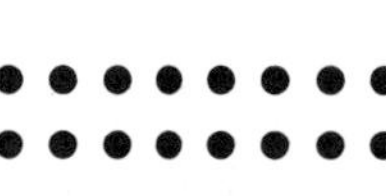 $16 \div 8 =$ ☐
3.  $24 \div 8 =$ ☐
4.  $32 \div 8 =$ ☐
5. 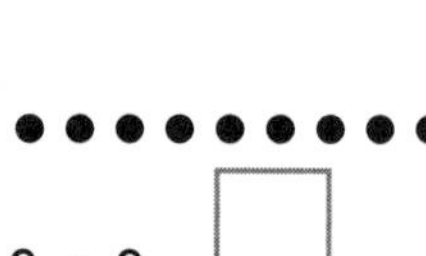 $9 \div 9 =$ ☐
6. 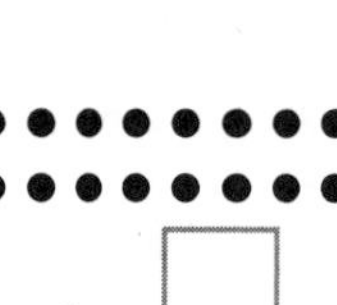 $18 \div 9 =$ ☐
7.  $27 \div 9 =$ ☐
8.  $45 \div 9 =$ ☐

Find the quotient.

9. $40 \div 8 =$ ☐
10. $56 \div 8 =$ ☐
11. $64 \div 8 =$ ☐
12. $72 \div 8 =$ ☐
13. $54 \div 9 =$ ☐
14. $63 \div 9 =$ ☐
15. $72 \div 9 =$ ☐
16. $81 \div 9 =$ ☐

Check

Find the quotient.

17. $18 \div 9 =$ ☐
18. $64 \div 8 =$ ☐
19. $36 \div 9 =$ ☐
20. $32 \div 8 =$ ☐

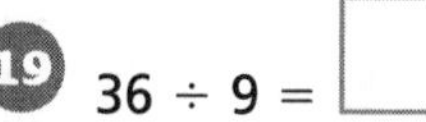

Meaning of Division

OBJECTIVE Form equal groups to divide

MATERIALS counters

Begin by asking students for different ways of dividing 6 counters into smaller equal size groups. Demonstrate with the counters as students suggest forming 2 groups of 3, 3 groups of 2, or 6 groups of 1. Point out that for each situation, the smaller groups each had the same number of counters.

Direct the students' attention to the first example.

Ask: **What does it mean to form equal groups?** The groups all have the same number in them. **How many triangles are there?** 8 **How many equal groups do you have to form?** 2 **How many triangles are in each group?** 4 **Can one group have 6 triangles and the other group have 2 triangles?** No, the groups have to be equal.

Continue to ask similar questions as you work through the next example.

TRY THESE Exercises 1–3 model the type of exercises students will find on the **Practice on Your Own** page.

- **Exercise 1** Form 2 groups of 3.
- **Exercise 2** Form 2 groups of 5.
- **Exercise 3** Form 3 groups of 3.

PRACTICE ON YOUR OWN Review the example at the top of the page. Ask students to explain each step in dividing the 8 squares into 4 equal groups.

CHECK Determine if students know how to form equal groups. Success is determined by 3 out of 3 correct responses.

Students who successfully complete the **Practice on Your Own** and **Check** are ready to move on to the next skill.

COMMON ERRORS

- Students may form unequal groups.
- Students may confuse the number of groups with how many are in each group.

Students who made more than two errors in the **Practice on Your Own**, or who were not successful in the **Check** section, may benefit from the **Alternative Teaching Strategy** on the next page.

Optional

Alternative Teaching Strategy

Model the Meaning of Division

OBJECTIVE Use counters to model the meaning of division

MATERIALS counters

Explain why it is necessary to be able to divide a group of objects equally. You may wish to relate it to a real life example:

Suppose you and 2 friends want to share equally 12 crackers with peanut butter.

Ask: **How many people will be sharing crackers?** 3

Distribute the counters. Have the students start by using the counters to model the 12 crackers.

Ask: **How many groups of crackers do you want to make?** 3

Have students draw three circles to represent the three groups. Then direct the students to start by putting one counter in each circle. Have them repeat this until all of the counters are gone.

Ask: **Does each group have an equal number of counters?** Yes **How many counters are in each group?** 4 counters

Repeat this activity with similar examples. Ask the students to tell you each time how many are in each group and how many groups there are. Have them use the word *equal* when discussing the number in each group.

○○○○○○○○○○○○

12 crackers

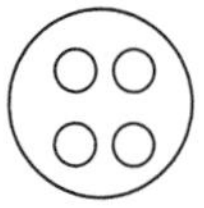 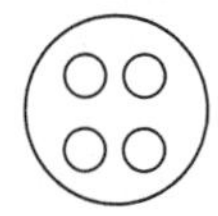 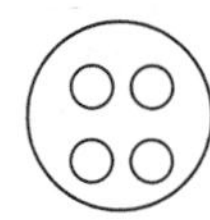

3 groups of 4

$12 \div 3 = 4$

Name ____________________ Skill ____________________

Grade 4 Skill 41

Meaning of Division

When you separate a large group into smaller equal groups, you are dividing.

Find the number in each group.
There are 8 triangles in all.

△△△△△△△△

Divide the 8 triangles into 2 equal groups.

[△△△△] [△△△△]

So, there are 4 triangles in each group.

Find the number of equal groups.
There are 12 triangles in all.

△△△△△△△△△△△△

Divide the 12 triangles into groups of 3 triangles.

[△△△] [△△△] [△△△] [△△△]

So, there are 4 groups.

Try These

Complete.

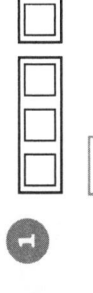

1. [□□□] [□□□]

   ☐ in all

   ☐ in each group

   ☐ groups

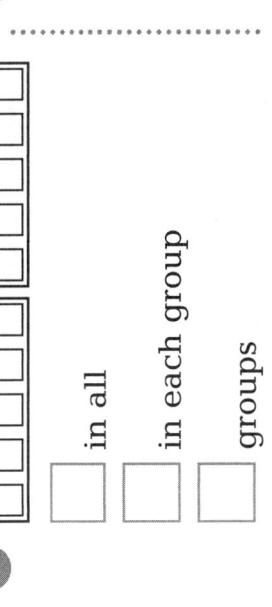

2. [□□□□□] [□□□□□]

   ☐ in all

   ☐ in each group

   ☐ groups

3. [□□□] [□□□] [□□□]

   ☐ in all

   ☐ in each group

   ☐ groups

Go to the next side.

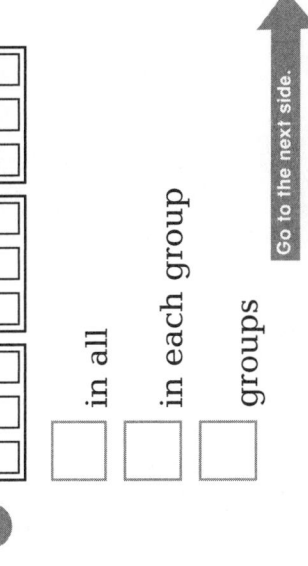

Name ______________________ Skill ______________________

Practice on Your Own

Skill 41

Think:
There are 8 squares in all.
There are 4 equal groups.
There are 2 squares in each group.

Complete.

1. ☐ in all
☐ in each group
☐ groups

2. ☐ in all
☐ in each group
☐ groups

3. ☐ in all
☐ in each group
☐ groups

Find the number in each group.

4. Divide 10 circles into 2 equal groups.
☐ in each group

5. Divide 12 circles into 4 equal groups.
☐ in each group

6. Divide 15 circles into 5 equal groups.
☐ in each group

Find the number of groups.

7. Divide 10 squares into groups of 5 squares.
☐ groups

8. Divide 14 squares into groups of 2 squares.
☐ groups

9. Divide 16 squares into groups of 4 squares.
☐ groups

Check

Complete.

10. ☐ in all ☐ groups
☐ in each group

11. Divide the 14 triangles into 2 equal groups.
☐ in each group

12. Divide the 14 circles into groups of 7 circles.
☐ groups

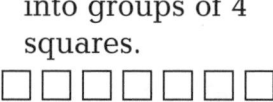

Skill 42 Grade 4 — Divide with Remainders

OBJECTIVE Divide by 1-digit divisors, with remainders in the quotient

MATERIALS counters

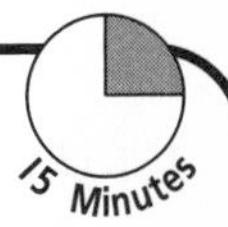

You may wish to have students model the example with counters.

When students have formed the 4 groups of 6, have them observe that they could form 4 equal groups, but 1 counter is left over. Note that, sometimes when they divide there will be some counters leftover. This is called the *remainder*.

Connect Steps 1–3 to the modeling just completed.

For Step 2, ask: **What number shows the total number of counters?** 25 **What number tells how many counters are in each group?** 6 **What number tells how many equal groups?** 4 **What is the remainder?** 1

TRY THESE In Exercises 1–4, students should circle groups to show each division.

- **Exercise 1** Circle groups of 4 with 2 left over.
- **Exercise 2** Circle groups of 3 with 2 left over.
- **Exercise 3** Circle groups of 5 with 1 left over.
- **Exercise 4** Circle groups of 4 with 2 left over.

PRACTICE ON YOUR OWN Review the example at the top of the page. Note that the example has a two-digit quotient. Emphasize that the "multiply, subtract" step is followed twice. Point out also that the remainder is always less than the divisor. You may wish to give another similar example, such as 44 ÷ 3.

Exercise 9 has a two-digit quotient; monitor students' work to provide help if necessary.

CHECK Success is determined by 3 out of 3 correct responses.

Students who successfully complete the **Practice on Your Own** and **Check** are ready to move on to the next skill.

COMMON ERRORS

- Students may forget to record the remainder.
- In divisions where the quotient is greater than 9, students may not finish the final step, and may record a remainder that is greater than the divisor.

Students who made more than three errors in the **Practice on Your Own**, or who were not successful in the **Check** section, may benefit from the **Alternative Teaching Strategy** on the next page.

Optional

Alternative Teaching Strategy

Model Division with Remainders

15 Minutes

OBJECTIVE Use tiles and arrays to understand division with remainders

MATERIALS square tiles, or counters; triangular flash cards

Students may need to take a step back in building understanding. Begin the activity by having students work with an exercise that does not have a remainder, for example 30 ÷ 5.

Distribute the tiles. Demonstrate how to form an array of rows and columns with the tiles.

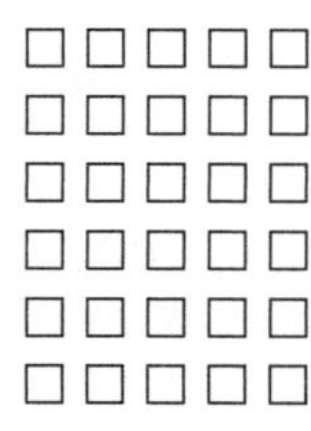

30 ÷ 5 = 6

Observe with the students that there are 6 rows of 5 tiles. All the rows have an equal number. If necessary, have students try several exercises similar to this one before moving on to modeling division with remainders.

Now ask students to model 27 ÷ 5. They will discover that they cannot complete an array. There are 5 rows of 5 tiles with two left over.

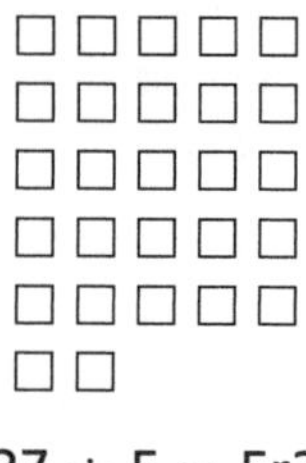

27 ÷ 5 = 5r2

Continue with other examples.

When students demonstrate understanding of equal groups and remainders, have students record the division symbolically.

If you find that students cannot remember division facts, give a short daily review. You may have pairs of students work together with flash cards. Triangular flash cards are simple to make and useful.

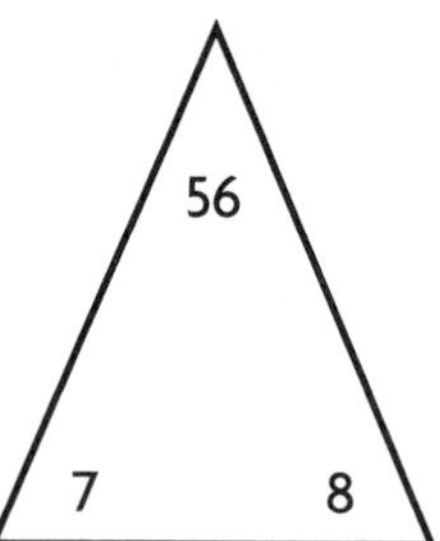

Name ____________________ Skill ____________________

Grade 4 Skill 

Divide with Remainders

Sometimes when you divide counters into equal groups, you may have some counters left over. The number left over is called the **remainder.**
Find 25 ÷ 6.

Divide 25 counters into groups of 6.

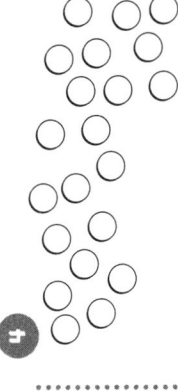

Think: The left over counter is the remainder.

Divide.

$$\begin{array}{r} 4 \\ 6\overline{)25} \\ -\,24 \\ \hline 1 \end{array}$$

4 ← number of groups
number of counters in a group → 6)25 ← number of counters
1 ← remainder

Write the remainder next to the quotient.

$$\begin{array}{r} 4\text{ r}1 \\ 6\overline{)25} \\ -\,24 \\ \hline 1 \end{array}$$

4 r1 ← quotient and remainder

Try These

Divide the counters into equal groups to find the quotient and remainder.

1. 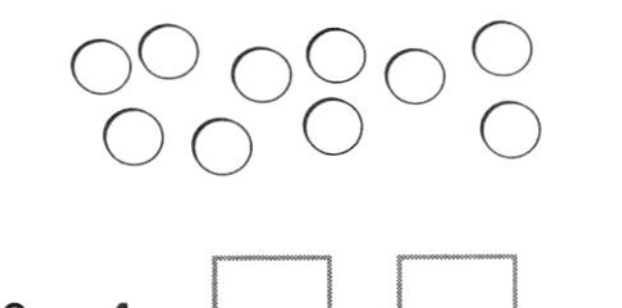
10 ÷ 4 = ☐ r ☐

2. 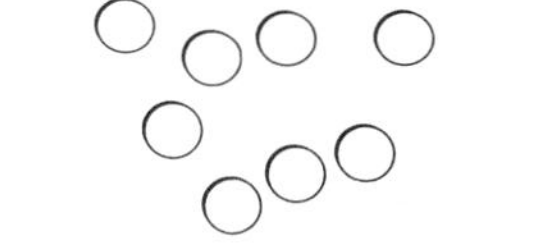
8 ÷ 3 = ☐ r ☐

3. 
16 ÷ 5 = ☐ r ☐

4. 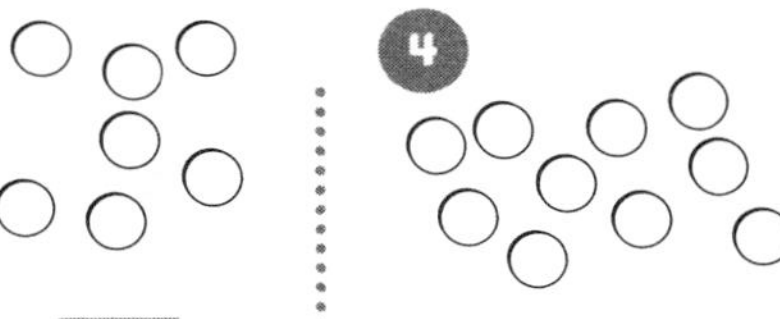
22 ÷ 4 = ☐ r ☐

Go to the next side.

Name ______________________ Skill ______________________

Practice on Your Own

Find 35÷ 2.

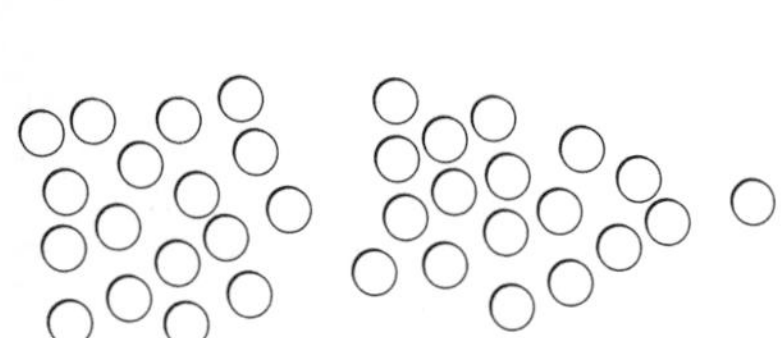

$$\begin{array}{r} 17 \text{ r1} \\ 2\overline{)35} \\ -\ 2 \\ \hline 15 \\ -\ 14 \\ \hline 1 \end{array}$$

Think:
Make 2 equal groups.

So, $35 \div 2 = 17$ r1.

Divide the counters into groups to find the quotient and remainder.

1

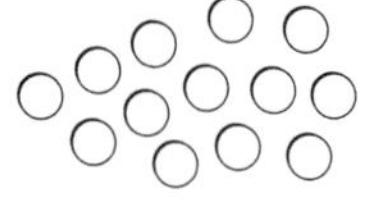

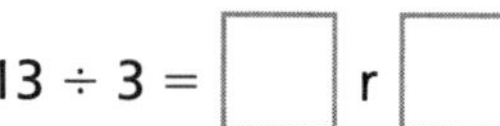

13 ÷ 3 = ☐ r ☐

2

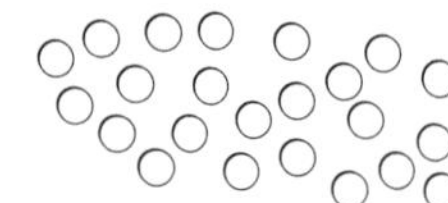
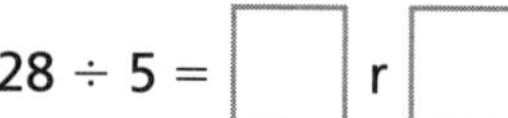

28 ÷ 5 = ☐ r ☐

3

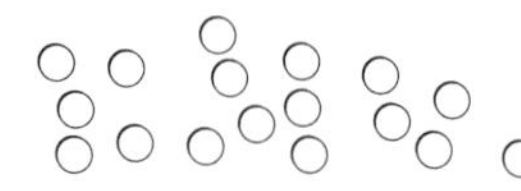

17 ÷ 7 = ☐ r ☐

4

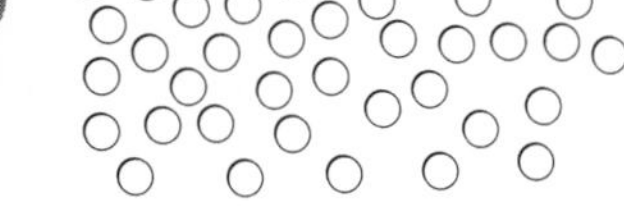

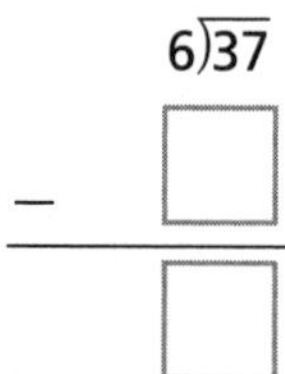

☐ r ☐
6)37
− ☐
☐

5

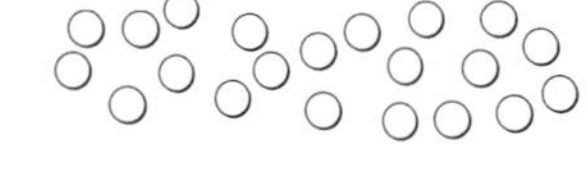
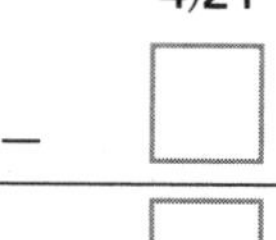

☐ r ☐
4)21
− ☐
☐

6

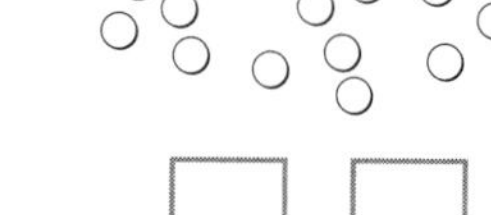
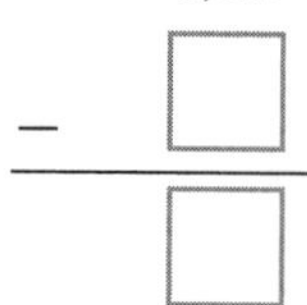

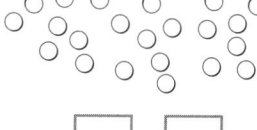

☐ r ☐
8)26
− ☐
☐

Find the quotient and remainder.

7 5)21

8 6)44

9 7)75

Check

10 9)28

11 3)32

12 9)56

Skill 43 Grade 4

Use Compatible Numbers

OBJECTIVE Estimate a quotient by using compatible numbers

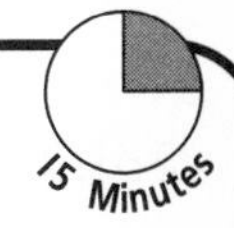

Before beginning, give a brief review of compatible numbers. Display this division: 28 ÷ 5.

Say: **Think of as many facts as you can that are close to this expression.** **Possible answers: 25 ÷ 5, 27 ÷ 3, 28 ÷ 4, and 30 ÷ 5**

Observe with students that they can find these quotients using mental math. Then have students decide which of the facts they listed might give an estimated quotient close to the actual quotient. Work with the students to find the actual quotient, then compare the quotients from the facts they chose to the result. Note that the facts students chose are compatible numbers.

Direct students' attention to the example. Note that numbers compatible with 4 are numbers that 4 divides evenly, such as 8, 12, or 16.

Ask: **What compatible number can you use to estimate the quotient 38 ÷ 4?** Try 36. **It is close to 38 and 4 divides 36 evenly. 36 ÷ 4 = 9. The compatible numbers are 36 and 4.**

How do compatible numbers help you estimate the quotient? **They are easy to divide.**

Students should know that they can use basic multiplication and division facts to help find compatible numbers.

TRY THESE Exercises 1–3 prompt students as they estimate quotients using compatible numbers.

- **Exercise 1** Compatible numbers 27, 3.
- **Exercise 2** Compatible numbers 48, 6.
- **Exercise 3** Compatible numbers 56, 9.

PRACTICE ON YOUR OWN Work through the example with students. Have them identify the compatible numbers. Explain that 90 is one number close to 94 that is divisible by 9.

CHECK Determine if students know how to use compatible numbers to estimate a quotient.

Success is determined by 3 out of 3 correct responses.

Students who successfully complete the **Practice on Your Own** and **Check** are ready to move on to the next skill.

COMMON ERRORS

- Students may not know basic multiplication and division facts.
- Students may not use the closest compatible numbers to the actual numbers, which results in an estimate that may be too high or too low.

Students who made more than two errors in the **Practice on Your Own**, or who were not successful in the **Check** section, may benefit from the **Alternative Teaching Strategy** on the next page.

Optional

Alternative Teaching Strategy

Choose Compatible Numbers

OBJECTIVE Estimate a quotient by using compatible numbers

MATERIALS large index cards

Prepare a card with a division sentence and at least four cards with division facts that are close to the division sentence.

Display a card, for example, $40 \div 6 = \square$. Then explain to the students that they are to estimate the quotient using compatible numbers. Remind them that compatible numbers are numbers close to the actual numbers. Compatible numbers are easier to divide because they divide evenly.

$40 \div 6 = \square$

$42 \div 6 = \square$

$40 \div 5 = \square$

$36 \div 6 = \square$

$48 \div 6 = \square$

Have students look at the cards with the close facts and choose the fact they think is closest to $40 \div 6 = \square$. Then have them divide 40 by 6 and compare the actual quotient to the quotients of the close facts. **$40 \div 6 = 6r4$ and $42 \div 6 = 7$ are the closest.** Point out that $42 \div 6 = 7$ is closest because the divisor remains the same and the dividend was increased by only 2.

Repeat this activity several times with other division sentences and close facts until the students are able to estimate without the fact cards. Have students note that although there could be more than one estimate, the closest estimate is preferable.

$40 \div 6 = 6r4$

$42 \div 6 = 7$ ←closest

$40 \div 5 = 8$

$36 \div 6 = 6$

$48 \div 6 = 8$

Name ______________________ Skill ______________

Grade 4 **Skill** 43

Use Compatible Numbers

You can estimate a quotient by using **compatible numbers**.

Compatible numbers are numbers close to the actual numbers and can be divided evenly.

Estimate $38 \div 4 = ■$.

Think: 36 is close to 38.
36 can be divided evenly by 4.

$38 \div 4 = ■$

$36 \div 4 = 9$ ← compatible numbers

So, $38 \div 4$ is about 9.

Estimate $34 \div 8 = ■$.

Think: 34 is close to 32.
32 can be divided evenly by 8.

$34 \div 8 = ■$

$32 \div 8 = 4$ ← compatible numbers

So, $38 \div 4$ is about 4.

Try These

Estimate the quotient. Use compatible numbers.

1. $28 \div 3 = ■$

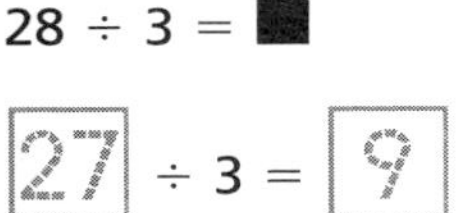

Think:
27 is close to 28.
27 can be divided
evenly by 3.

27 ÷ 3 = 9

So, $28 \div 3$ is about ☐.

2. $47 \div 6 = ■$

Think: 48 is close to 47. 48 can be divided evenly by 6.

☐ ÷ ☐ = ☐

So, $47 \div 6$ is about ☐.

3. $52 \div 9 = ■$

Think: 52 is close to 54. 54 can be divided evenly by 9.

☐ ÷ ☐ = ☐

So, $52 \div 9$ is about ☐.

Go to the next side.

Name ______________________ Skill ______________________

Practice on Your Own

Estimate: 94 ÷ 9 = ■.

90 ÷ 9 = 10 (compatible numbers)

Think:
90 is close to 94,
90 can be divided evenly by 9.

So, 94 ÷ 9 is about 10.

Estimate. Use compatible numbers.

1. 25 ÷ 6 = ■.

Think:
24 is close to 25,
24 can be divided evenly by 6.

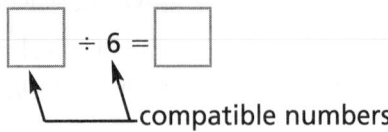

☐ ÷ 6 = ☐ (compatible numbers)

So, 25 ÷ 6 is about ☐.

2. 70 ÷ 8 = ■.

Think:
72 is close to 70,
72 can be divided evenly by 8.

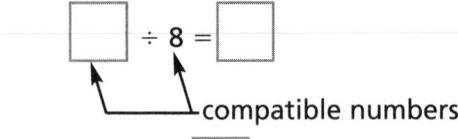

☐ ÷ 8 = ☐ (compatible numbers)

So, 70 ÷ 8 is about ☐.

3. 52 ÷ 5 = ■

☐ ÷ ☐ = ☐

So, 52 ÷ 5 is about ☐.

4. 47 ÷ 7 = ■

☐ ÷ ☐ = ☐

So, 47 ÷ 7 is about ☐.

5. 66 ÷ 8 = ■

☐ ÷ ☐ = ☐

So, 66 ÷ 8 is about ☐.

6. 84 ÷ 9 is about ☐.
7. 61 ÷ 7 is about ☐.
8. 41 ÷ 6 is about ☐.

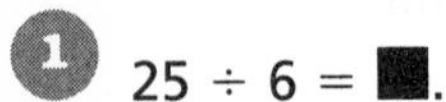

Check

Estimate. Use compatible numbers.

9. 32 ÷ 6 is about ☐.
10. 58 ÷ 9 is about ☐.
11. 73 ÷ 8 is about ☐.

Skill 44 Grade 4 Divide by 1-Digit Numbers

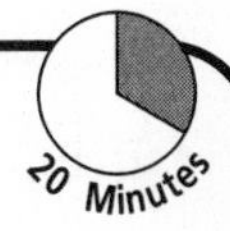

OBJECTIVE Divide 2-digit and 3-digit numbers by 1-digit numbers

Begin by recalling that the answer to a division problem is the *quotient*. Point out the place-value chart above the quotient in the division problems.

Ask: **What number is being divided?** 128 **What number will you divide by?** 8 **If you begin dividing with the left-most digit in 128, what will you divide first?** 1 **Is that digit greater than the divisor?** No, 1< 8

Point out that since 1 < 8, students can use 12 tens to begin dividing.

How many whole groups of 8 are in 12? 1 **How do you place the 1 in the quotient?** In the tens place.

Work through Steps 2 and 3, emphasizing the "divide, multiply, subtract, compare" steps used in dividing. Have students pay special attention to the compare step, noting that if the remainder is greater than the divisor, then continue dividing.

TRY THESE Exercises 1–4 model the type of exercises students will find on the **Practice on Your Own** page.

- **Exercises 1–3** 2-digit dividends.
- **Exercise 4** 3-digit dividends.

PRACTICE ON YOUR OWN Review the example at the top of the page. Ask students to explain why the first digit is placed in the tens place.

CHECK Determine if students can set up the division correctly, aligning digits they divide and placing the quotient in the correct place. Success is determined by 2 out of 3 correct responses.

Students who successfully complete the **Practice on Your Own** and **Check** are ready to move on to the next skill.

COMMON ERRORS

- Students may write the digits of the quotient in the wrong place.
- Students may forget to subtract before bringing down the next digit in the dividend.
- Students may multiply or subtract incorrectly.

Students who made more than two errors in the **Practice on Your Own**, or were not successful in the **Check** section, may benefit from the **Alternative Teaching Strategy** on the next page.

Optional

Alternative Teaching Strategy
Model Dividing by 1-Digit Numbers

OBJECTIVE Use base-ten blocks to model division of 1-digit numbers

MATERIALS base-ten blocks

You may wish to begin by asking students to suggest a possible context for dividing 34 by 2. **Possible responses: number of students in each of 2 buses, cost of each of 2 shirts, etc.**

Write the division on the board for the students to copy on their papers.

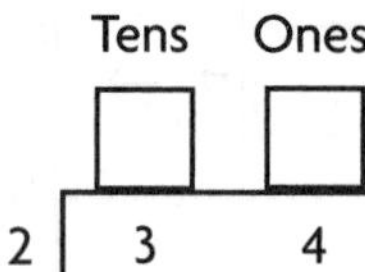

Distribute the base-ten blocks.

Ask: **Into how many groups are you dividing 34?** 2 groups

How many tens blocks can you put in each of 2 groups? 1 each

Where do you record the 1 ten in each group in the division? **Above the 3 in the tens place.** **What do you need to do next to the ten block to divide the blocks evenly between the two groups?** **Regroup the leftover ten as 10 ones.**

How many ones do you have now? **14** **Can you divide 14 ones evenly between the 2 groups?** **Yes.** **How many ones does each group have?** **7** **Where do you record the 7 ones?** **Above the 4 in the ones place.**

Repeat for exercises with 3-digit dividends. When the students show understanding of the division process, remove the base-ten blocks and have them try an exercise using only paper and pencil. Ask students to tell you each step as they complete the division.

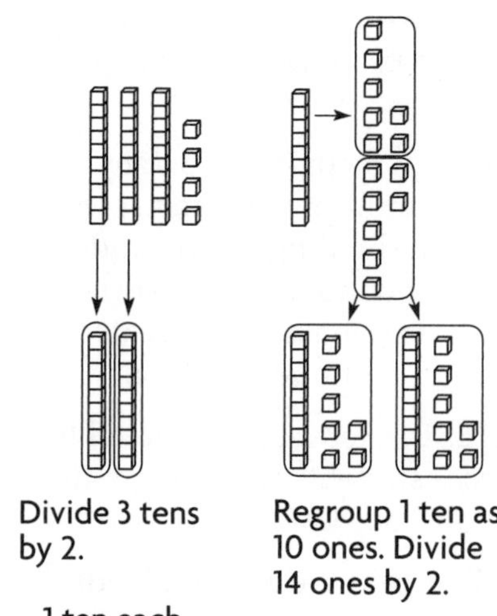

Divide 3 tens by 2.

1 ten each

Regroup 1 ten as 10 ones. Divide 14 ones by 2.

7 ones each

So, $34 \div 2 = 17$.

Name ______________________ Skill ______________________

Grade 4
Skill 44

Divide by 1-Digit Numbers

Find 128 ÷ 8.

Step 1
Decide where to place the first digit in the quotient.

Hundreds	Tens	Ones
quotient →	☐	☐
8)1	2	8

$12 > 8$, so use 12 tens. Place the first digit in the tens place.

Think: $1 < 8$, so look at the tens.

Step 2
Divide the 12 tens.

Hundreds	Tens	Ones	
	1	☐	Divide 12 ÷ 8.
8)1	2	8	Multiply 8 × 1.
−	8		Subtract 12 − 8.
	4		Compare 4 < 8.

Think: The remainder must be less than the divisor.

Step 3
Bring down the ones. Divide.

Hundreds	Tens	Ones	
	1	6	
8)1	2	8	
−	8	↓	Divide 48 ÷ 8.
	4	8	Multiply 8 × 6.
	− 4	8	Subtract 48 − 48.
		0	Compare 0 < 8.

There is no remainder.
So, 128 ÷ 8 = 16.

Try These

Complete the division.

1. 16 ÷ 2 = ■

Tens	Ones
	☐
2)1	6
− ☐	☐
	☐

2. 45 ÷ 3 = ■

Tens	Ones
☐	☐
3)4	5
− ☐	↓
☐	☐
− ☐	☐
	☐

3. 64 ÷ 4 = ■

Tens	Ones
☐	☐
4)6	4
− ☐	↓
☐	☐
− ☐	☐
	☐

4. 108 ÷ 3 = ■

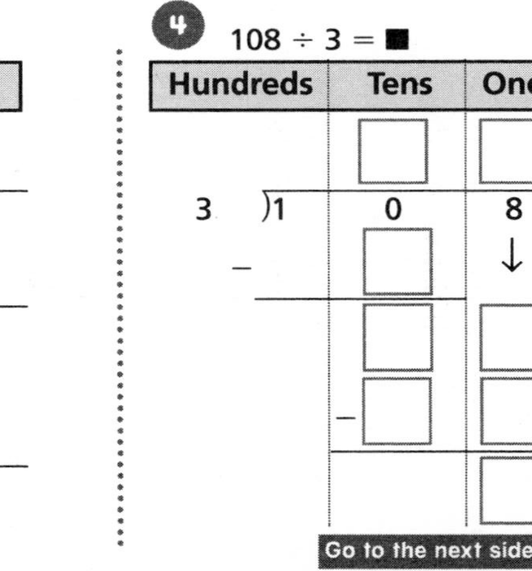

Hundreds	Tens	Ones
	☐	☐
3)1	0	8
−	☐	↓
	☐	☐
	− ☐	☐
		☐

Go to the next side.

Name ______________________ Skill ______________________

Practice on Your Own

Skill 44

Find 108 ÷ 9.

Think:
Decide where to place the first digit in the quotient. 1 < 9, so look at the tens place.

Hundreds	Tens	Ones
quotient →	1	2
divisor → 9)1	0	8
−	9	↓
	1	8
	− 1	8
		0

Remember:
Bring down the ones.
Multiply: 9 × **2** = 18.
Subtract: 18 − 18.
Compare: 0 < 8.

So, 108 ÷ 9 = 12.

Complete the division.

1. 36 ÷ 3 = ■

Tens	Ones
□	□
3)3	6
− □	↓
□	□
	− □
	□

2. 72 ÷ 6 = ■

Tens	Ones
□	□
6)7	2
− □	↓
□	□
− □	□
	□

3. 120 ÷ 6 = ■

Hundreds	Tens	Ones
	□	□
6)1	2	0
− □	□	↓
	□	□
		− □
		□

4. 60 ÷ 5 = ■ ______

5. 75 ÷ 3 = ■ ______

6. 324 ÷ 3 = ■ ______

Check

Find the quotient.

7. 33 ÷ 3 = ■ ______

8. 216 ÷ 3 = ■ ______

9. 256 ÷ 8 = ■ ______

SKILL 37

TRY THESE

1. 2
2. 2
3. 1

PRACTICE

1. 1
2. 2
3. 3
4. 4
5. 1
6. 2
7. 3
8. 4
9. 6
10. 7
11. 8
12. 9
13. 6
14. 7
15. 8
16. 9

CHECK

17. 3
18. 1
19. 4
20. 8

SKILL 38

TRY THESE

1. 2
2. 4
3. 5

PRACTICE

1. 1
2. 2
3. 3
4. 4
5. 1
6. 2
7. 3
8. 4
9. 6
10. 7
11. 8
12. 9
13. 6
14. 7
15. 8
16. 9

CHECK

17. 7
18. 6
19. 9
20. 3

SKILL 39

TRY THESE

1. 1
2. 2
3. 3

PRACTICE

1. 1
2. 2
3. 3
4. 4
5. 1
6. 2
7. 3
8. 4
9. 6
10. 7
11. 8
12. 9
13. 6
14. 7
15. 8
16. 9

CHECK

17. 8
18. 8
19. 7
20. 7

Answer Card

Division

Grade 4

Answer Card

Division

Grade 4

SKILL 40

TRY THESE

1. 1
2. 2
3. 4

PRACTICE

1. 1
2. 2
3. 3
4. 4
5. 1
6. 2
7. 3
8. 5
9. 5
10. 7
11. 8
12. 9
13. 6
14. 7
15. 8
16. 9

CHECK

17. 2
18. 8
19. 4
20. 4

SKILL 41

TRY THESE

1. 6, 3, 2
2. 10, 5, 2
3. 9, 3, 3

PRACTICE

1. 6, 3, 2
2. 10, 5, 2
3. 12, 6, 2
4. 5
5. 3
6. 3
7. 2
8. 7
9. 4

CHECK

10. 9, 3, 3
11. 7
12. 2

SKILL 42

TRY THESE

1. 2 r2
2. 2 r2
3. 3 r1
4. 5 r2

PRACTICE

1. 4 r1
2. 5 r3
3. 2 r3
4. 6 r1, 36, 1
5. 5 r1, 20, 1
6. 3 r2, 24, 2
7. 4 r1
8. 7 r2
9. 10 r5

CHECK

10. 3 r1
11. 10 r2
12. 6 r2

Answer Card
Division
Grade 4

SKILL 43

TRY THESE

1. 9
2. 48, 6, 8, 8
3. 54, 9, 6, 6

PRACTICE

Answers may vary. Possible answers are given.

1. 24, 4, 4
2. 72, 9, 9
3. 50, 5, 10, 10
4. 49, 7, 7, 7
5. 64, 8, 8, 8
6. 9
7. 9
8. 7

CHECK

9. 5
10. 6
11. 9

SKILL 44

TRY THESE

1.
```
   8
2)16
 −16
   0
```

2.
```
  15
3)45
 −3↓
  15
 −15
   0
```

3.
```
  16
4)64
 −4↓
  24
 −24
   0
```

4.
```
   36
3)108
  −9↓
   18
  −18
    0
```

PRACTICE

1.
```
  12
3)36
 −3↓
  06
  −6
   0
```

2.
```
  12
6)72
 −6↓
  12
 −12
   0
```

SKILL 44 (continued)

3.
```
   20
6)120
 −12↓
   00
   −0
    0
```

4. 12
5. 25
6. 108

CHECK

7. 11
8. 72
9. 32

Number Sense

Fractions

Model Parts of a Whole

OBJECTIVE Build fraction skills by counting parts of a whole

MATERIALS paper fraction models

Begin the lesson by displaying 2 squares: one divided into 4 unequal parts, the other into 4 equal parts. Point out that a fraction names equal parts of a whole. Ask students to compare the parts in each figure to decide which shows equal parts.

Direct the students' attention to Model A on the page.

Say: **Into how many equal parts is the whole figure divided?** 8

How many parts of the whole are shaded? 1

Explain that a fraction can represent the part of the figure that is shaded.

Remind students that a fraction has a numerator and a denominator. Demonstrate how the fraction is formed:

$\frac{1}{8}$ 1 ←number of shaded parts; 8 ←number of equal parts in the whole

Point out that fractions are named using ordinal numbers.

Say: **You read this fraction as 1 eighth.**

Continue with Model B. Have the students name the fraction that represents the shaded part of the hexagon.

Language Note: Students may need help saying and spelling *numerator* and *denominator*. You may wish to post the ordinal numbers for students reference.

TRY THESE In Exercises 1–3, students model parts of a whole step-by-step.

- **Exercise 1** Count thirds.
- **Exercise 2** Count fourths.
- **Exercise 3** Count eighths.

PRACTICE ON YOUR OWN Review the example at the top of the page. Have students identify the numerator and denominator.

CHECK Make sure students can distinguish the parts from the whole, and can represent them correctly as a fraction.

Success is determined by 3 out of 3 correct responses.

Students who successfully complete the **Practice on Your Own** and **Check** are ready to move on to the next skill.

COMMON ERRORS

- Students may compare shaded and unshaded parts to each other, and record the fraction as a part to part relationship.

Students who made more than four errors in the **Practice on Your Own**, or were not successful in the **Check** section, may benefit from the **Alternative Teaching Strategy** on the next page.

Optional

Alternative Teaching Strategy
Model Parts of a Whole

OBJECTIVE Use models to count parts of a whole

MATERIALS large paper circles, crayons or markers

Distribute paper circles. Have students fold the circles in half, fourths, and eighths, respectively.

Have students note that when they unfold the circles, the parts are all equal size. Ask students to color in any number of equal parts they wish.

Choose one model as an illustration, and hold it up. Have students name the number of parts in the whole, and the number of parts shaded. Help them write the fraction.

For example:

3 ←number of shaded parts
—
4 ←number of equal parts in the whole

Explain that $\frac{3}{4}$ or 3 fourths of the circle is shaded. Ask students to identify the numerator **3** and denominator **4**.

Repeat the activity several times, having students write the fraction and name the whole and shaded parts.

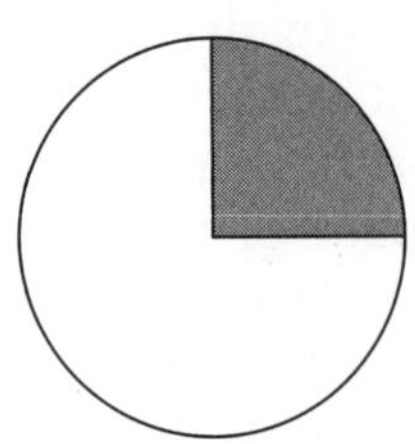

1 fourth is shaded

___ is shaded

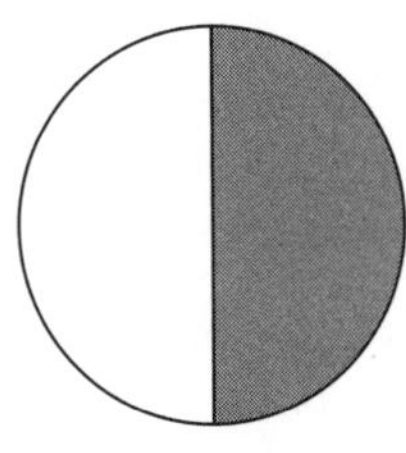

1 half is shaded

___ is shaded

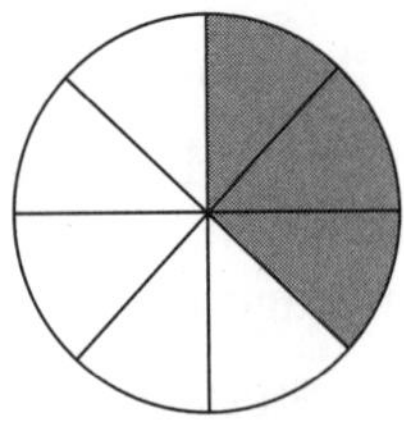

3 eighths are shaded

___ are shaded

Name ______________________ Skill ______________________

Grade 4 Skill 45

Model Parts of a Whole

You can write a fraction to name part of a whole.

Model A
The **whole** is divided into **8** equal parts.
1 out of **8** parts is shaded.

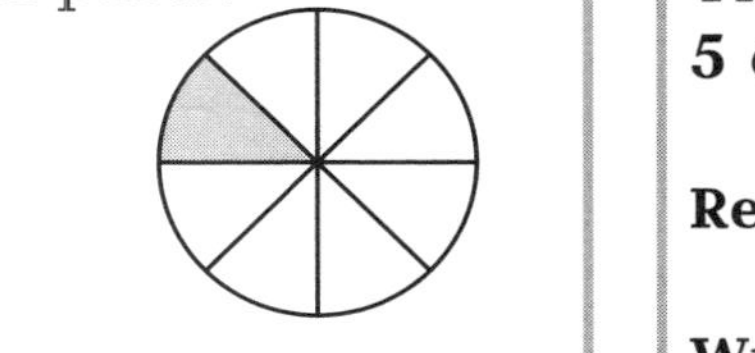

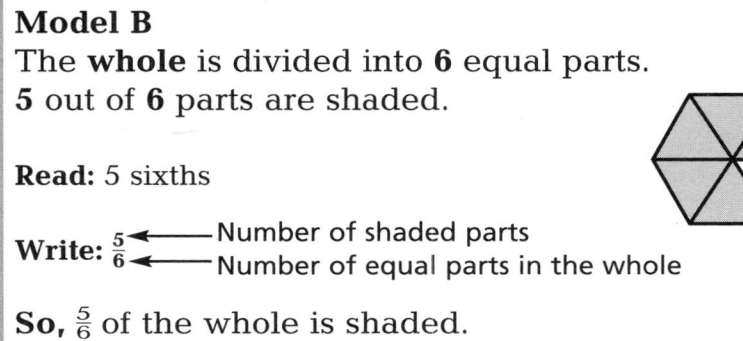

Read: 1 eighth

Write: $\frac{1}{8}$ ← Number of shaded parts / ← Number of equal parts in the whole

So, $\frac{1}{8}$ of the whole is shaded.

Model B
The **whole** is divided into **6** equal parts.
5 out of **6** parts are shaded.

Read: 5 sixths

Write: $\frac{5}{6}$ ← Number of shaded parts / ← Number of equal parts in the whole

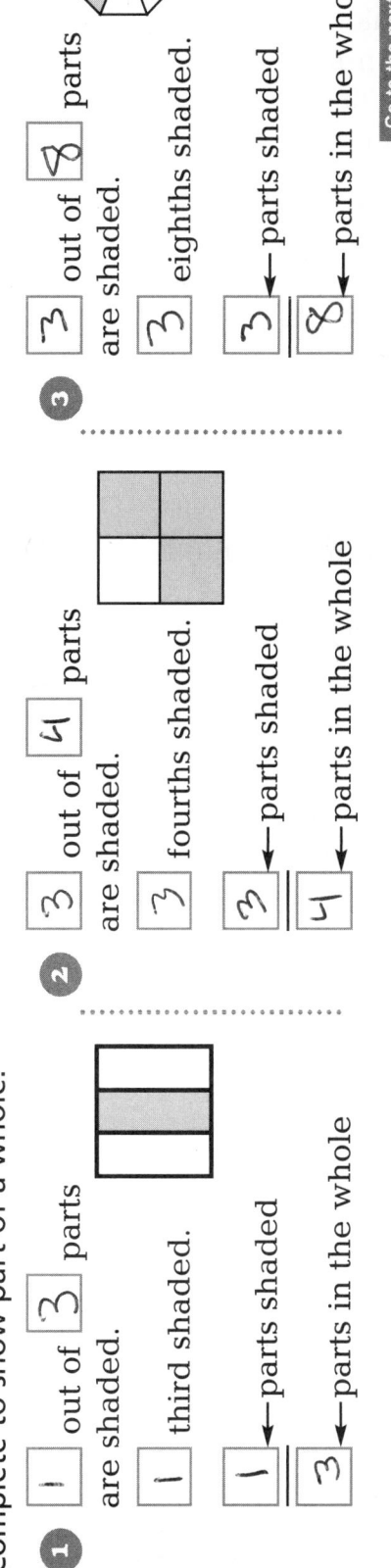

So, $\frac{5}{6}$ of the whole is shaded.

Try These

Complete to show part of a whole.

1. [1] out of [3] parts are shaded.

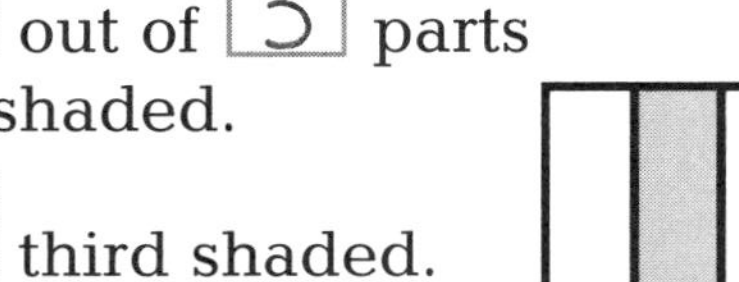

[1] third shaded.

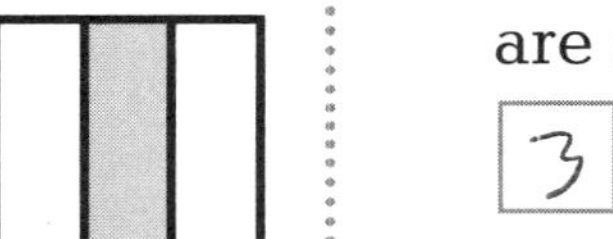

[1] ← parts shaded
[3] ← parts in the whole

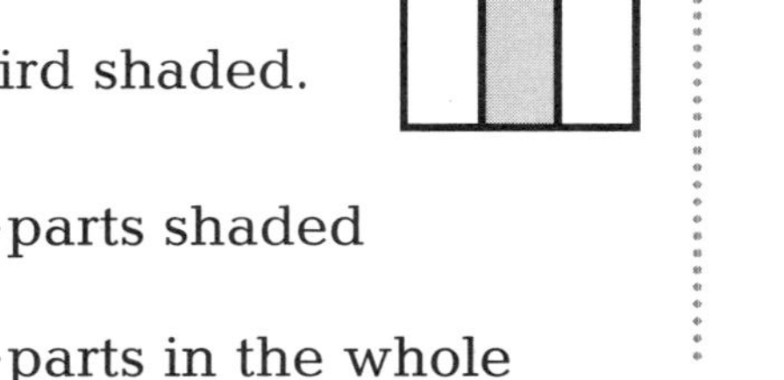

2. [3] out of [4] parts are shaded.

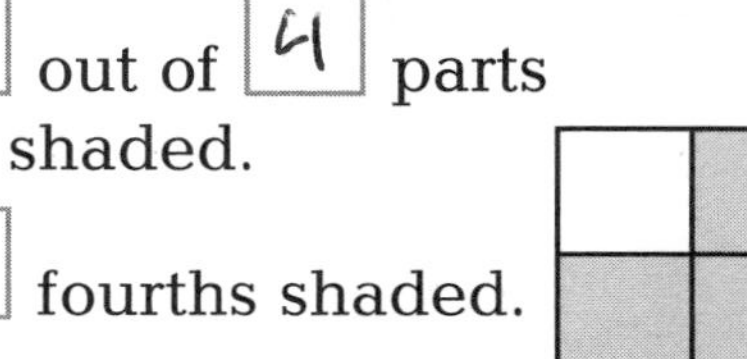

[3] fourths shaded.

[3] ← parts shaded
[4] ← parts in the whole

3. [3] out of [8] parts are shaded.

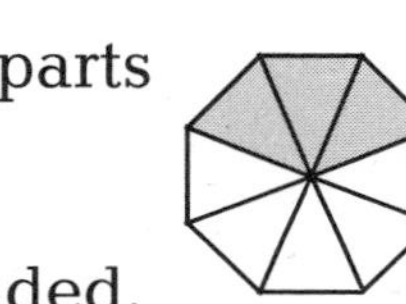

[3] eighths shaded.

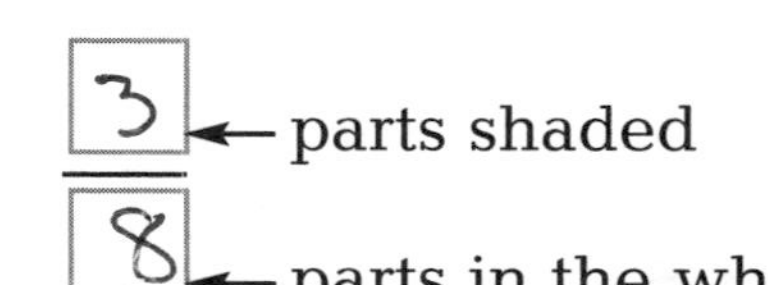

[3] ← parts shaded
[8] ← parts in the whole

Name ______________________ Skill ______________________

Practice on Your Own

Skill 45

Think:
The **numerator** tells how many parts are shaded.

The **denominator** tells how many equal parts are in the whole.

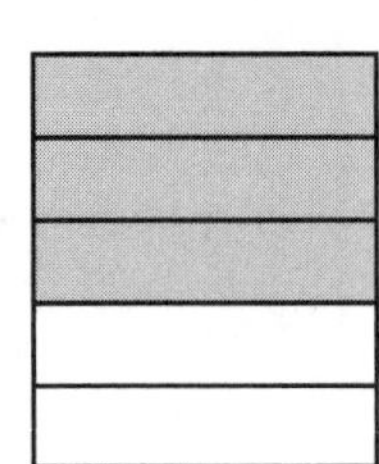

3 out of **5** parts are shaded.

3 fifths are shaded.

$\frac{3}{5}$ 3 ← Number of shaded parts; 5 ← Number of equal parts in the whole

Complete to show part of a whole.

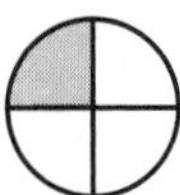
1 fourth shaded.

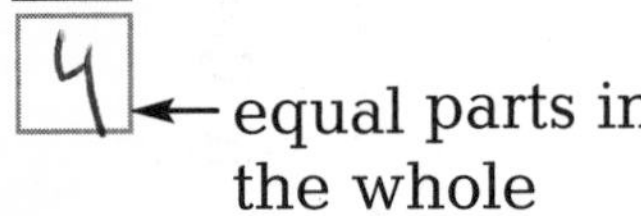
1 ← parts shaded
4 ← equal parts in the whole

2 fifths shaded.

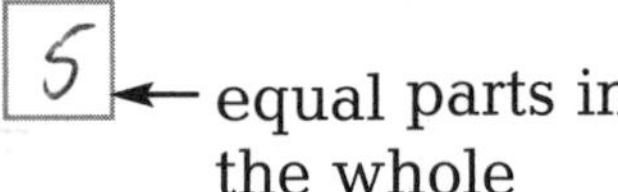
2 ← parts shaded
5 ← equal parts in the whole

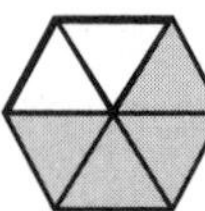
4 sixths shaded.

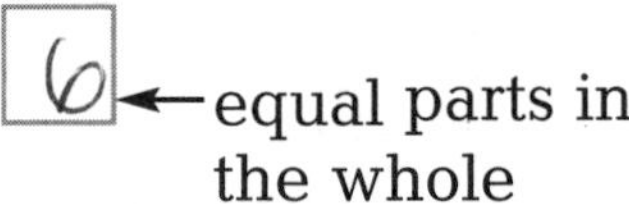
4 ← parts shaded
6 ← equal parts in the whole

4 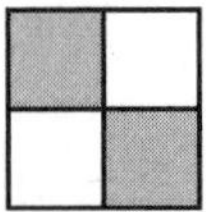
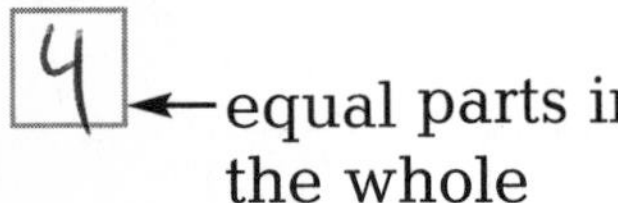
2 ← parts shaded
4 ← equal parts in the whole

5 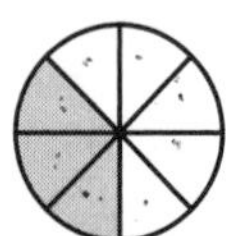
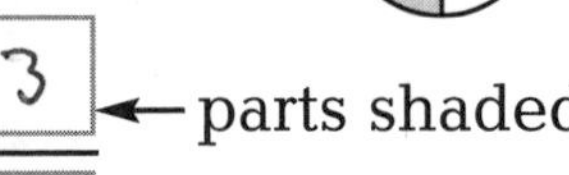
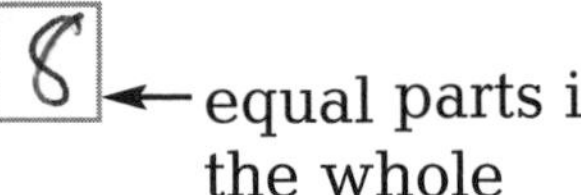
3 ← parts shaded
8 ← equal parts in the whole

6 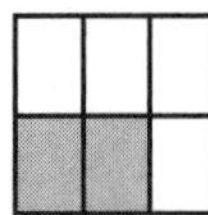
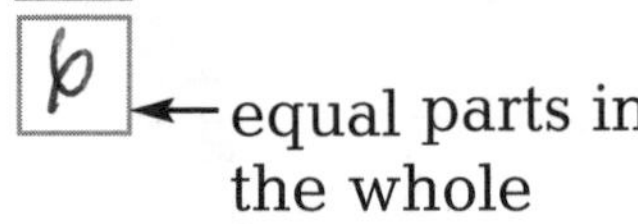
2 ← parts shaded
6 ← equal parts in the whole

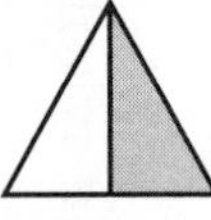
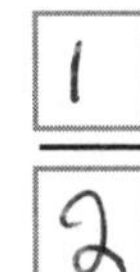
7 $\frac{1}{2}$

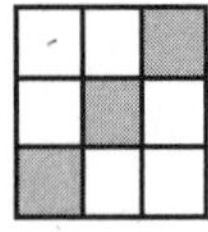
8 $\frac{3}{9}$

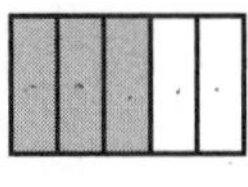
9 $\frac{3}{5}$

Check

Complete to show part of a whole.

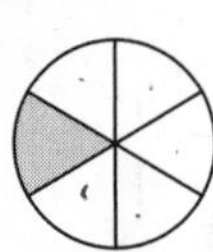

10 $\frac{1}{6}$

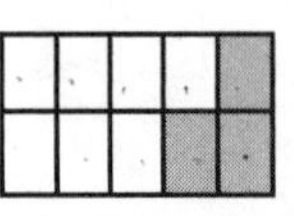

11 $\frac{3}{10}$

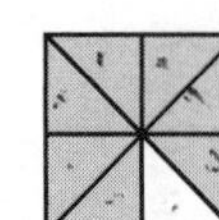
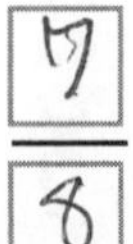
12 $\frac{7}{8}$

Skill 46 Grade 4

Model Parts of a Group (or Set)

OBJECTIVE Model parts of a group

MATERIALS (optional) two-color counters, pieces of yarn

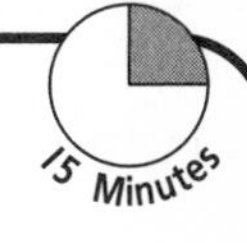

You may wish to have students use counters and circles of yarn to represent Model A, Model B, and the exercises.

Recall with students how they represented part of a whole using a geometric region. In that situation, the whole was divided into equal-size parts. Point out that the models on the card show groups of figures separated into parts. In this situation, each part of the group must have the same number of figures. It is not necessary for the figures in the part to be the same size.

Ask: **In Model A, how many parts are in the whole group?** 3 **So, each part is 1 third of the group. How many thirds are in the whole group?** 3

In Model B, how many figures are in each part? 3 **Are the parts equal?** Yes **How can you tell?** Each part has the same number of figures. **If 2 parts in the group are shaded, what do the shaded parts equal?** 2 thirds

TRY THESE Exercises 1–3 are sequenced to reinforce the part-to-group relationship.

- **Exercise 1** Students can count to recognize that the group has 5 parts. One figure represents 1 part or 1 fifth of the group.
- **Exercises 2–3** The next step in conceptual development is represented by these exercises. Students may need help recognizing parts of the whole.

PRACTICE ON YOUR OWN Have students explain the example at the top of the page.

Ask: **If I shade another part in the group, would that change the number of parts in the group?** No. **Explain.** There are still 3 parts. Only the number of parts shaded has changed. So the numerator changes.

The exercises begin by reinforcing the language of part-to-group comparison. Exercises 1–6 provide cued practice in writing the fraction for the pictorial model.

CHECK Determine if students understand that the shaded part represents part of the group and is the numerator of the fraction. Note if they accurately record the number of parts of the group as the denominator.

Success is determined by 3 out of 3 correct responses.

Students who successfully complete the **Practice on Your Own** and the **Check** are ready to move on to the next skill.

COMMON ERRORS

- Students may write the numerator correctly, but may write the number of figures in one part as the denominator.
- Students may write any part of a group or any part that has 3 figures as $\frac{1}{3}$; e.g., for Ex. 4 students may write $\frac{1}{3}$ instead of $\frac{3}{7}$.

Students who made more than three errors in **Practice on Your Own**, or were not successful in the **Check** section, may benefit from the **Alternative Teaching Strategy** on the next page.

Optional

Alternative Teaching Strategy

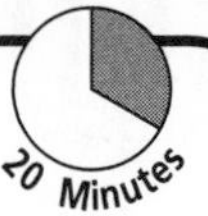

Partitioning a Group of Objects

OBJECTIVE Use counters to understand how to form and partition a group

MATERIALS 2-color counters (or construction paper cut into small squares), notebook paper, colored pencils to match the counters

Provide students with a group of 6 two-color counters. Ask students to display the counters so they are all the same color. Have them organize the counters into a 3 × 2 array, then partition the group into two equal parts by pushing one group of 3 counters to the side, and turn the counters so they are the second color.

Ask: **How many parts did you make?** 2 **How many counters are in each part?** 3 **Are the parts equal?** Yes **How can you tell?** Because they each have the same number of counters

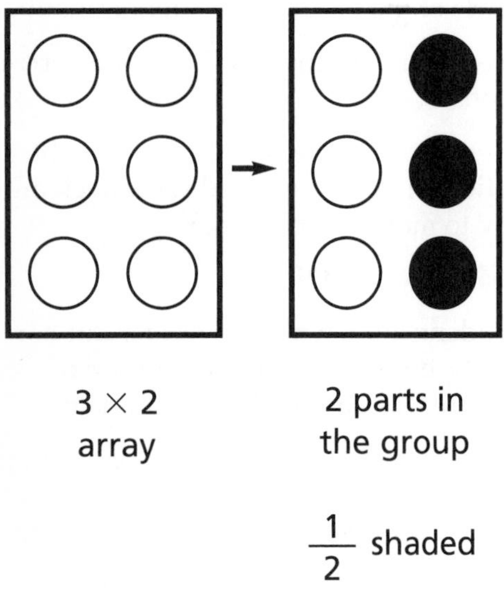

Provide additional counters and have students partition 8 counters, 12 counters, 16 counters, and so on, into 4 equal groups. Have students turn the counters so 3 parts are the second color. Have students draw pictures of what they do each time and record the fraction.

What fraction can you write to represent 3 of 4 equal parts? 3 fourths

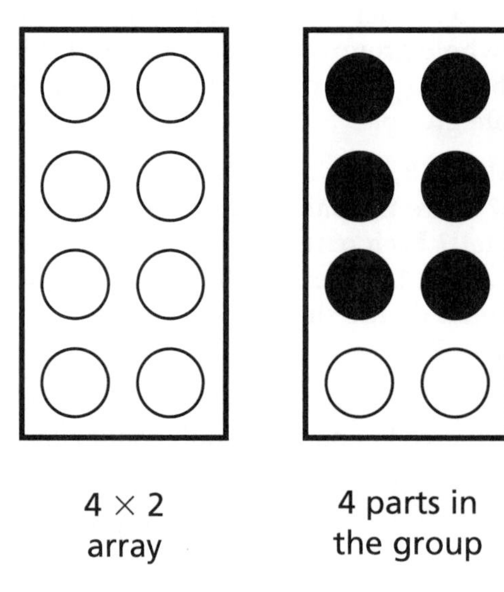

Model Parts of a Group (or Set)

Grade 4 Skill 

A fraction can name part of a whole. It can also name part of a group.

Model A
There are 3 figures in this group.

1 of **3** figures is shaded.
So, 1 of **3** parts is shaded.
1 third is shaded.

part shaded ⟶ 1
number of parts in the group ⟶ 3

$\frac{1}{3}$

Model B
There are 9 figures in this group.
There are 3 equal parts.

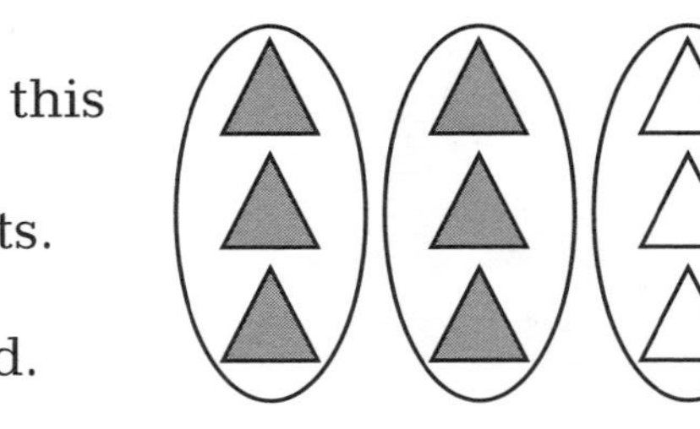

2 of **3** parts are shaded.
2 thirds are shaded.

part shaded ⟶ 2
number of parts in the group ⟶ 3

$\frac{2}{3}$

Try These

Complete to show part of a group.

1. 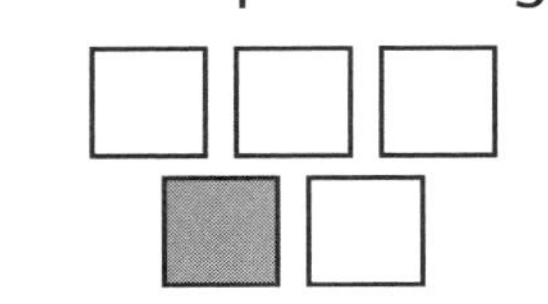

There are ☐ parts in the group.
☐ parts are shaded.
☐ fifth is shaded.

2. 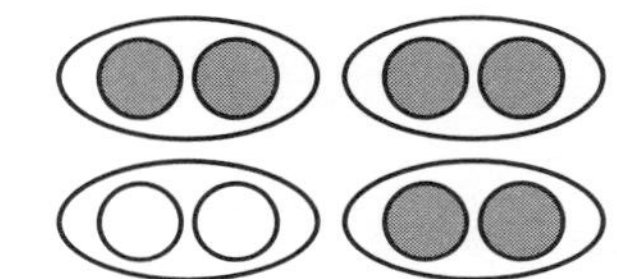

There are ☐ parts in the group.
☐ parts are shaded.
☐ fourths are shaded.

3. 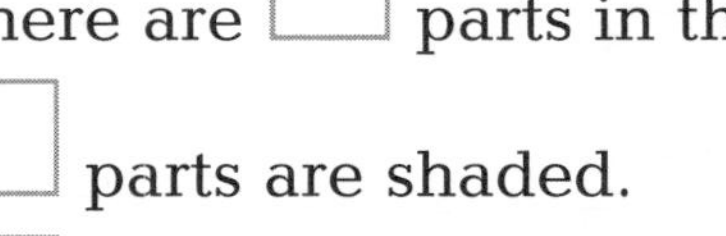

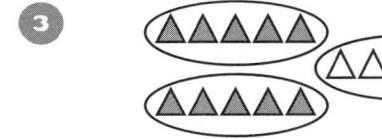

There are ☐ parts in the group.
☐ parts are shaded.
☐ thirds are shaded.

Go to the next side.

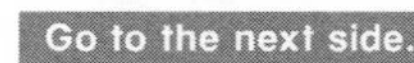

Name ______________________ Skill ______________________

Skill 46

Practice on Your Own

Think:
There are 12 figures in this group.
There are 3 parts.

1 of **3** parts is shaded.

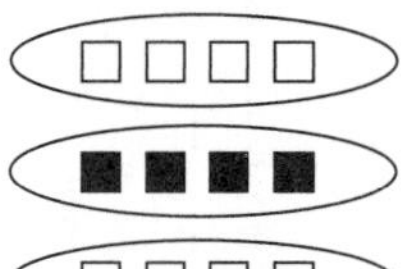

part shaded → 1
number of parts → 3

Complete to show part of a group.

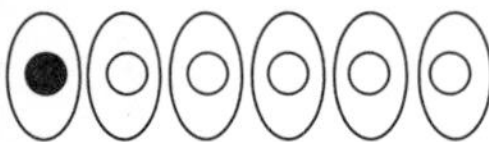

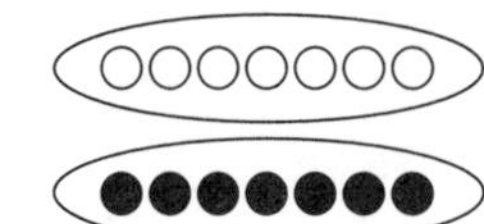

1. ☐ of ☐ parts is shaded.

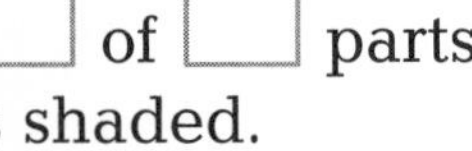
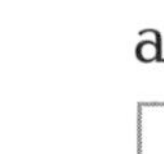

2. ☐ of ☐ parts are shaded.

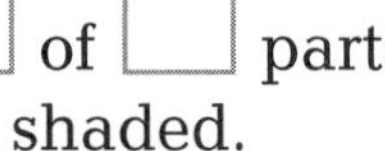

3. ☐ of ☐ parts is shaded.

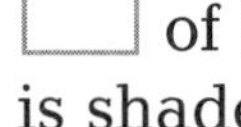

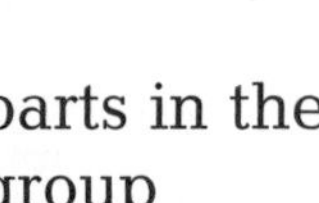
☐ ← parts shaded
☐ ← parts in the group

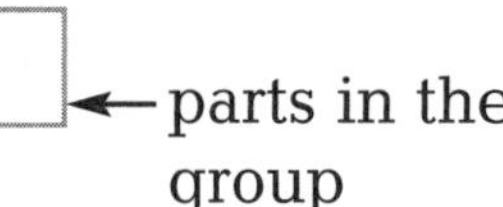
☐ ← parts shaded
☐ ← parts in the group

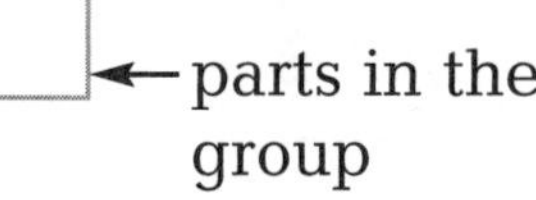
☐ ← parts shaded
☐ ← parts in the group

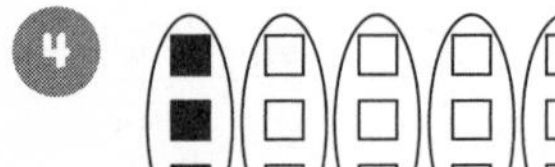

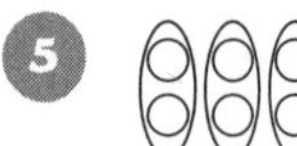
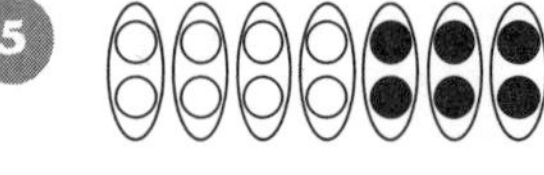

4. ☐ ← parts shaded
☐ ← parts in the group

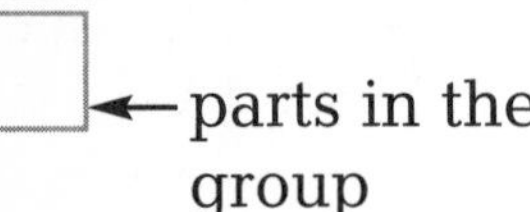

5. ☐ ← parts shaded
☐ ← parts in the group

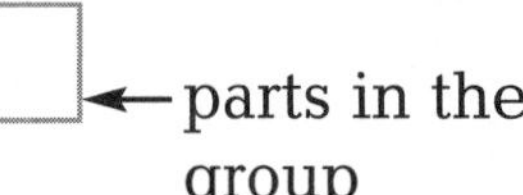

6. ☐ ← parts shaded
☐ ← parts in the group

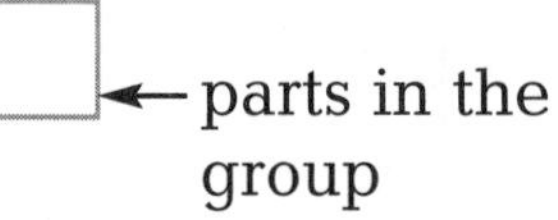

7. 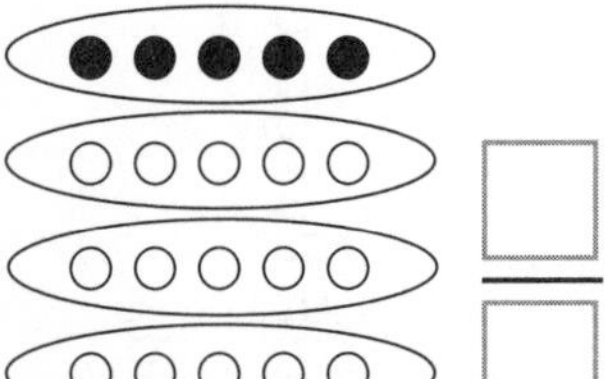

8. 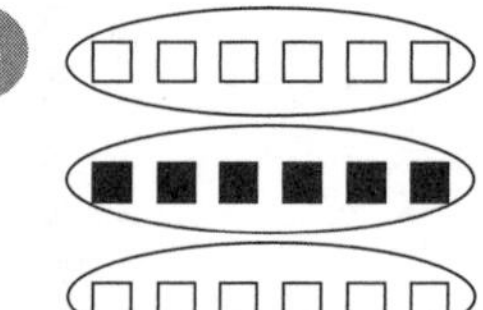
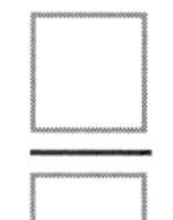

9. 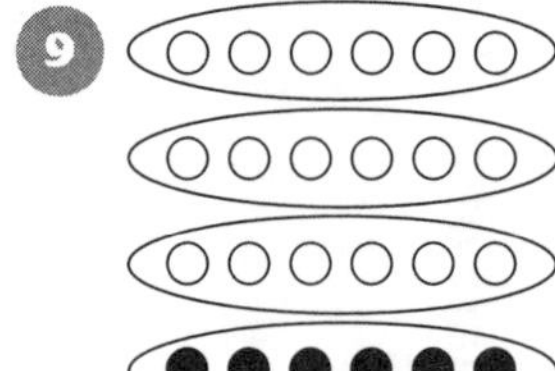
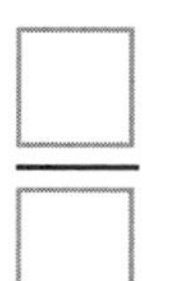

Check

Complete to show part of a group.

10. 

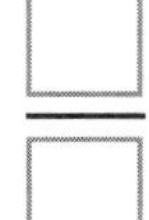

11. 
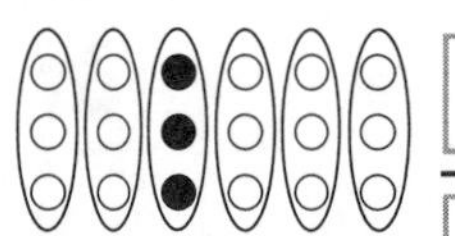

12. 
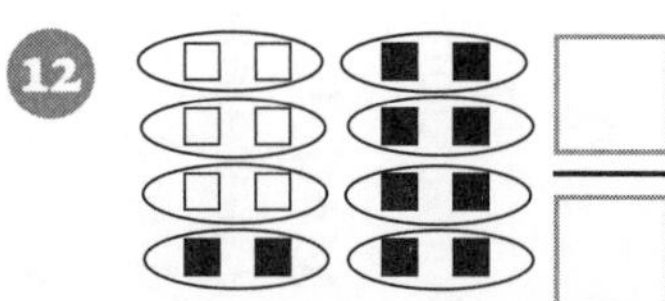

Skill 47 Grade 4

Count Parts of a Whole

OBJECTIVE Count parts of a whole

MATERIALS (optional) fraction strips

You may wish to begin the lesson by reviewing how to model a fraction. Display a fraction strip for 1 third and have students tell you the fraction and write it as shown below.

$\frac{1}{3}$ ← 1: parts shaded; 3: number of equal parts in the whole

Review the meaning of numerator and denominator. Then, direct the students' attention to the rectangle in Model A.

Ask: **Into how many equal parts is whole figure divided?** 4 **How many parts are shaded?** 1 **What fraction shows that just one part is shaded?** $\frac{1}{4}$ **What if 4 out of 4 parts were shaded, then what fraction shows 4 out of 4 parts shaded?** $\frac{4}{4}$

Direct students' attention to Model B.

Ask: **As the number of shaded parts increases, what do you notice about the number of equal parts or the denominator?** The number stays the same.

Then have students name each fraction, explaining that they are counting by fourths.

TRY THESE In these three exercises students count parts of a whole step-by-step.

- **Exercise 1** One part shaded.
- **Exercise 2** Two parts shaded.
- **Exercise 3** Three parts shaded.

PRACTICE ON YOUR OWN Review the example at the top of the page. Have students identify the numerator 3 and denominator 5.

CHECK Make sure students can distinguish the parts from the whole, and can represent the fraction correctly.

Success is determined by 3 out of 3 correct responses.

Students who successfully complete the **Practice on Your Own** and **Check** are ready to move on to the next skill.

COMMON ERRORS

- Students may confuse the numerator and the denominator.
- Students may compare shaded and unshaded parts to each other, and record the fraction as part-to-part relationship.

Students who made more than three errors in the **Practice on Your Own**, or were not successful in the **Check** section, may benefit from the **Alternative Teaching Strategy** on the next page.

Optional

Alternative Teaching Strategy

Count Parts of a Whole

OBJECTIVE Count parts of a whole

MATERIALS Fraction strips for the fractions: $\frac{0}{5}, \frac{1}{5}, \frac{2}{5}, \frac{3}{5}, \frac{4}{5}, \frac{5}{5}$

Make corresponding fraction flash cards. Prepare sets of flash cards for other fractions.

Before beginning, review with the class how to model a fraction. Review the meaning of numerator and denominator.

Then distribute a set of shaded figures to the students. Have them arrange the figures in order, from no parts shaded to all parts shaded.

Distribute the corresponding fractions. Have them match fractions and figures.

Repeat the activity with other fraction sets. Then take out some of the figures or flash cards and have students identify which are missing.

Finally, use only the flash cards to encourage students to count fractions in symbolic form only.

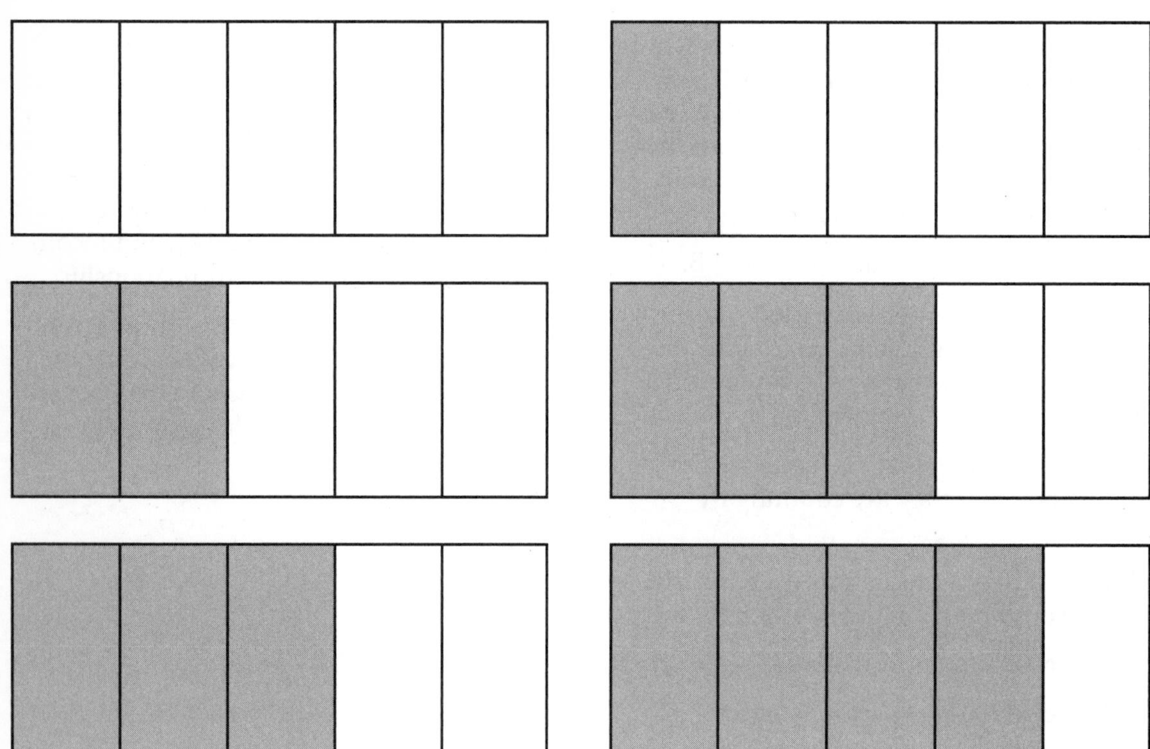

Name ____________________ Skill ____________________

Grade 4 **Skill 47**

Count Parts of a Whole

You can write a fraction to show how many parts of a whole are shaded.

Model A

The whole figure is divided into 4 equal-size parts.

1 of **4** parts are shaded.

So, 1 fourth of the whole is shaded.

part shaded → 1
parts in the whole → 4

Model B

You can use a pattern to write fractions for the shaded part of a whole.

Model					
Parts Shaded	0	1	2	3	4
Number of Equal Parts	4	4	4	4	4
Fraction	$\frac{0}{4}$	$\frac{1}{4}$	$\frac{2}{4}$	$\frac{3}{4}$	$\frac{4}{4}$

Try These

Complete to show part of a whole.

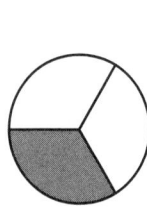

1. 

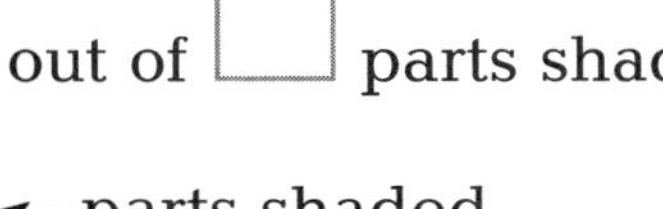

☐ out of ☐ parts shaded.

☐ ← parts shaded
☐ ← parts in the whole

2. ☐ out of ☐ parts shaded.

☐ ← parts shaded
☐ ← parts in the whole

3. ☐ out of ☐ parts shaded.

☐ ← parts shaded
☐ ← parts in the whole

Go to the next side.

Name ______________________ Skill ______________________

Practice on Your Own

Skill 47

Think:
The whole is divided into **5** equal parts.

3 out of **5** parts are shaded.

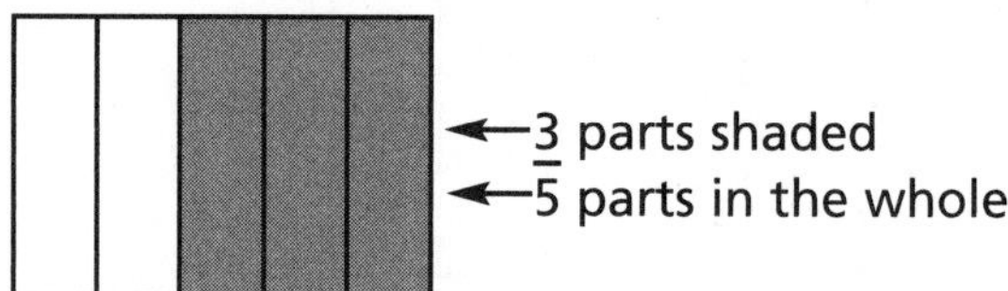

Complete. Show the part of the whole that is shaded.

Model						
1 Parts Shaded						
2 Number of Equal Parts						
3 Fraction						

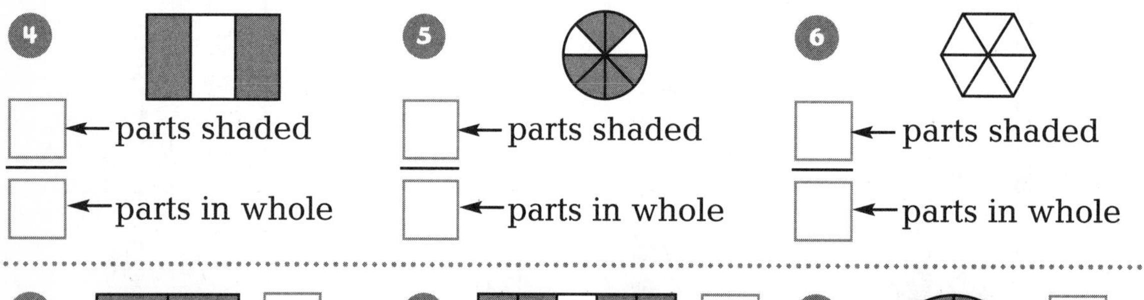

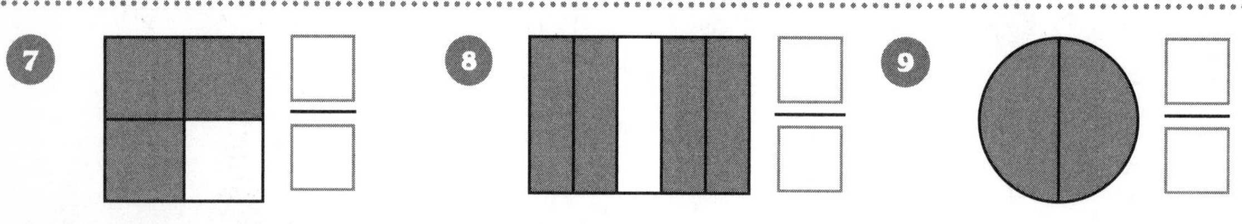

Check

Show the part of the whole that is shaded.

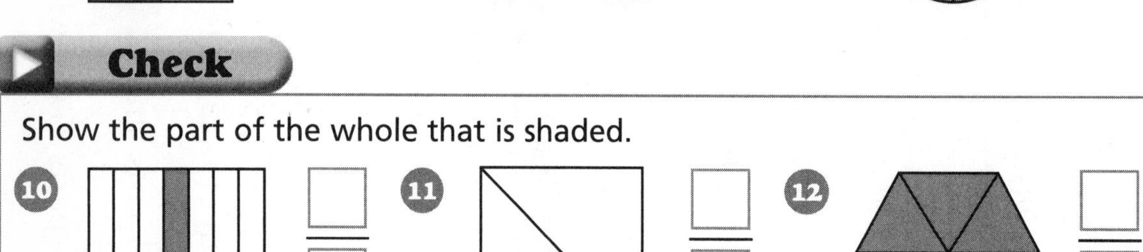

Skill 48 Grade 4

Count Parts of a Group

OBJECTIVE Count parts of a group

MATERIALS (optional) two-color counters

You may wish to have students represent Model A and Model B with two-color counters.

Ask students to count the number of parts in Model A. 5 Verify that they recognize the parts as the encircled figures. Have students confirm that each part is 1 fifth of the group. Then have them count by fifths: 1 fifth, 2 fifths, and so forth.

Ask: **How many fifths make a whole group?** 5 **If no parts of the group are shaded, how many fifths is that?** 0 fifths **If one part is shaded, how many fifths is that?** 1 fifth

Recall that the numerator of the fraction represents the number of parts being considered, and the denominator represents the number of parts in the whole group.

As you work through Model B with the students,

Ask: **In the pattern, what number in the fraction stays the same?** The denominator **Why?** Because the number of parts in the whole group does not change **What number changes?** The numerator, because the number of parts being considered changes

TRY THESE Exercises 1–3 show a pattern of thirds of the group. The sentences, for example,

____ out of ____ parts are shaded,

help students keep track of the number of parts and reinforce part-to-group concepts.

PRACTICE ON YOUR OWN Review the example at the top of the page with students. You may wish to have them represent the parts with counters. Ask student to count from 1 sixth to 6 sixths.

Exercises 1–3 provide a pattern for the students to complete. Exercises 4–6 are not patterned, however, students are supported with cued fraction boxes. Exercises 7–9 provide an opportunity for students to record fractions without written cues.

CHECK Determine if students recognize that the numerator counts the number of parts being considered.

Success is determined by 3 out of 3 correct responses.

Students who successfully complete the **Practice on Your Own** and the **Check** are ready to move on to the next skill.

COMMON ERRORS

- Students may write the numerator correctly, but may write the number of figures in one part as the denominator.
- Students may count parts correctly, but record the number of parts being considered as the denominator and name the numerator as 1.

Students who made more than 3 errors in **Practice on Your Own**, or were not successful in the **Check** section, may benefit from the **Alternative Teaching Strategy** on the next page.

Optional

Alternative Teaching Strategy

Count Parts of a Group

20 Minutes

OBJECTIVE Use concrete models to represent and count parts of a group

MATERIALS 2-color counters (or construction paper cut into small squares), construction paper student-made workmats

Have students use construction paper or notebook paper to make workmats for 3, 4, 5, 6, . . . parts in a group.

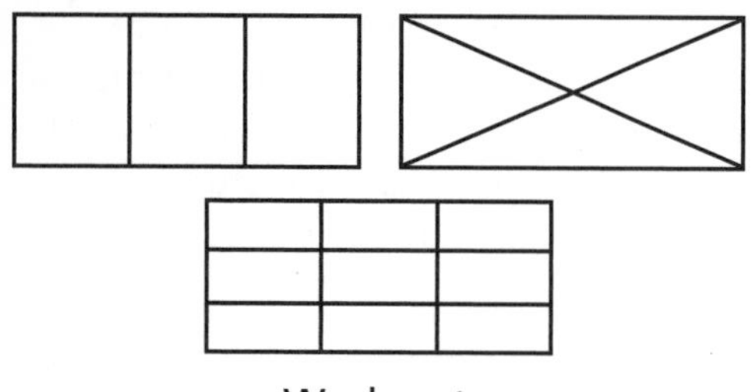

Workmats

Begin with the mat for thirds. Distribute a group of counters to each student and have the students use 6 of the counters.

Ask the students to make 3 equal groups using the counters and the workmats. Have them tell you the number of parts and orally count: 1 third, 2 thirds, 3 thirds.

Next, have students remove the counters, and now partition 12 counters into 3 equal groups. Have them count by thirds again.

Ask students to show $\frac{1}{3}$ by turning the counters in 1 part to the second color.

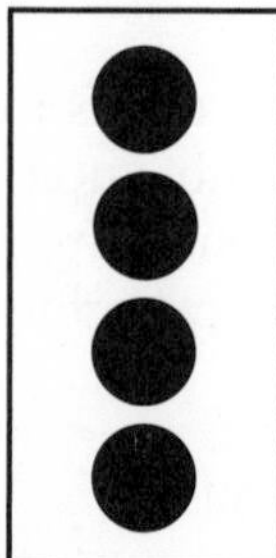  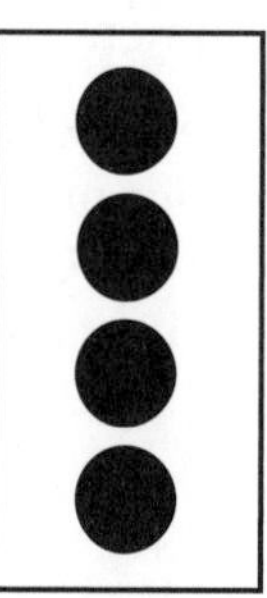

Count: 1 third, 2 thirds, 3 thirds

Repeat the activity for fourths or fifths. Each time have students count using the fraction name: 1 fourth, 2 fourths, 3 fourths, 4 fourths.

After students have completed several examples in this way, have them draw a picture of a group of 15 figures separated into 5 parts. Ask students to write the fractions to represent 1 of 5 parts, 2 of 5 parts, and so on.

As a culminating activity, show these fractions:

$\frac{\square}{5}$ $\frac{\square}{8}$ $\frac{\square}{9}$

Have students tell you the number of parts in a group with each denominator, then count to tell the fractions for each group. (e.g., 1 fifth, 2 fifths, 3 fifths, . . .)

Name ______________________ Skill ______________________

Grade 4 **Skill 48**

Count Parts of a Group

You can write a fraction to show how many parts of a group are shaded.

Model A
There are 5 parts in this group.

0 of **5** parts are shaded.

parts shaded → $\frac{0}{5}$ ← number of parts in the group

Model B
You can use a pattern to write fractions for the shaded parts.

Model						
Parts Shaded	0	1	2	3	4	5
Number of Parts	5	5	5	5	5	5
Fraction	$\frac{0}{5}$	$\frac{1}{5}$	$\frac{2}{5}$	$\frac{3}{5}$	$\frac{4}{5}$	$\frac{5}{5}$

Try These

Complete to show how many parts are shaded.

1. ☐ out of ☐ parts are shaded.

☐ ← parts shaded
☐ ← number of parts in the group

2. 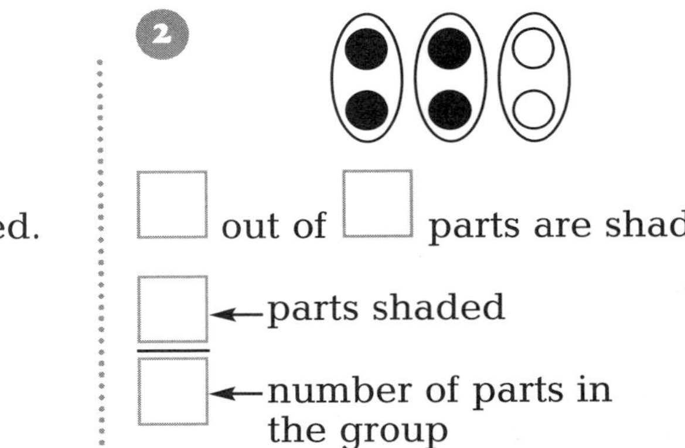

☐ out of ☐ parts are shaded.

☐ ← parts shaded
☐ ← number of parts in the group

3. ☐ out of ☐ parts are shaded.

☐ ← parts shaded
☐ ← number of parts in the group

Go to the next side.

Name ______________________ Skill ______________________

Practice on Your Own

Skill 48

Think: There are 6 parts in this group.

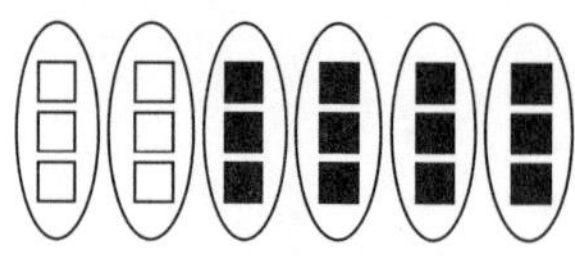

4 out of **6** parts are shaded.

parts shaded ⟶ 4
number of parts ⟶ 6

Complete to show how many parts are shaded.

Model					
1 Parts Shaded					
2 Number of Parts					
3 Fraction					

Complete to show how many parts are shaded.

4 ← parts shaded ← parts in the group

5 ← parts shaded ← parts in the group

6 ← parts shaded ← parts in the group

7

8

9

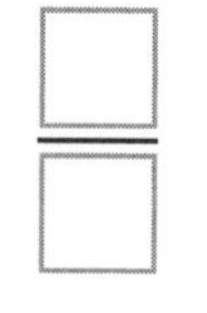

Check

Complete to show how many parts are shaded.

10 

11 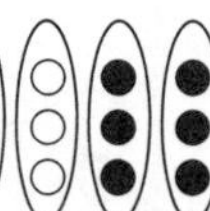

12 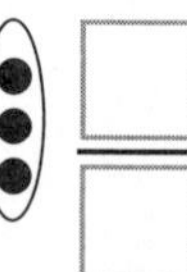

Skill 49 Grade 4

Compare Parts of a Whole

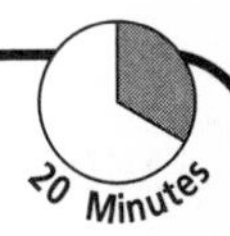

OBJECTIVE Compare fractions of a whole with the same denominator

MATERIALS (optional) fraction strips

Some students may be able to work with just pictorial models to compare two fractions; however, you may find that other students need to use fraction strips or paper folding to represent each situation on the front of the Skill Card.

Review the idea that for comparing two fractions the whole and the parts must be the same size and shape, and each whole must have the same number of parts. Help students see that the fractions that represent each figure will have the same denominator.

Draw attention to the models on the card.

Ask: **In Step 1, what can you tell me about the size and the shape of the rectangles that represent each fraction?** They are the same size and shape. **How many equal parts does each rectangle have?** 5 **What do you notice about the denominators of both fractions?** They are the same.

Have students look at the figure that represents $\frac{2}{5}$ and confirm that the numerator tells how many parts of the whole are shaded. Have students do the same for $\frac{3}{5}$.

In Step 2, ask students to compare the parts shaded. Some students will compare the lengths of the shaded parts and others will count the number of parts shaded. Students should recognize since 2 is less than 3, then $\frac{2}{5} < \frac{3}{5}$. If necessary, review the comparison symbols.

Help students understand that if two fractions have the same denominator, then the fraction with the greater part shaded—or the one with the greater numerator—is the greater fraction.

TRY THESE The dashed vertical line on both figures should help students see that the parts to the left of the line "match" and the parts to the right show how much greater one fraction is than the other.

PRACTICE ON YOUR OWN Review the examples at the top of the page.

For Exercises 4–6, students are asked to shade the models first, then compare the fractions. Check that students are shading accurately.

CHECK Determine if students understand that the fraction with the greater numerator is the greater fraction.

Success is determined by 3 out of 3 correct responses.

Students who successfully complete the **Practice on Your Own** and the **Check** are ready to move on to the next skill.

COMMON ERROR

- Students may accurately compare fractions, but use the wrong inequality symbol.

Students who made more than three errors in **Practice on Your Own**, or were not successful in the **Check** section, may benefit from the **Alternative Teaching Strategy** on the next page.

Optional

Alternative Teaching Strategy

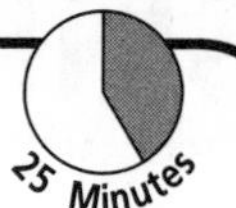

Use Models to Compare Fractions

OBJECTIVE Use a model to understand how to compare fractions with the same denominators

MATERIALS fraction strips or small pieces of paper of equal size, crayons or markers

Review the meaning of models that represent parts of a whole. Note that the whole is divided into equal size parts. Each part represents a fraction of the whole.

Then have students use blank fraction strips or fold a paper rectangle into equal size parts.

Take 2 blank strips for sixths. Have students shade 1 sixth on one strip and 2 sixths on the other.

Ask: **What fraction represents the strip with one part shaded?** 1 sixth **How can you tell?** There are 6 parts in the whole, 1 part is shaded.

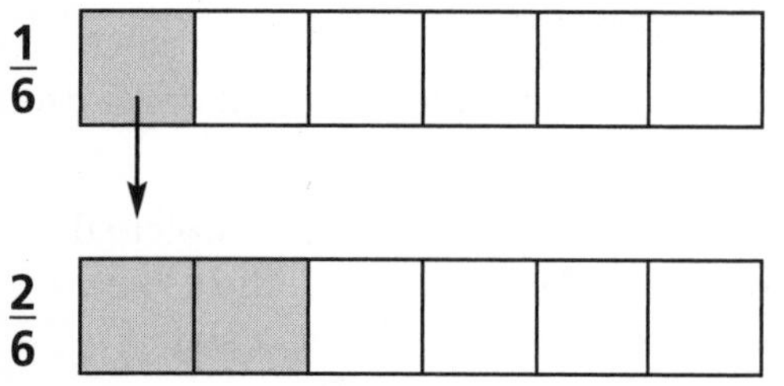

Repeat for the 2-sixths strip. Have students align the strips as shown below, then use one-to-one correspondence to match shaded parts. Have students tell you that 2 sixths has more shaded parts, so 2 sixths is greater than 1 sixth. Ask students to write the comparison two ways:

$\frac{2}{6} > \frac{1}{6}$ 2 sixths is greater than 1 sixth

$\frac{1}{6} < \frac{2}{6}$ 1 sixth is less than 2 sixths

Continue with other fraction strips, each time having students verify the number of parts in the whole, compare the shaded parts, and record the comparison 2 ways. Confirm that for fractions with the same denominator, you compare the numerators.

Name ______________________ Skill ______________________

Grade 4 Skill 49

Compare Parts of a Whole

You can use models to compare fractions. Compare: $\frac{2}{5} \bigcirc \frac{3}{5}$

Step 1 Make a model of each fraction. The denominators are the same. **So,** draw equal-size parts in both models. Shade parts to show the numerator.

$\frac{2}{5}$ →

$\frac{3}{5}$ →

Step 2 To compare the numerators, compare the number of equal-size shaded parts.

$\frac{2}{5}$ →

$\frac{3}{5}$ →

$\frac{2}{5}$ has fewer parts shaded than $\frac{3}{5}$.

Step 3 Record the result using >, <, or =.

Remember:
> means *is greater than*
< means *is less than*

$\frac{2}{5}$ is less than $\frac{3}{5}$

$\frac{2}{5} < \frac{3}{5}$

Try These

Complete to show the greater part shaded.

1. 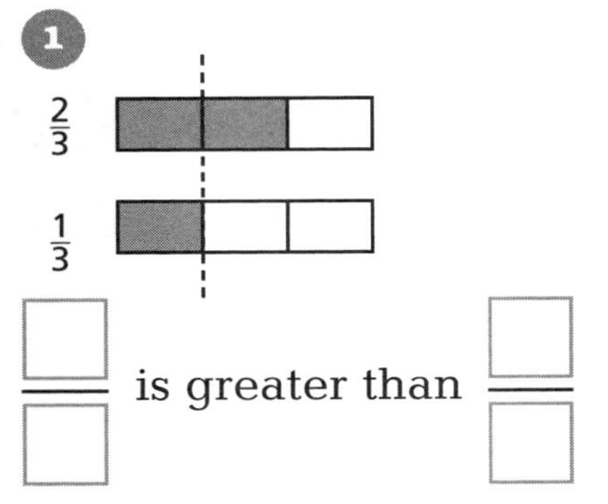

$\frac{\square}{\square}$ is greater than $\frac{\square}{\square}$

2. $\frac{4}{6}$

$\frac{5}{6}$

$\frac{\square}{\square}$ is greater than $\frac{\square}{\square}$

3. 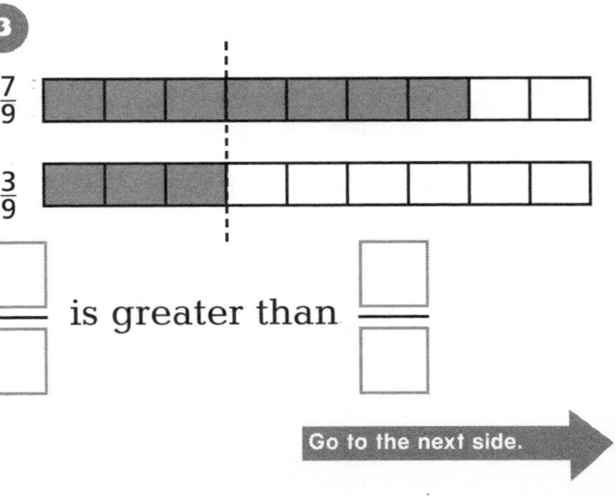

$\frac{\square}{\square}$ is greater than $\frac{\square}{\square}$

Go to the next side.

Name ______________________ Skill ______________________

Practice on Your Own

Skill 49

Think:
Compare the numbers of equal parts that are shaded.

$\frac{5}{8}$ 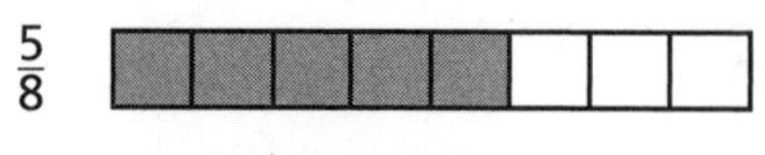

$\frac{4}{8}$ 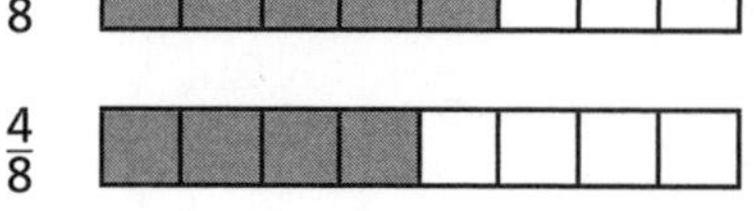

5 eighths is greater than **4** eighths. $\frac{5}{8} > \frac{4}{8}$

$\frac{3}{3}$ 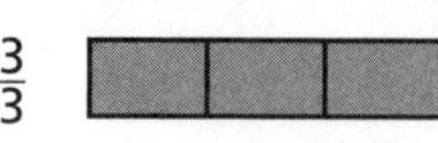

$\frac{3}{3}$ 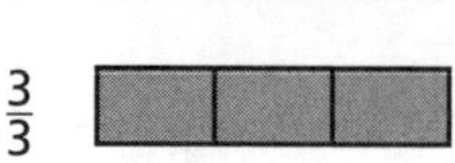

$\frac{3}{3} = \frac{3}{3}$

Compare. Choose >,<, or =.

1. $\frac{5}{6}$ 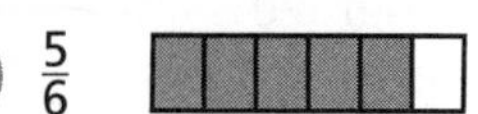

   $\frac{3}{6}$

   $\frac{5}{6} \bigcirc \frac{3}{6}$

2. $\frac{4}{4}$ 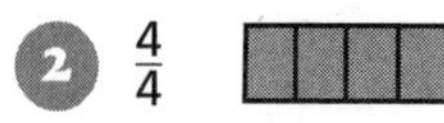

   $\frac{4}{4}$

   $\frac{4}{4} \bigcirc \frac{4}{4}$

3. $\frac{1}{10}$ 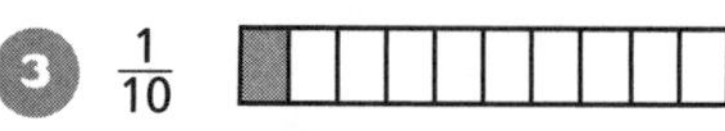

   $\frac{2}{10}$

   $\frac{1}{10} \bigcirc \frac{2}{10}$

Shade the model to show the fractions. Compare.

4. $\frac{3}{3}$ 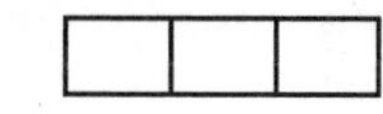

   $\frac{2}{3}$

   $\frac{3}{3} \bigcirc \frac{2}{3}$

5. $\frac{2}{5}$ 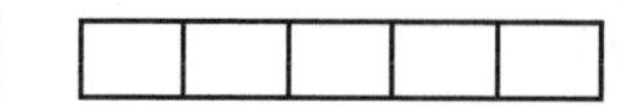

   $\frac{4}{5}$

   $\frac{2}{5} \bigcirc \frac{4}{5}$

6. $\frac{1}{4}$ 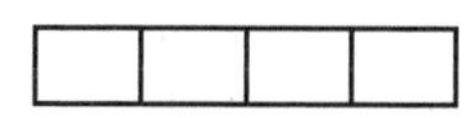

   $\frac{0}{4}$

   $\frac{1}{4} \bigcirc \frac{0}{4}$

Compare. Choose >,<, or =.

7. $\frac{2}{2} \bigcirc \frac{2}{2}$
8. $\frac{9}{12} \bigcirc \frac{8}{12}$
9. $\frac{3}{6} \bigcirc \frac{3}{6}$

Check

Compare. Choose >,<, or =.

10. $\frac{6}{7} \bigcirc \frac{2}{7}$
11. $\frac{4}{10} \bigcirc \frac{9}{10}$
12. $\frac{11}{12} \bigcirc \frac{10}{12}$

Skill 50 Grade 4 Fractions with Denominators of 10 and 100

OBJECTIVE Write fractions with denominators of 10 and 100.

MATERIALS decimal squares

Begin by reminding students that when a whole is divided into equal parts, each part is a fraction of the whole.

Direct students' attention to the first example.

Ask: **Into how many equal parts is the square divided?** 10 **What fraction does one part represent?** $\frac{1}{10}$ **How many parts are shaded?** 2 **What fraction of the square is shaded?** $\frac{2}{10}$

Continue to ask similar questions as you work through the next two examples. You may wish to display decimal squares as you work through the next examples.

Point out that for each example, the same size square was divided into equal parts. Note that the tenths parts are larger than the hundredths parts.

TRY THESE Exercises 1–3 model the type of exercises students will find on the **Practice on Your Own** page.

- **Exercise 1** Fraction with a denominator of 10.
- **Exercise 2** Fraction with a denominator of 100.
- **Exercise 3** Equivalent fractions.

PRACTICE ON YOUR OWN Review the example at the top of the page. Ask students to explain how they know the models have the same amount shaded. **The region or area for $\frac{5}{10}$ is the same size as the area for $\frac{50}{100}$.**

CHECK Determine if the students can identify the correct name for each denominator, can write fractions for tenths and hundredths, and can recognize equivalent fractions. Success is determined by 3 out of 3 correct responses.

Students who successfully complete the **Practice on Your Own** and **Check** are ready to move on to the next skill.

COMMON ERRORS

- Students may count the wrong number of shaded parts or count the parts that are not shaded.
- Students may write the correct numerator, but write the number of unshaded parts as the denominator.
- Students may write all denominators as 10 or as 100 without regard to the number of the parts in the model.

Students who made more than three errors in the **Practice on Your Own**, or who were not successful in the **Check** section, may benefit from the **Alternative Teaching Strategy** on the next page.

Optional

Alternative Teaching Strategy

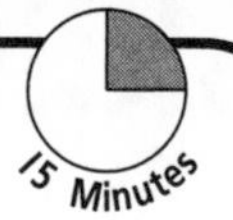

Model Fractions with Denominators of 10 and 100

OBJECTIVE Use decimal squares to represent fractions with denominators of 10 and 100.

MATERIALS decimal squares, paper

Distribute the decimal squares.

Have students separate them into two piles. One pile has squares divided into 10 equal parts and the other pile has squares that are divided into 100 equal parts.

Say: **Find the square that would represent the fraction $\frac{3}{10}$.**

Ask: **Into how many equal parts is the square divided?** 10 **Does this number represent the numerator or denominator of the fraction?** denominator **How many parts out of 10 are shaded?** 3 out of 10 parts

Say: **Then this number represents the numerator of the fraction.**

Have students record the fraction.

Continue with the square for 30 hundredths.

Say: **Find a square divided into 100 equal parts with the same area shaded as $\frac{3}{10}$.**

After the students have found the square, Ask: **How many parts out of 100 are shaded?** 30 out of 100 parts

What fraction does this square show? $\frac{30}{100}$

Have students record the fraction to the right of $\frac{3}{10}$.

Explain that the two fractions are equal. Ask students to show this by writing an equal sign between $\frac{3}{10}$ and $\frac{30}{100}$.

Repeat this activity with similar examples. As the students select squares, have them tell you the fraction name.

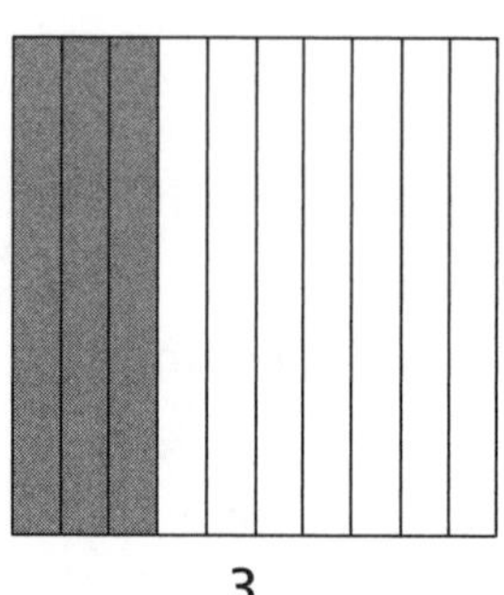

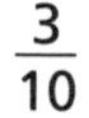

$\frac{3}{10}$

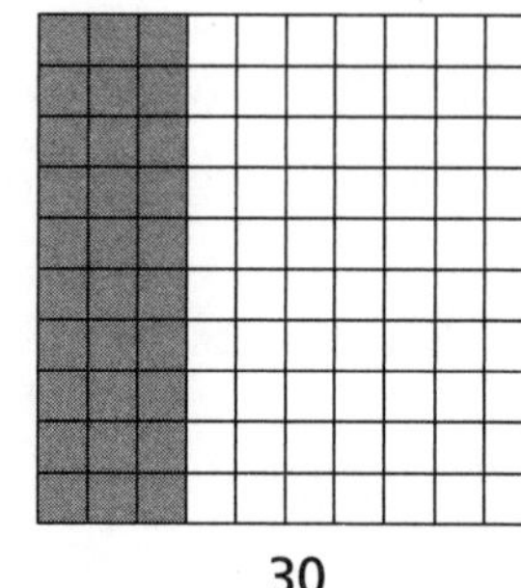

$\frac{30}{100}$

Name ______________________ Skill ______________________

Grade 4 Skill 50

Fractions with Denominators of 10 and 100

A **fraction** is a number that names part of a whole.

Fractions with Denominators of 10

The whole is divided into 10 equal parts. Each part is one tenth.

2 out of 10 parts are shaded.

Read: two tenths

Write: $\frac{2}{10}$ ← parts shaded / ← parts in the whole

Fractions with Denominators of 100

The whole is divided into 100 equal parts. Each part is one hundredth.

20 out of 100 parts are shaded.

Read: twenty hundredths

Write: $\frac{20}{100}$ ← parts shaded / ← parts in the whole

Equivalent Fractions

Equivalent fractions name the same amount.

Compare the shaded parts of the models. They are the same size.

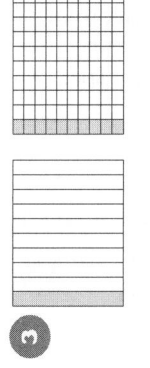

$\frac{2}{10} = \frac{20}{100}$

So, $\frac{2}{10}$ and $\frac{20}{100}$ are equivalent fractions.

Try These

Complete.

1. Read: ☐ tenth

   Write: $\frac{\square}{\square}$ ← parts shaded / ← parts in the whole

2. Read: ☐ hundredths

   Write: $\frac{\square}{\square}$ ← parts shaded / ← parts in the whole

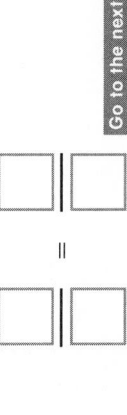

3. Complete to show equivalent fractions.

   $\frac{\square}{\square} = \frac{\square}{\square}$

Go to the next side.

Name ______________________ Skill ______________________

Practice on Your Own

Skill 50

Think:
The models have the same amount shaded. So, the fractions are equivalent.

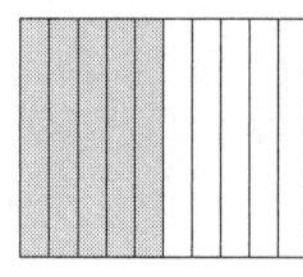
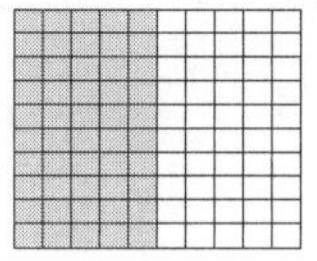

$\frac{5}{10}$ = $\frac{50}{100}$

five tenths — fifty hundredths

Write a fraction for each.

1. three tenths $\frac{\square}{\square}$
2. 9 tenths $\frac{\square}{\square}$
3. twenty-five hundredths $\frac{\square}{\square}$
4. ninety hundredths $\frac{\square}{\square}$

Complete to show equivalent fractions.

5. 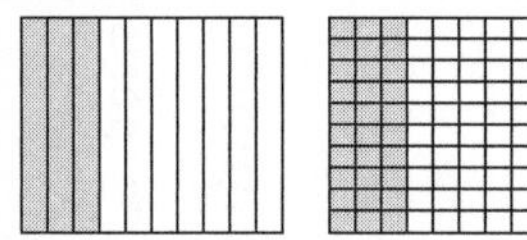 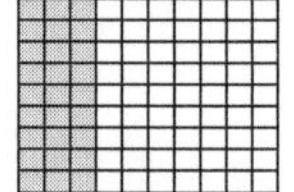 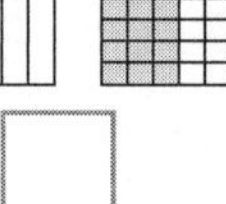

$\frac{3}{10} = \frac{\square}{100}$

6.  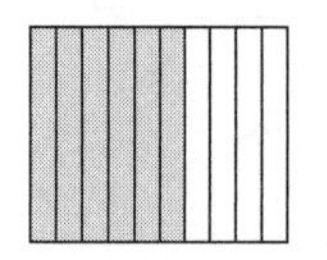

$\frac{6}{10} = \frac{\square}{100}$

7.  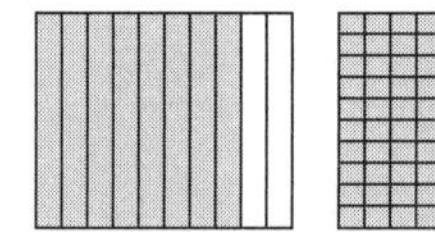

$\frac{8}{10} = \frac{\square}{100}$

8. 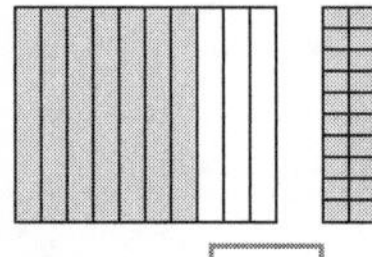 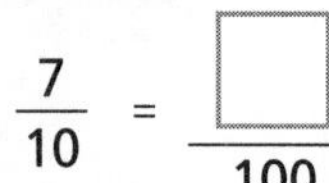

$\frac{7}{10} = \frac{\square}{100}$

9.  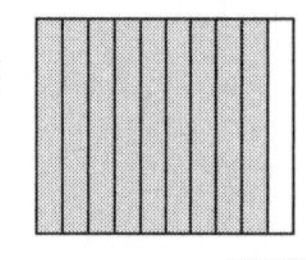 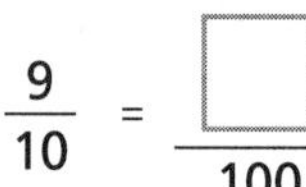

$\frac{9}{10} = \frac{\square}{100}$

10.  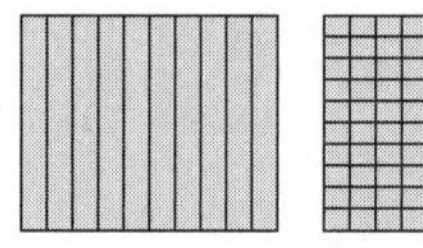

$\frac{10}{10} = \frac{\square}{100}$

Check

11. Write a fraction for seven tenths.

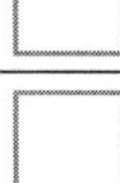

$\frac{\square}{\square}$

12. Write a fraction for seventy-five hundredths.

$\frac{\square}{\square}$

13. Complete to show equivalent fractions.

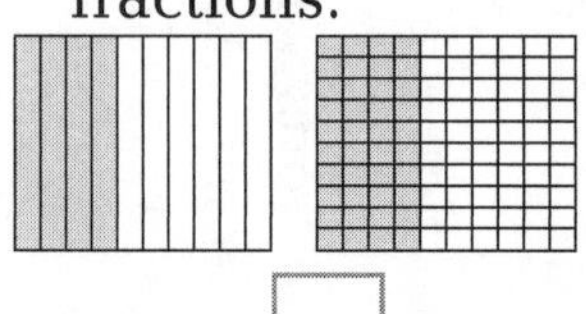

$\frac{4}{10} = \frac{\square}{100}$

Answer Card

Fractions

Grade 4

Skill 45

TRY THESE

1. 1, 3, 1, $\frac{1}{3}$
2. 3, 4, 3, $\frac{3}{4}$
3. 3, 8, 3, $\frac{3}{8}$

PRACTICE

1. 1, $\frac{1}{4}$
2. 2, $\frac{2}{5}$
3. 4, $\frac{4}{6}$
4. $\frac{2}{4}$
5. $\frac{3}{8}$
6. $\frac{2}{6}$
7. $\frac{1}{2}$
8. $\frac{3}{9}$
9. $\frac{3}{5}$

CHECK

10. $\frac{1}{6}$
11. $\frac{3}{10}$
12. $\frac{7}{8}$

Skill 46

TRY THESE

1. 5, 1, 1
2. 4, 3, 3
3. 3, 2, 2

PRACTICE

1. 1, 6, $\frac{1}{6}$
2. 2, 5, $\frac{2}{5}$
3. 1, 2, $\frac{1}{2}$
4. $\frac{1}{6}$
5. $\frac{3}{7}$
6. $\frac{3}{4}$
7. $\frac{1}{4}$
8. $\frac{1}{3}$
9. $\frac{1}{4}$

CHECK

10. $\frac{1}{7}$
11. $\frac{1}{6}$
12. $\frac{5}{8}$

Skill 47

TRY THESE

1. 1, 3, $\frac{1}{3}$
2. 2, 3, $\frac{2}{3}$
3. 3, 3, $\frac{3}{3}$

PRACTICE

1. 0, 1, 2, 3, 4, 5
2. 5, 5, 5, 5, 5, 5
3. $\frac{0}{5}$, $\frac{1}{5}$, $\frac{2}{5}$, $\frac{3}{5}$, $\frac{4}{5}$, $\frac{5}{5}$
4. $\frac{2}{3}$
5. $\frac{6}{8}$
6. $\frac{0}{6}$
7. $\frac{3}{4}$
8. $\frac{4}{5}$
9. $\frac{2}{2}$

CHECK

10. $\frac{1}{7}$
11. $\frac{0}{2}$
12. $\frac{6}{6}$

Answer Card

Fractions

Grade 4

SKILL 48

TRY THESE

1. 1, 3, $\frac{1}{3}$
2. 2, 3, $\frac{2}{3}$
3. 3, 3, $\frac{3}{3}$

PRACTICE

1. 0, 1, 2, 3, 4
2. 4, 4, 4, 4, 4
3. $\frac{0}{4}, \frac{1}{4}, \frac{2}{4}, \frac{3}{4}, \frac{4}{4}$
4. $\frac{4}{8}$
5. $\frac{2}{6}$
6. $\frac{1}{2}$
7. $\frac{3}{7}$
8. $\frac{3}{6}$
9. $\frac{4}{5}$

CHECK

10. $\frac{1}{3}$
11. $\frac{1}{5}$
12. $\frac{3}{6}$

SKILL 49

TRY THESE

1. $\frac{2}{3}, \frac{1}{3}$
2. $\frac{5}{6}, \frac{4}{6}$
3. $\frac{7}{9}, \frac{3}{9}$

PRACTICE

1. $>$
2. $=$
3. $<$
4. Shade 3 parts, shade 2 parts, $>$
5. Shade 2 parts, shade 4 parts, $<$
6. Shade 1 part, shade 0 parts, $>$
7. $=$
8. $>$
9. $=$

CHECK

10. $>$
11. $<$
12. $>$

SKILL 50

TRY THESE

1. one, $\frac{1}{10}$
2. ten, $\frac{10}{100}$
3. $\frac{1}{10}, \frac{10}{100}$

PRACTICE

1. $\frac{3}{10}$
2. $\frac{9}{10}$
3. $\frac{25}{100}$
4. $\frac{90}{100}$
5. 30
6. 60
7. 80
8. 70
9. 90
10. 100

CHECK

11. $\frac{7}{10}$
12. $\frac{75}{100}$
13. 40

Number Sense

Decimals

Model Decimals (Tenths and Hundredths)

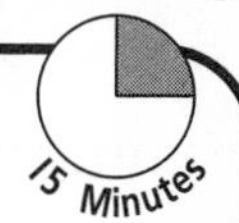

OBJECTIVE Model decimals for tenths and hundredths

MATERIALS decimal squares

Begin by pointing to the place-value labels on the place-value chart. Explain that any place value to the right of the decimal point ends in *ths*. Also recall that places to the right of the decimal point are less than 1.

Direct students' attention to the first example. Explain that the square represents a whole. It is not divided into parts. Point out that 1 can be written as a decimal: 1.0. The zero in the tenths place shows that there are no tenths.

Have students look at the square for tenths and compare it to the square for 1. Although they are the same size and shape, the tenths square is divided into 10 equal parts, one part is written as 0.1 to show that there are no ones and only 1 tenth. Continue in the same way for the example for hundredths.

TRY THESE Exercises 1–4 model the type of exercises students will find on the **Practice on Your Own** page.

- **Exercises 1–2** Write decimals for tenths.
- **Exercises 3–4** Write decimals for hundredths.

PRACTICE ON YOUR OWN Review the example at the top of the page. Work with students to compare the two models. Help students understand that 8 hundredths is a very small part of the whole and that 8 tenths is close to 1.

CHECK Determine if students know the difference between tenths and hundredths. Success is determined by 3 out of 3 correct responses.

Students who successfully complete the **Practice on Your Own** and **Check** are ready to move on to the next skill.

COMMON ERRORS

- Students may write tenths in the tens place or may write tenths two places to the right of the decimal point.
- Students may write a decimal such as 3 hundredths as 0.300.

Students who made more than three errors in the **Practice on Your Own**, or who were not successful in the **Check** section, may benefit from the **Alternative Teaching Strategy** on the next page.

Optional

Alternative Teaching Strategy

15 Minutes

Model Decimals (Tenths and Hundredths)

OBJECTIVE Use decimal squares to model decimals

MATERIALS decimal squares

Begin by writing the following on the board for the students.

ones	.	tenths
0	.	6

Ask: **What does the zero in the ones place mean?** there are no ones **What does the digit 6 represent?** 6 tenths

Explain that the decimal separates ones from tenths and hundredths.

In the number six tenths, what does the tenths mean? The whole is divided into 10 equal parts

What does the six represent in the number six tenths? There are 6 out of 10 parts being considered.

Distribute the decimal squares. Ask students to find a decimal square that matches 0.6. Have students confirm that the square they chose represents 6 tenths. Students should be able to say that the whole is divided into 10 parts and 6 out of 10 parts are shaded.

Repeat this activity with similar examples. Each time, have the students use the place-value names when reading the numbers.

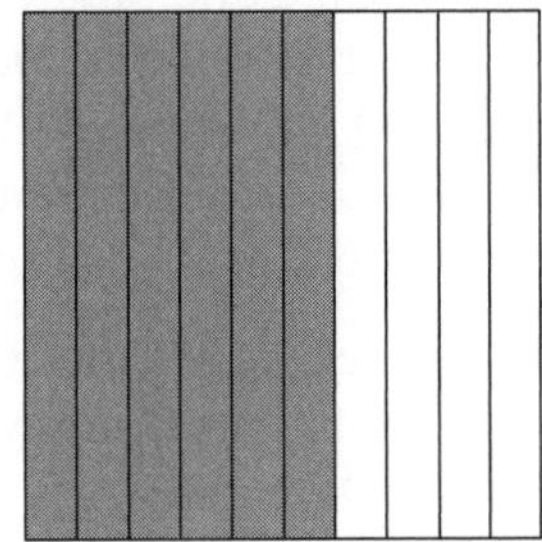

Name ______________________ Skill ______________________

Grade 4 Skill 51

Model Decimals (Tenths and Hundredths)

Use decimal squares to model decimal numbers.

A decimal uses place value and a decimal point to show values less than one.

This model represents one whole or 1.

ones	.	tenths
1	.	0

decimal point

Read: one
Write: 1.0

The whole is divided into 10 equal parts. 1 of 10 equal parts of a whole is one tenth.

1 out of **10** equal parts is shaded. **So,** one tenth is shaded.

ones	.	tenths
0	.	1

Read: one tenth
Write: 0.1

The whole is divided into 100 equal parts. 1 of 100 equal parts of a whole is one hundredth.

1 out of **100** equal parts is shaded. **So,** one hundredth is shaded.

ones	.	tenths	hundredths
0	.	0	1

Read: one hundredth
Write: 0.01

Try These

Write how many parts are shaded. Write the decimal two ways.

1. 3 out of 10 parts shaded
Read: ☐ tenths
Write: ☐ . ☐ ☐

2. ☐ out of ☐ parts shaded
Read: ☐ tenths
Write: ☐ . ☐ ☐

3. ☐ out of ☐ parts shaded
Read: ☐ hundredths
Write: ☐ . ☐ ☐

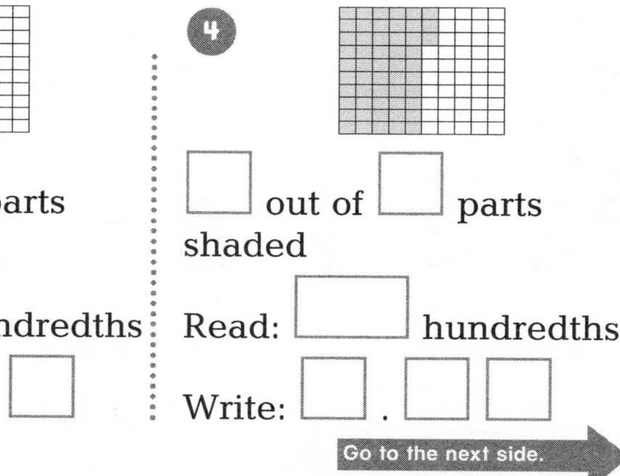

4. ☐ out of ☐ parts shaded
Read: ☐ hundredths
Write: ☐ . ☐ ☐

Go to the next side.

Name ______________________ Skill ______________________

Practice on Your Own

8 out of 10 equal parts are shaded.

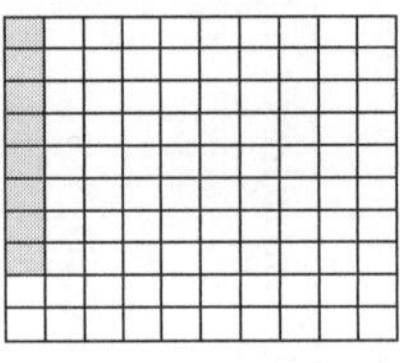

ones	.	tenths
0	.	8

Read: eight tenths

Write: 0.8

8 out of 100 equal parts are shaded.

ones	.	tenths	hundredths
0	.	0	8

Read: eight hundredths

Write: 0.08

Write how many parts are shaded. Write the decimal two ways.

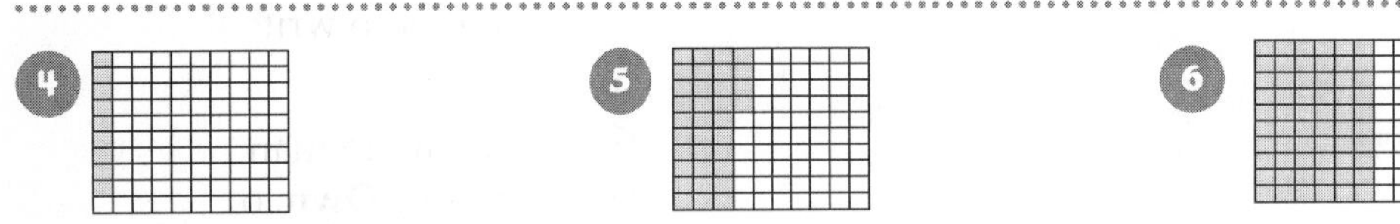

1. ☐ out of ☐ parts shaded.
   Read: ☐ tenths
   Write: ☐ . ☐
2. ☐ out of ☐ parts shaded.
   Read: ☐ tenths
   Write: ☐ . ☐
3. ☐ out of ☐ parts shaded.
   Read: ☐ tenths
   Write: ☐ . ☐
4. Read: ☐ hundredths
   Write: ☐ . ☐ ☐
5. Read: ☐ hundredths
   Write: ☐ . ☐ ☐
6. Read: ☐ hundredths
   Write: ☐ . ☐ ☐

Write the decimal for the shaded part.

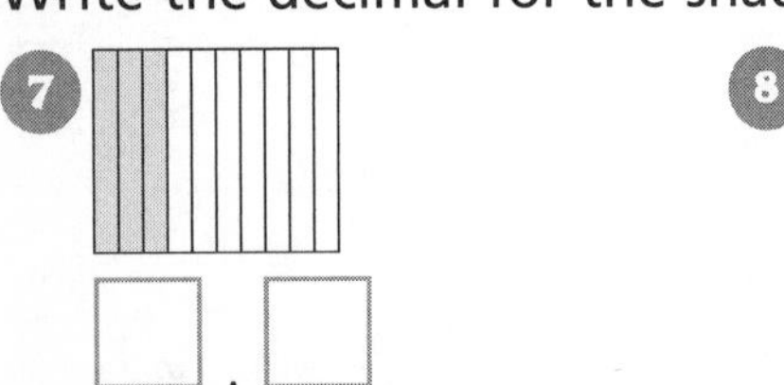

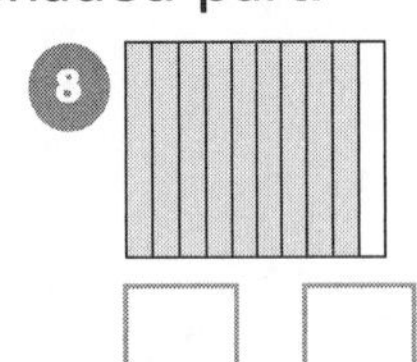

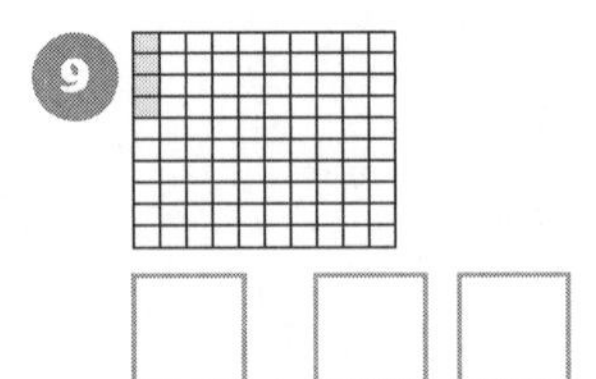

7. ☐ . ☐
8. ☐ . ☐
9. ☐ . ☐ ☐

Check

Write the decimal for the shaded part.

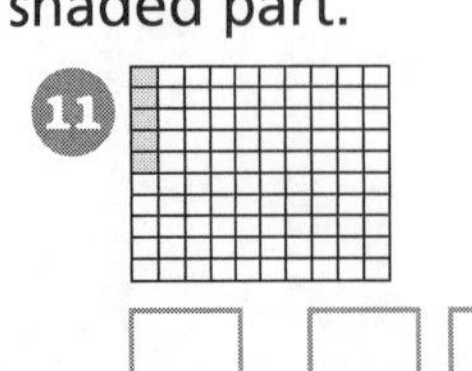

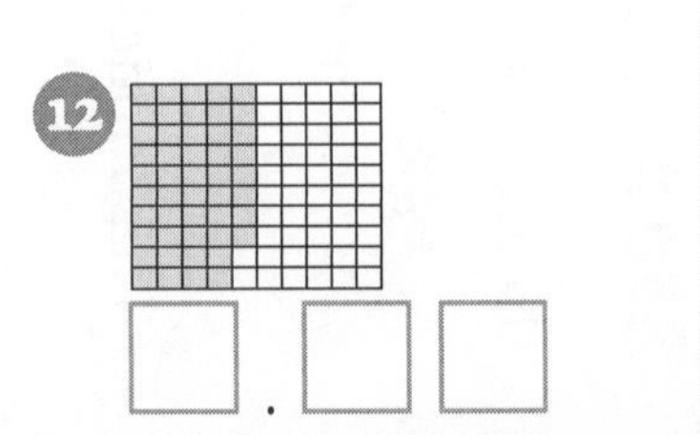

10. ☐ . ☐
11. ☐ . ☐ ☐
12. ☐ . ☐ ☐

Skill 52 Grade 4

Relate Decimals to Money

OBJECTIVE Relate decimals to money

You may wish to begin the lesson by discussing with students why money is written using decimals. Explain that the amounts shown to the right of the decimal point represent parts of a dollar.

Direct students' attention to the first example.

Ask: **How many dimes are in 1 dollar?** 10 **What part or fraction of a dollar is 1 dime?** $\frac{1}{10}$ **You write** $\frac{1}{10}$ **of a dollar as $0.10.**

How many pennies are in 1 dollar? 100 **What fraction of a dollar is 1 penny?** $\frac{1}{100}$ **How do you write** $\frac{1}{100}$ **of a dollar as a decimal?** $0.01

Point out the relationship between decimal notation and money notation in the next example.

You may wish to show students the expanded form of 15 hundredths and 15 cents so students understand the meaning of the digits in each place.

$$0.15 = 0.10 + 0.05$$

$$\$0.15 = \$0.10 + \$0.05$$

TRY THESE Exercises 1–3 model the type of exercises students will find on the **Practice on Your Own** page.

- **Exercises 1–3** Write amounts written in money notation as a number of dimes and pennies, then as tenths and hundredths.

PRACTICE ON YOUR OWN Review the example at the top of the page. Ask students to explain how dimes and tenths are related.

CHECK Determine if students can write amounts of money using money notation. Success is determined by 3 out of 3 correct responses.

Students who successfully complete the **Practice on Your Own** and **Check** are ready to move on to the next skill.

COMMON ERRORS

- Students may write 1 penny as $0.1 thinking of it as 1 tenth of a dime.
- Students may forget to write the dollar sign.

Students who made more than two errors in the **Practice on Your Own**, or who were not successful in the **Check** section, may benefit from the **Alternative Teaching Strategy** on the next page.

Optional

20 Minutes

Alternative Teaching Strategy
Model Relating Decimals to Money

OBJECTIVE Model amounts of money and use money notation

MATERIALS play money (dollars, dimes, and pennies)

Distribute the play money. Display the dollar and ask the students to show you an equivalent amount using the dimes. Observe that 10 dimes or 10 tenths equal 1 dollar. Have students set aside 9 of the dimes.

Ask: **What part of a dollar is still showing?** $\frac{1}{10}$ Record the amount as $0.10.

Have students add 2 more dimes to the 1 on their desks.

Ask: **Now, what part of a dollar is represented?** $\frac{3}{10}$ **How would you write that using money notation?** $0.30 Recall that this amount is read as 30 cents.

Now have students show an amount equivalent to 1 dollar using pennies. As students count the pennies, suggest that they arrange them in rows of 10. Observe that 100 pennies or 100 hundredths equal 1 dollar.

Have students move one penny aside.

Ask: **What part of a dollar does 1 penny represent?** $\frac{1}{100}$ of a dollar

Ask: **How would you write that using money notation?** $0.01

If students have trouble showing this, display a place value table and record the amount in the table.

Continue building the connection between pennies and hundredths through amounts to 9 cents. When students have demonstrated understanding, work with amounts such as $0.24, $0.85, $1.10, $1.05, and $1.17.

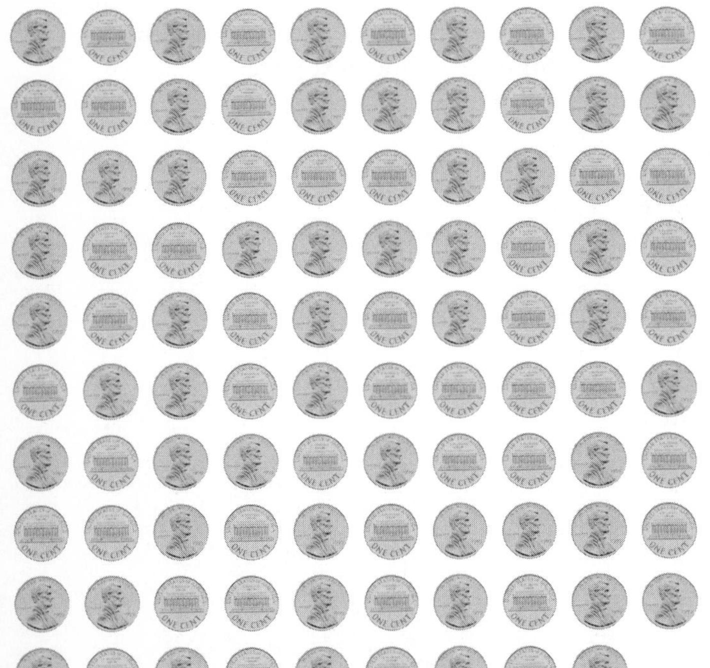

1 hundredth of 1 dollar
$0.01

Name ______________________ Skill ______________________

Relate Decimals to Money

Grade 4 Skill 52

Write amounts of money using decimals.

Dimes and Pennies

Here are some ways to think about dimes and pennies.

10 dimes = 1 dollar
1 dime = 1 tenth of a dollar
↓
$\frac{1}{10}$ of a dollar
↓
0.1 of a dollar
↓
$0.10

100 pennies = 1 dollar
1 penny = 1 hundredth of a dollar
↓
$\frac{1}{100}$ of a dollar
↓
0.01 of a dollar or $0.01

You can see that a dime and a tenth show $\frac{1}{10}$ of 100. A penny and a hundredth show $\frac{1}{100}$ of 100.

Use a Place Value Chart

Use the place value chart to understand the meaning of $0.15 and 0.15.

Dollars		Dimes	Pennies
0	.	1	5

$0.15 = 15 hundredths of a dollar
= 15 pennies
= 1 dime 5 pennies

Ones		Tenths	Hundredths
0	.	1	5

0.15 = 15 hundredths
= 1 tenth 5 hundredths

Try These

Complete.

1.

Dollars		Dimes	Pennies
0	.	4	7

$0.47 = ☐ dimes ☐ pennies
= ☐ tenths ☐ hundredths

2.

Dollars		Dimes	Pennies
0	.	6	2

$0.62 = ☐ dimes ☐ pennies
= ☐ tenths ☐ hundredths

3.

Dollars		Dimes	Pennies
0	.	2	3

$0.23 = ☐ dimes ☐ pennies
= ☐ tenths ☐ hundredths

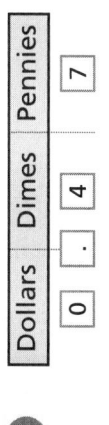

Go to the next side.

Name ______________________ Skill ______________________

Practice on Your Own

Skill 52

Think:
Use a place-value chart to write the money amount.

three dollars and forty-seven cents

Dollars	Dimes	Pennies
3	4	7

$3.47

Complete.

1. one dollar and eighty-four cents

Dollars	Dimes	Pennies
☐	☐	☐

$☐ . ☐ ☐

2. two dollars and twenty-seven cents

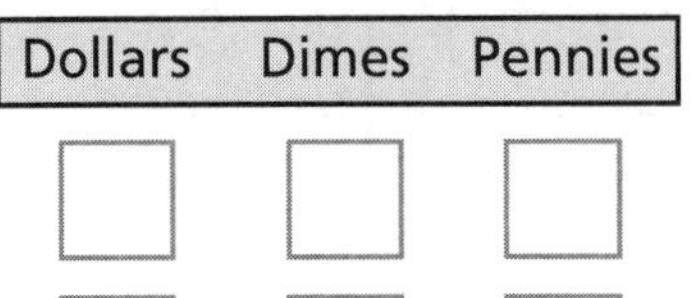

Dollars	Dimes	Pennies
☐	☐	☐

$☐ . ☐ ☐

3. two dollars and fifty-two cents

Dollars	Dimes	Pennies
☐	☐	☐

$☐ . ☐ ☐

4. three dollars and thirty-six cents

Dollars	Dimes	Pennies
☐	☐	☐

$☐ . ☐ ☐

5. five dollars and eight cents

Dollars	Dimes	Pennies
☐	☐	☐

$☐ . ☐ ☐

6. three cents

Dollars	Dimes	Pennies
☐	☐	☐

$☐ . ☐ ☐

Write the decimal.

7. one dollar and nineteen cents

$☐ . ☐ ☐

8. two dollars and five cents

$☐ . ☐ ☐

9. thirty-nine cents

$☐ . ☐ ☐

Check

Write the decimal.

10. two dollars and eighty-two cents

$☐ . ☐ ☐

11. one dollar and eighteen cents

$☐ . ☐ ☐

12. fourteen cents

$☐ . ☐ ☐

Answer Card

Decimals

Grade 4

SKILL 51

TRY THESE

1. three, 0.3
2. 6, 10, six, 0.6
3. 7, 100, seven, 0.07
4. 53, 100, fifty-three, 0.53

PRACTICE

1. 4, 10, four, 0.4
2. 5, 10, five, 0.5
3. 9, 10, nine, 0.9
4. nine, 0.09
5. thirty-four, 0.34
6. sixty, 0.60
7. 0.3
8. 0.9
9. 0.04

CHECK

10. 0.2
11. 0.05
12. 0.48

SKILL 52

TRY THESE

1. 4, 7, 4, 7
2. 6, 2, 6, 2
3. 2, 3, 2, 3

PRACTICE

1. 1, 8, 4, $1.84
2. 2, 2, 7, $2.27
3. 2, 5, 2, $2.52
4. 3, 3, 6, $3.36
5. 5, 0, 8, $5.08
6. 0, 0, 3, $0.03
7. $1.19
8. $2.05
9. $0.39

CHECK

10. $2.82
11. $1.18
12. $0.14

Algebra and Functions

Skill 53 Grade 4

Missing Addends

OBJECTIVE Find missing addends

MATERIALS counters

You may wish to have students use counters to represent mathematical sentences.

Display an addition sentence such as 7 + ___ = 11. Have students identify the known addend and sum. Suggest a story that fits the addition sentence: For example, there are 7 geese resting on the grass and some geese are swimming. There are 11 geese in all. How many geese are swimming?

Have students model the sentence using counters. Then ask one student to write the addition sentence on the board **7 + 4 = 11.**

Write the subtraction sentence 11 − 7 = ___. Ask students to retell the story so that it fits the subtraction sentence: There are 11 geese in all; 7 geese are resting on the grass. How many geese are swimming? Have students model the sentence using counters. Then ask one student to write the subtraction sentence on the board **11 − 7 = 4.**

Ask: **What happens to the sum when the addition sentence is rewritten as a subtraction sentence? The sum becomes the number to subtract from.**

Link the oral activity to the steps in the lesson. If necessary, have students model the steps with counters.

TRY THESE Exercises 1–3 prepare students for the types of exercises they will encounter in the **Practice on Your Own** section.

- **Exercises 1 and 3** Second addend is missing.
- **Exercises 2 and 4** First addend is missing.

PRACTICE ON YOUR OWN Review the example at the top of the page. Have students identify the addends and the sum in each number sentence. **7 + 13 = 20, 20 − 7 = 13**

In Exercises 1–7 students are given a subtraction sentence to help them find the missing addend. In Exercises 8–11 students are expected to find the missing addend without such prompts.

CHECK Determine if students understand they can use subtraction to find a missing addend.

Success is indicated by 4 out of 4 correct responses.

Students who successfully complete the **Practice on Your Own** and **Check** are ready to move on to the next skill.

COMMON ERRORS

- Students may not know basic facts for addition and subtraction, and thus write an incorrect addend.
- Students may have difficulty changing the order of numbers when they write subtraction sentences from addition sentences.

Students who made more than 4 errors in the **Practice on Your Own**, or who were not successful in the **Check** section, may benefit from the **Alternative Teaching Strategy** on the next page.

Optional

Alternative Teaching Strategy

Model Missing Addends with Cubes

OBJECTIVE Find missing addends

MATERIALS connecting centimeter cubes, index cards prepared with addition and subtraction sentences written on each

Distribute at least 10 of each color of the connecting cubes, and index cards with addition sentences and subtraction sentences such as 9 + ___ = 13 and 13 − ___ = 4 written on them.

Have students model the known addend with cubes of one color. Then have them count on with cubes of a second color to find the missing addend.

9 + ___ = 13

9 + 4 = 13

Ask students to use the cubes to represent the subtraction sentence they would write to find the missing addend. 13 − 9 = 4

Repeat the activity several times. Encourage students to use the connecting cubes to model each addition and subtraction sentence on the index cards.

Name ______________________ Skill ______________________

Grade 4 Skill 53

Missing Addends

Use subtraction to find a missing addend. Find 6 + ■ = 15.

Step 1
Rewrite the addition sentence as a subtraction sentence.

6 + ■ = 15	addend + addend = sum
15 − 6 = ■	sum − addend = addend

Step 2
Subtract the known addend from the sum.

15	−	6	=	■
sum		known addend		missing addend

Step 3
Write the missing addend.

15 − 6 = 9

So, 6 + 9 = 15.

Try These

Find the missing addend.

1. 7 + ■ = 11

11	−	7	=	4
sum		known addend		missing addend

7 + ☐ = 11

2. ■ + 8 = 17

17	−	8	=	☐
sum		known addend		missing addend

☐ + 8 = 17

3. 6 + ■ = 14

14	−	6	=	☐
sum		known addend		missing addend

6 + ☐ = 14

4. ■ + 3 = 12

12	−	3	=	☐
sum		known addend		missing addend

☐ + 3 = 12

Go to the next side.

Name ______________________ Skill ______________________

Practice on Your Own

Find 7 + ■ = 20. addend + addend = sum

20	−	7	=	■
sum		known addend		missing addend

20 − 7 = 13

Think: You can use subtraction to find a missing addend.

So, 7 + **13** = 20.

Find the missing addend.

1. 8 + ■ = 15

   15 − 8 = 7
   sum known addend missing addend

   8 + □ = 15

2. ■ + 3 = 14

   14 − 3 = □
   sum known addend missing addend

   □ + 3 = 14

3. 7 + ■ = 16

   16 − 7 = □
   sum known addend missing addend

   7 + □ = 16

4. 6 + ■ = 19

   19 − 6 = □

   6 + □ = 19

5. ■ + 5 = 14

   14 − 5 = □

   □ + 5 = 14

6. 4 + ■ = 21

   21 − 4 = □

   4 + □ = 21

7. ■ + 7 = 18

   18 − 7 = □

   □ + 7 = 18

8. 5 + □ = 11
9. □ + 9 = 16
10. 7 + □ = 13
11. □ + 11 = 15

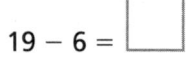

Check

Find the missing addend.

12. □ + 8 = 12
13. 6 + □ = 18
14. □ + 7 = 10
15. 2 + □ = 16

Skill 54 Grade 4

Number Patterns for Addition

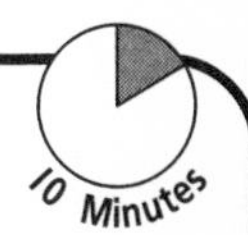

OBJECTIVE Find and use addition patterns to extend a sequence of numbers

Begin the skill by having students look at the sequence of numbers at the top of the skill page. Explain that the task is to examine the series of numbers to see if they can discover the pattern. Then they will use the pattern to find the next number.

Explain that a number line can help them find the pattern.

Say: **Look at the number line. Are the numbers increasing or decreasing?** increasing **When the numbers increase, the pattern may be an addition pattern. What number do you add to 5 to get 7?** 2 **What number do you add to 7 to get 9?** 2 **What do you think is the rule for the pattern?** Add 2.

Point out that once they know the rule, they can extend the pattern to find the next number. **You know that the rule is add 2. To what number in the sequence should you add 2 to get the next number?** the last number, 15 **What is the next number?** 17

Students may note that they can use skip counting to find the rule and the next number. Discuss how the number line illustrates the pattern and makes it easier to find the rule.

TRY THESE In Exercises 1–3, the students find the rule and the next number step by step.

- **Exercise 1** Rule: + 5.
- **Exercise 2** Rule: + 3.
- **Exercise 3** Rule: + 2.

PRACTICE ON YOUR OWN Review the example with students. Ask students if the next number in the pattern will be greater than or less than 14. Remind them that since addition patterns increase, the next number is greater than 14.

CHECK Determine if students know how to find and complete a number pattern for addition. Success is determined by 3 out of 3 correct responses.

Students who successfully complete the **Practice on Your Own** and **Check** are ready to move on to the next skill.

COMMON ERRORS

- Students may not understand how to find the rule and may think they can add 1 to get the next number.
- Students may understand the pattern, but add incorrectly.

Students who made more than two errors in the **Practice on Your Own**, or who were not successful in the **Check** section, may benefit from the **Alternative Teaching Strategy** on the next page.

Optional

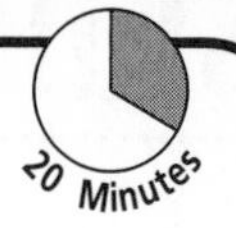

Alternative Teaching Strategy
Model Finding an Addition Pattern

OBJECTIVE Use counters to model patterns for addition

MATERIALS counters

Students may benefit from working in pairs or small groups. Distribute counters and explain to students that they will use the counters to find a number pattern.

Have students arrange counters in groups of 2, 5, 8, 11, 14, and 17. Ask students to look carefully at the groups. Students may note that the number of counters in the groups increases. Suggest that since the number of counters increases, the pattern may be an addition pattern.

Say: **There are two counters in the first group. How many more counters did you need to get 5 counters?** 3

How many more counters than 5 did you need to get 8 counters? 3 **11 counters?** 3 **14 counters?** 3 **17 counters?** 3

Each group seems to have 3 more counters than the last group. What rule do you see for the pattern? Add 3.

Discuss how the students can find the number of counters for the group that comes next. Explain that they can find out by taking the number of counters in the last group, 17, and adding 3 more counters. The next group will have 20 counters.

Repeat the activity several times with the counters. Have students identify the pattern and find the next number.

When students show an understanding of number patterns with counters, have them find number patterns using numbers only.

2
○○

5
○○○
○○

8
○○○○
○○○○

11
○○○○○
○○○○○
○

14
○○○○○
○○○○○
○○○○

17
○○○○○
○○○○○
○○○○○
○○

Name ______________________ Skill ______________________

Grade 4 Skill 54

Number Patterns for Addition

Find the pattern in: 5, 7, 9, 11, 13, 15, ■. Find the next number.

Step 1 You can look for a pattern in the numbers on a number line.

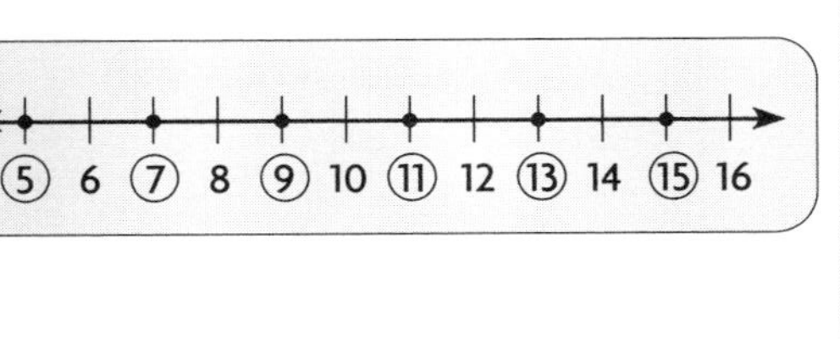

Step 2 Find a rule for the pattern. You can add 2 to each number to get the next number.

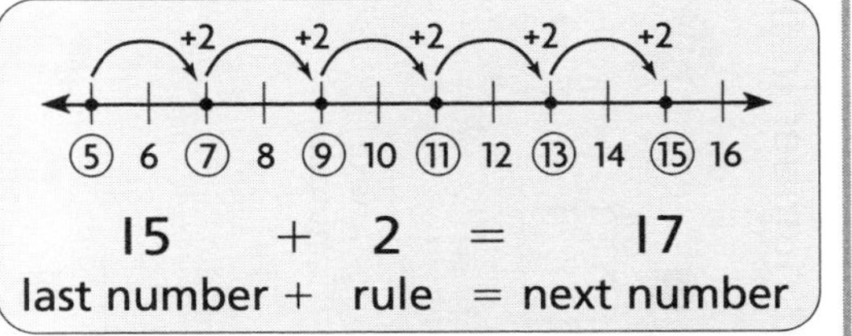

Step 3 Use the rule to extend the pattern. Find the next number in the pattern.

5, 7, 9, 11, 13, 15, ■

So, the next number is 17.

Try These

Find the pattern and the rule. Write the next number in the pattern.

1 10, 15, 20, 25, 30, ■

10 15 20 25 30

The rule is ☐.

☐ + ☐ = ☐

last number + rule = next number

10, 15, 20, 25, 30, ☐

2 1, 4, 7, 10, 13, ■

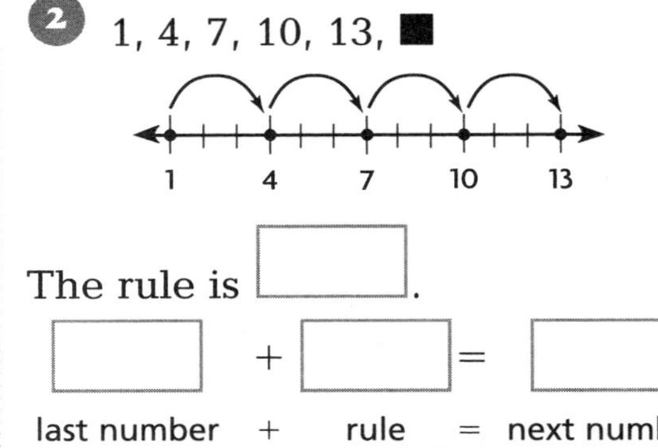

The rule is ☐.

☐ + ☐ = ☐

last number + rule = next number

1, 4, 7, 10, 13, ☐

3 20, 22, 24, 26, 28, 30, ■

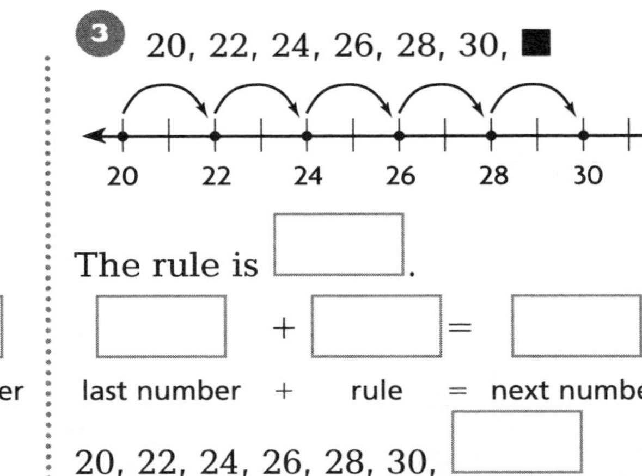

The rule is ☐.

☐ + ☐ = ☐

last number + rule = next number

20, 22, 24, 26, 28, 30, ☐

Go to the next side.

Name ______________________ Skill ______________________

Practice on Your Own

Skill 54

Find the pattern.
Write the next number.
2, 6, 10, 14, ■.

Think: What number can I add to get the next number?

The rule is add 4.

14	+	4	=	18
last number	+	rule	=	next number

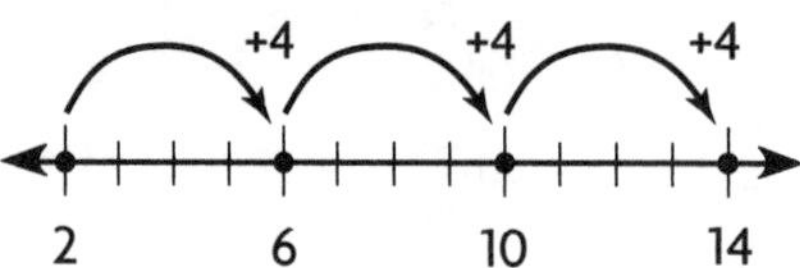

So, the next number is 18.

Find the pattern and the rule. Write the next number in the pattern.

1. 2, 5, 8, 11, ■

   The rule is ☐.

   ☐ + ☐ = ☐
   last number + rule = next number

   2, 5, 8, 11, ☐

2. 3, 9, 15, 21, ■

   The rule is ☐.

   ☐ + ☐ = ☐
   last number + rule = next number

   3, 9, 15, 21, ☐

3. 5, 13, 21, 29, ■

   The rule is ☐.

   ☐ + ☐ = ☐
   last number + rule = next number

   5, 13, 21, 29, ☐

4. 4, 9, 14, 19, ■

   The rule is ☐.

   4, 9, 14, 19, ☐

5. 1, 9, 17, 25, ■

   The rule is ☐.

   1, 9, 17, 25, ☐

6. 15, 19, 23, 27, ■

   The rule is ☐.

   15, 19, 23, 27, ☐

Write the next number in the pattern.

7. 10, 20, 30, 40, ☐
8. 32, 35, 38, 41, ☐
9. 17, 26, 35, 44, ☐

Write the next number in the pattern.

10. 15, 25, 35, 45, ☐
11. 16, 20, 24, 28, ☐
12. 21, 29, 37, 45, ☐

Skill 55 Grade 4 Number Patterns for Subtraction

OBJECTIVE Find number patterns using subtraction

MATERIALS number lines

Demonstrate counting back from a number on the number line. Remind students that in this activity they are subtracting and that the numbers in the pattern will decrease.

Direct students' attention to Steps 1–3.

Ask: **How is the number line labeled?** from 5 to 15 **What numbers are in the pattern?** 15, 13, 11, 9, 7, 5 **If you begin at 15 on the number line, and move left to 13, how many spaces do you move?** 2 **How many spaces is it from 13 to 11?** 2 **From 11 to 9?** 2 **If the pattern holds, how many spaces should you move from 5 to get the next number in the pattern?** 2 **What is the number when you move 2 spaces left from 5?** 3

Work through Steps 2 and 3 of the skill. Note how the rule for the pattern is expressed and why subtraction is the operation to use for this pattern.

TRY THESE Exercises 1–3 provide number lines for students to use for each pattern. Remind students to verify the pattern by counting spaces or subtracting from one number to the next.

- **Exercise 1** Use the rule: Subtract 5.
- **Exercise 2** Use the rule: Subtract 3.
- **Exercise 3** Use the rule: Subtract 6.

PRACTICE ON YOUR OWN Review the example with students. Point out that Exercises 1–3 provide number lines, but Exercises 4–9 do not. Remind students that they can use subtraction to find these patterns.

CHECK Determine if students know how to find a pattern involving subtraction and if they understand how to use the rule to find a missing number. Success is indicated by 3 out of 3 correct responses.

Students who successfully complete the **Practice on Your Own** and **Check** are ready to move on to the next skill.

COMMON ERRORS

- Students may understand the pattern, but subtract incorrectly.
- Students may use an incorrect number as a starting point.

Students who made more than three errors in the **Practice on Your Own**, or who were not successful in the **Check** section, may also benefit from the **Alternative Teaching Strategy** on the next page.

Optional

Alternative Teaching Strategy
Model Finding a Subtraction Pattern

OBJECTIVE Use number lines to count backward and state a pattern

MATERIALS number lines or lined paper turned sideways

Provide blank number lines, or have students turn lined paper sideways to make their own number lines. Have students number the interval from zero through twenty. Then, starting with 20, have students skip count back by 2s and circle each number they say. Next have the students write the numbers as a pattern in the order that they counted. As a group, discuss the different ways the pattern might be described. **Possible responses: multiples of 2; numbers decrease by 2 each time; all the even numbers between 20 and 0.** Together develop the rule for the pattern using subtraction.

(0) 1 (2) 3 (4) 5 (6) 7 (8) 9 (10) 11 (12) 13 (14) 15 (16) 17 (18) 19 (20)

Pattern:
20, 18, 16, 14, 12, 10, 8, 6, 4, 2

The rule for the pattern is: Subtract 2 to get the next number.

Repeat the activity for multiples of 3 from 0 to 21. Then ask students to label a number line from 36 to 18. Students skip count back by 4s, write the pattern, and write the rule.

When you feel that students have understood these patterns, provide this sequence:

44, 41, 38, 35, 32, □

Have them find the pattern, write the rule, and find the next number. Provide help when needed through questions such as:

Do the numbers increase or decrease? decrease What operation do you use to decrease a number? subtraction When you begin subtracting, what numbers do you use? 44 and 41 What is the difference? 3 What should you do next? Check if the difference between each pair of consecutive numbers is the same.

Name ______________________ Skill ______________________

Grade 4
Skill 55

Number Patterns for Subtraction

Extend the pattern. 15, 13, 11, 9, 7, 5, ■

Step 1 Start at 15. Move to the left since the numbers in the pattern decrease.

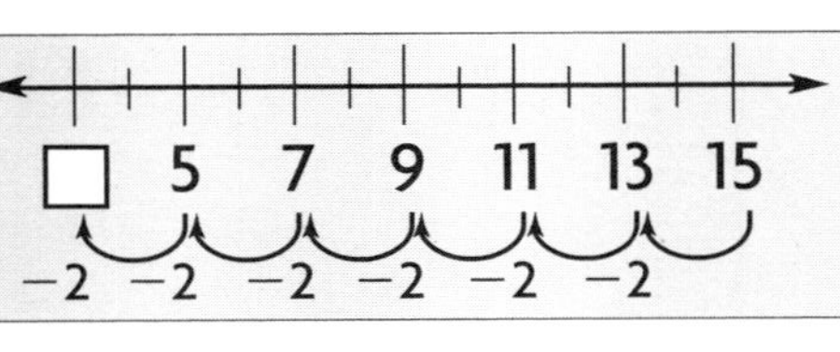

Step 2 Note that you subtracted 2 from each number to get the next number. State a rule for the pattern you find.

□ 5 7 9 11 13 15

−2 −2 −2 −2 −2 −2

The rule is: Subtract 2.

Step 3 Use the rule to find the next number in the pattern.

15, 13, 11, 9, 7, 5, ■

5	−	2	=	3
last number	−	rule	=	next number

So, the next number is 3.

Try These

State the rule. Then write the next number in the pattern.

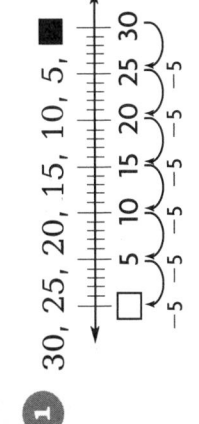

1. 30, 25, 20, 15, 10, 5, ■

□ 5 10 15 20 25 30

−5 −5 −5 −5 −5 −5

The rule is □.

□	−	□	=	□
last number	−	rule	=	next number

30, 25, 20, 15, 10, 5, □

2. 18, 15, 12, 9, 6, ■

□ 6 9 12 15 18

−3 −3 −3 −3 −3

The rule is □.

□	−	□	=	□
last number	−	rule	=	next number

18, 15, 12, 9, 6, □

3. 26, 20, 14, 8, ■

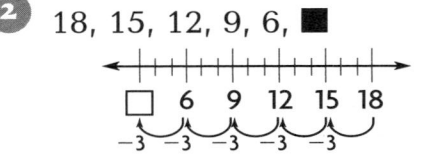

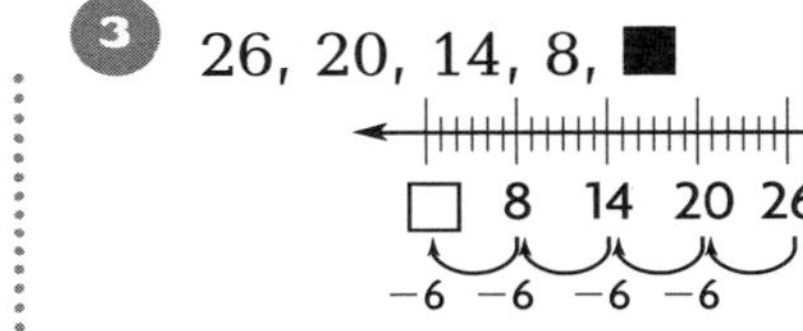

The rule is □.

□	−	□	=	□
last number	−	rule	=	next number

26, 20, 14, 8, □

Go to the next side.

Name ______________________ Skill ______________________

Practice on Your Own

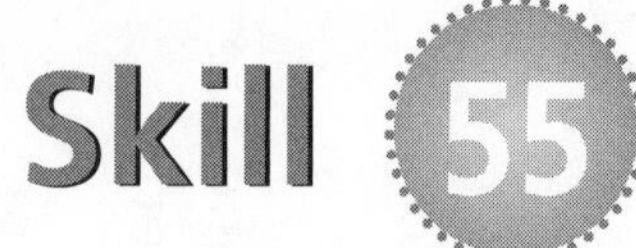

Find the pattern. Write the next number.

18, 14, 10, 6, ■

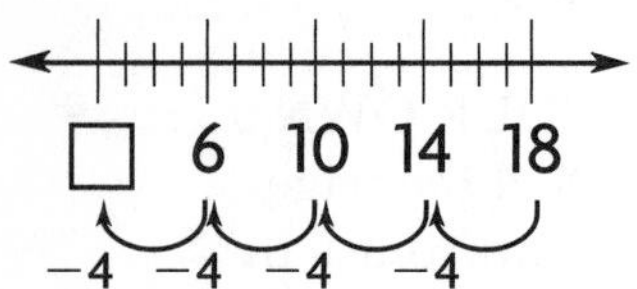

The rule is: subtract 4.

6 − 4 = 2

last number − rule = next number

Think: What number can I subtract to get the next number?

So, the next number is 2. 18, 14, 10, 6, **2**

State the rule.

1. 14, 11, 8, 5, ■

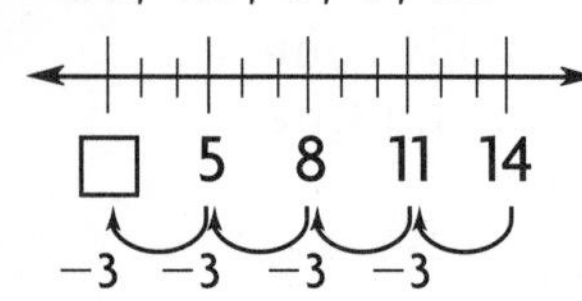

The rule is □.

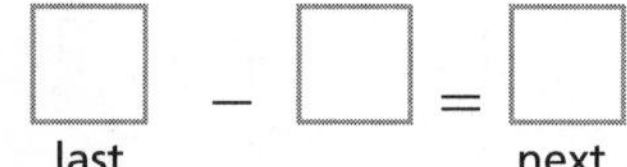

14, 11, 8, 5, □

2. 23, 21, 19, 17, ■

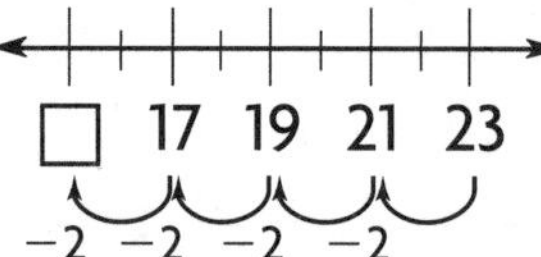

The rule is □.

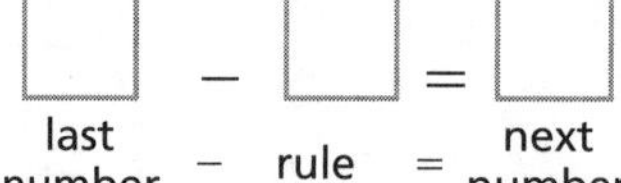

23, 21, 19, 17, □

3. 27, 21, 15, 9, ■

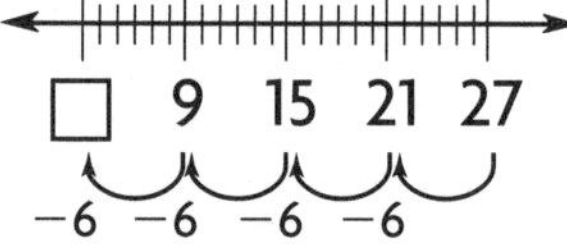

The rule is □.

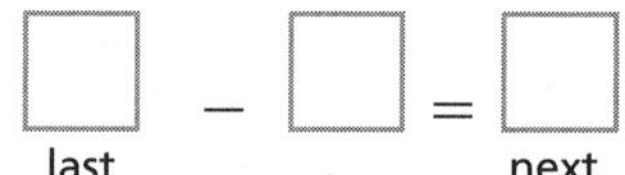

27, 21, 15, 9, □

4. 19, 15, 11, 7, ■

The rule is □.

19, 15, 11, 7, □

5. 31, 26, 21, 16, ■

The rule is □.

31, 26, 21, 16, □

6. 42, 35, 28, 21, ■

The rule is □.

42, 35, 28, 21, □

Write the next number.

7. 25, 21, 17, 13, □

8. 22, 20, 18, 16, □

9. 40, 32, 24, 16, □

Check

Write the next number.

10. 42, 37, 32, 27, □

11. 20, 17, 14, 11, □

12. 33, 29, 25, 21, □

Skill 56 Grade 4

Missing Factors

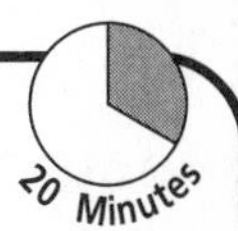

OBJECTIVE Find missing factors

Explain to the students that they are going to find the missing factor in a multiplication sentence. Point out the multiplication table. Suggest that when they do not remember a multiplication fact, they can use the multiplication table to help them find missing factors.

You may wish to familiarize the students with the table. Explain that there are factors in the left-most column, and factors in the top row. Where the row of one factor meets the column of another, they can find the product of those factors.

Then direct students' attention to the steps in the skill.

Say: **What do you already know about the multiplication fact?** **One factor is 4 and the product is 20.**

Ask students to find the row for the factor, 4 and look across until they find the product, 20.

Ask: **When you look up the column to the top, what factor do you find?** **5** **The multiplication table shows that 4 × 5 = 20. So, what is the missing factor?** **5**

TRY THESE In Exercises 1–3, students find the missing factor step by step.

- **Exercise 1** Multiplication fact for 3.
- **Exercise 2** Multiplication fact for 7.
- **Exercise 3** Multiplication fact for 8.

PRACTICE ON YOUR OWN Review the example at the top of the page. Ask a volunteer to explain how to use the multiplication table to find the missing factor.

CHECK Determine if students understand the relationship between the factors and the product, and can use one factor and the product to find a missing factor. Success is determined by 2 out of 3 correct responses.

Students who successfully complete the **Practice on Your Own** and **Check** are ready to move on to the next skill.

COMMON ERRORS

- Students may not understand how to use the multiplication table.
- Students may have trouble tracking a row or column, and thus write an incorrect factor as a missing factor.

Students who made more than two errors in the **Practice on Your Own**, or who were not successful in the **Check** section, may benefit from the **Alternative Teaching Strategy** on the next page.

Optional

Alternative Teaching Strategy

Model Finding Missing Factors

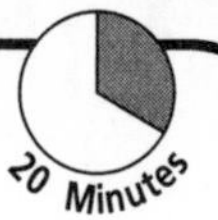

OBJECTIVE Find missing factors using arrays

MATERIALS counters or tiles

Prepare multiplication sentences on cards that show missing factors. Begin with products of 30 or less. Be sure to provide missing factors in both positions in the multiplication sentence.

Distribute the counters. Have students choose a card. Point out that the number sentences they have chosen show a missing factor. Explain that they already know one factor and the product. Suggest that they can use what they know to discover the missing factor.

Demonstrate how to use an array to find the missing factor for the multiplication sentence.

4 × □ = 24.

○○○○○○
○○○○○○
○○○○○○
○○○○○○

4 × 6 = 24

Explain that the product represents the total number of counters to use. Have students count out 24 counters.

Then point out that they can make 4 rows of counters using all 24 counters. The number of counters in each row will be the other factor. There are 6 counters in a row. So, 4 × 6 = 24.

Repeat the activity for □ × 6 = 18. Suggest they make rows of 6 with the 18 counters. The number of rows will be the missing factor. There are 3 rows. So, 3 × 6 = 18.

□ × 6 = 18.

○○○○○○
3 ○○○○○○
○○○○○○

3 × 6 = 18

Continue the activity by having students choose other cards. When the students understand how to find a missing factor using arrays, give them more counters and have them work with greater products.

Name ______________________ Skill ______________________

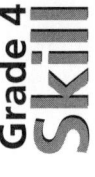

Grade 4 Skill 56

Missing Factors

Use a multiplication table to find missing factors.

Find $4 \times ■ = 20$.

factors: 4, ■ — product: 20

Think: 4 times what number is 20?

Step 1 Find the row for the factor, 4.
Look **across** the row. Find the product, 20.

Step 2 Look **up** the column.
Find the missing factor. It is 5.

Step 3 Complete the multiplication sentence.

$4 \times \mathbf{5} = 20$

factor row — factor column — product row × column

Multiplication Table

×	0	1	2	3	4	5	6	7	8	9
0	0	0	0	0	0	0	0	0	0	0
1	0	1	2	3	4	5	6	7	8	9
2	0	2	4	6	8	10	12	14	16	18
3	0	3	6	9	12	15	18	21	24	27
4	0	4	8	12	16	20	24	28	32	36
5	0	5	10	15	20	25	30	35	40	45
6	0	6	12	18	24	30	36	42	48	54
7	0	7	14	21	28	35	42	49	56	63
8	0	8	16	24	32	40	48	56	64	72
9	0	9	18	27	36	45	54	63	72	81

Factors → row, column

Try These

Use the multiplication table to find the missing factors.

1. $3 \times ■ = 9$

   Row Factor: ☐ Product: ☐

   Missing Factor: ☐

   $3 \times$ ☐ $= 9$

2. $7 \times ■ = 42$

   Row Factor: ☐ Product: ☐

   Missing Factor: ☐

   $7 \times$ ☐ $= 42$

3. $8 \times ■ = 24$

   Row Factor: ☐ Product: ☐

   Missing Factor: ☐

   $8 \times$ ☐ $= 24$

Go to the next side.

Name ______________________ Skill ______________________

Practice on Your Own

Skill 56

Find ■ × 7 = 21.

Think:
Find the column for 7.
Look **down** the column to 21.
Look **left across** the row to 3.
The missing factor is 3.

×	0	1	2	3	4	5	6	7	8	9
0	0	0	0	0	0	0	0	0	0	0
1	0	1	2	3	4	5	6	7	8	9
2	0	2	4	6	8	10	12	14	16	18
3	0	3	6	9	12	15	18	21	24	27
4	0	4	8	12	16	20	24	28	32	36
5	0	5	10	15	20	25	30	35	40	45
6	0	6	12	18	24	30	36	42	48	54
7	0	7	14	21	28	35	42	49	56	63
8	0	8	16	24	32	40	48	56	64	72
9	0	9	18	27	36	45	54	63	72	81

3 × 7 = 21
row × column = product

Use the multiplication table to find the missing factors.

1. ■ × 2 = 8
   Column Factor: ☐
   Product: ☐
   Missing Factor: ☐
   ☐ × 2 = 8

2. ■ × 5 = 30
   Column Factor: ☐
   Product: ☐
   Missing Factor: ☐
   ☐ × 5 = 30

3. ■ × 9 = 54
   Column Factor: ☐
   Product: ☐
   Missing Factor: ☐
   ☐ × 9 = 54

4. ☐ × 4 = 20
5. 6 × ☐ = 30
6. ☐ × 4 = 36

7. 6 × ☐ = 18
8. ☐ × 7 = 28
9. 8 × ☐ = 48

Check

Use the multiplication table to find the missing factors.

10. 8 × ☐ = 64
11. ☐ × 7 = 63
12. 8 × ☐ = 56

Number Patterns (for Multiplication)

15 Minutes

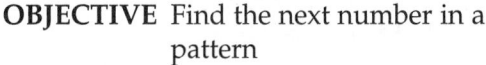

OBJECTIVE Find the next number in a pattern

Begin by reviewing how to skip count. Have students skip count by two from 20 to 34. You may want to record each number on the board so students can reflect on the pattern created by skip counting.

Ask: **What would be the next number?** 36 **What is another way to describe the pattern in the numbers using one of the four operations-addition, subtraction, multiplication, or division?** Possible answer: You add 2 to each number to get the next number.

Direct students' attention to the number line in Step 1.

Ask: **What numbers are circled on the number line?** 21, 24, 27, 30, 33, and 36

Now redirect them to the number line in Step 2: **What is added to each of the circled numbers on the number line in Step 2?** 3

Invite students to state the rule for the pattern in their own words.

For Step 3, refer to the equation and mention to students that they can use a number sentence to state some mathematical rules.

TRY THESE Exercises 1–3 provide prompts to help student state rules for patterns.

- **Exercise 1** Patterns with 2.
- **Exercise 2** Patterns with 5.
- **Exercise 3** Patterns with 10.

PRACTICE ON YOUR OWN Review the example at the top of the page. Ask students to explain how they know that the rule is to add 4 to each number. Possible response: I use the number line to count the spaces between the numbers. That way I can tell what to add to get the next number.

- **Exercises 1–3** provide help with identifying the rule and writing the equation to find the next number in the pattern.
- **Exercises 4–6** The patterns involve multiples of 3, 4, and 6. Students continue to receive a cue for stating a rule, but the prompt for writing the equation-form is removed.
- **Exercises 7–9** involve multiples of 7, 8, and 9. Students have an opportunity to work without cues before moving on to the **Check.**

CHECK Determine if students can state the rule for finding the next number in the pattern. Success is indicated by 3 out of 3 correct responses.

Students who successfully complete the **Practice on Your Own** and **Check** are ready to move on to the next skill.

COMMON ERRORS

- Students may not recognize the pattern when larger numbers are used in sequence.
- Students may add or skip count incorrectly.

Students who made more than four errors in the **Practice on Your Own** section, or who were not successful in the **Check** section, may benefit from the **Alternative Teaching Strategy** on the next page.

Optional

Alternative Teaching Strategy

Model Number Patterns

20 Minutes

OBJECTIVE Use a number line to state the rule for finding the next number in a pattern

MATERIALS number line

You might start the activity with students working individually. When students have demonstrated competence finding patterns with small numbers on a number line, have them work in pairs to find number patterns with larger numbers.

Use this activity to build students' competence in multiplication by helping them recognize number patterns. For example, if you have students skip count by four on a number line, make sure they begin on a number that is a multiple of 4, such as 20.

Before students use their number lines, model skip counting on the number line.

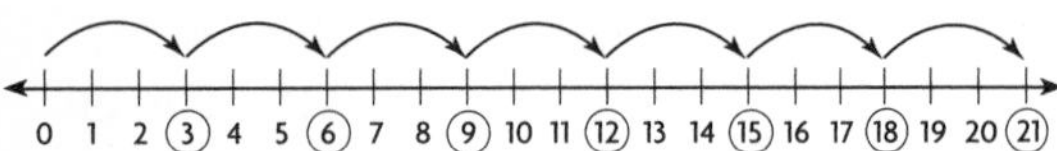

Distribute copies of the number lines. Instruct students to label one of the number lines from 1 through 21. Tell students that they will use the number line to skip count by 3. Have them begin at 0 and then skip count by 3 as far as they can on the number line. Tell them to circle each number they land on.

When students have completed skip counting, ask them to tell you what numbers they have circled on their number lines. You may want to write these numbers on the board: 3, 9, 12, 15, 18, 21.

Have students compare their findings. If any student found a different set of numbers, ask that student to skip count on the number line again while you monitor the work.

Talk to students about the numbers they found by skip counting. Remind them that these numbers show a pattern. Tell them that you are thinking of a number that fits this pattern; in fact, it is the next number in the pattern.

Ask: **What would be the next number after 21?** 24 **How do you find that number?** **Possible response: If you skip count by 3, the next number is 24.**

Help students state a rule for pattern. **Add 3.**

Continue with this activity by having students skip count on the other number lines. Have students share their findings each time and find the rule for each pattern. You might have students work their way up gradually, skip counting by 2, 4, 5, and then 10. For each number line, students must state the rule for the pattern and the next number in the pattern.

Name ______________________ Skill ______________________

Grade 4
Skill 57

Number Patterns

Find the next number in the pattern. 21, 24, 27, 30, 33, 36, ■

Step 1 Decide if you can use skip-counting to find the pattern.

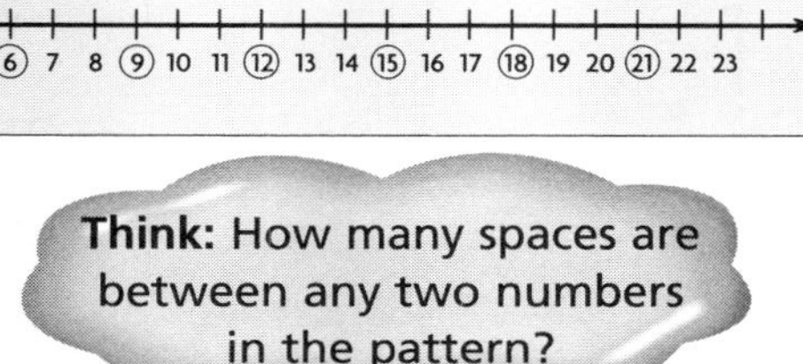

Step 2 You skip-count by 3s to find the next number in the pattern. State a rule for the pattern.

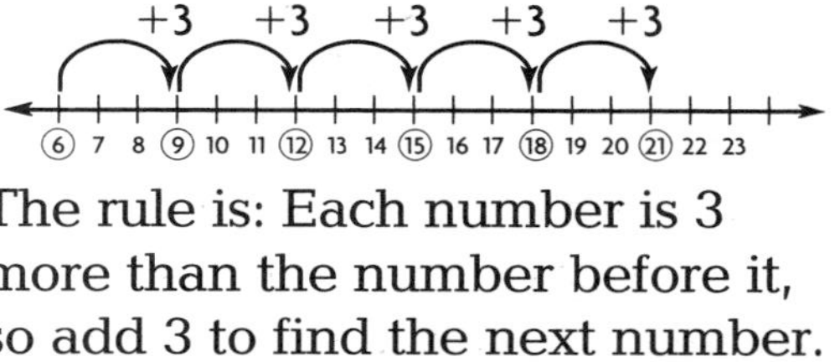

The rule is: Each number is 3 more than the number before it, so add 3 to find the next number.

Step 3 Use the rule to find the next number in the pattern.

21, 24, 27, 30, 33, 36, ■

36	+	3	=	39
last number	+	rule	=	next number

So, the next number is 39.

Try These

State the rule. Then write the next number.

1 12, 14, 16, 18, 20, 22, ■

12 14 16 18 20 22

The rule is add 2 to find the next number.

22	+	2	=	24
last number	+	rule	=	next number

12, 14, 16, 18, 20, 22, ____

2 65, 70, 75, 80, 85, 90, ■

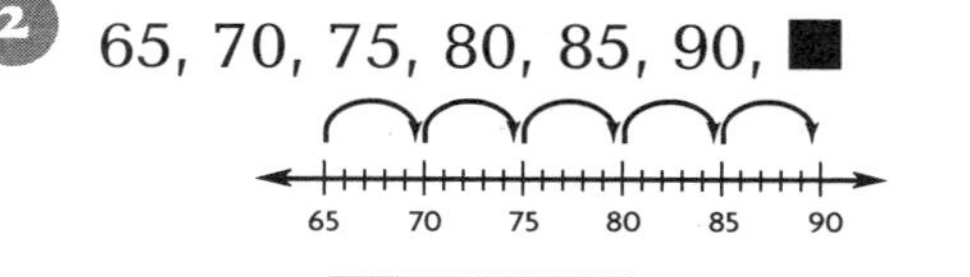

The rule is ____.

____	+	____	=	____
last number	+	rule	=	next number

65, 70, 75, 80, 85, 90, ____

3 30, 40, 50, 60, 70, 80, ■

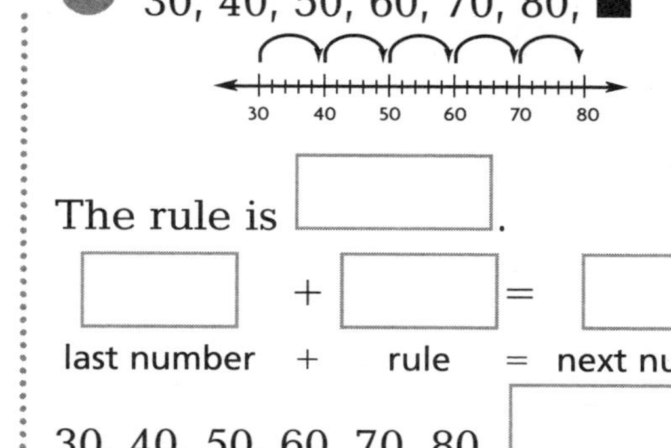

The rule is ____.

____	+	____	=	____
last number	+	rule	=	next number

30, 40, 50, 60, 70, 80, ____

Go to the next side.

Name ______________________ Skill ______________________

Practice on Your Own

Skill 57

Find the pattern. Write the next number.

20, 24, 28, 32, 36, 40, ■

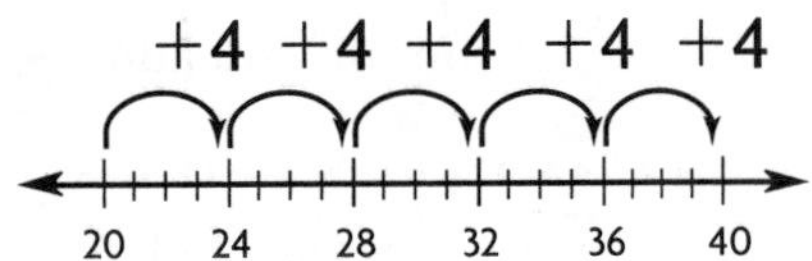

Think:
What number can I add to get the next number?

The rule is: Add 4 to find the next number.

40 + 4 = 44

last number + rule = next number

The next number is 44. 20, 24, 28, 32, 36, 40, **44**

State the rule. Write the next number.

1. 8, 10, 12, 14, ■

   The rule is ☐.

   ☐ + ☐ = ☐

   last number + rule = next number

   8, 10, 12, 14, ☐

2. 15, 20, 25, 30, ■

   The rule is ☐.

   ☐ + ☐ = ☐

   last number + rule = next number

   15, 20, 25, 30, ☐

3. 20, 30, 40, 50, ■

   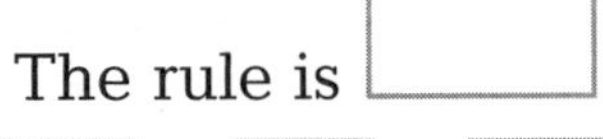

   The rule is ☐.

   ☐ + ☐ = ☐

   last number + rule = next number

   20, 30, 40, 50, ☐

4. 12, 15, 18, 21, ■

   The rule is ☐.

   12, 15, 18, 21, ☐

5. 32, 36, 40, 44, ■

   The rule is ☐.

   32, 36, 40, 44, ☐

6. 30, 36, 42, 48, ■

   The rule is ☐.

   30, 36, 42, 48, ☐

Write the next number.

7. 14, 21, 28, 35, ☐
8. 40, 48, 56, 64, ☐
9. 36, 45, 54, 63, ☐

Check

Write the next number.

10. 33, 36, 39, 42, ☐
11. 12, 18, 24, 30, ☐
12. 54, 63, 72, 81, ☐

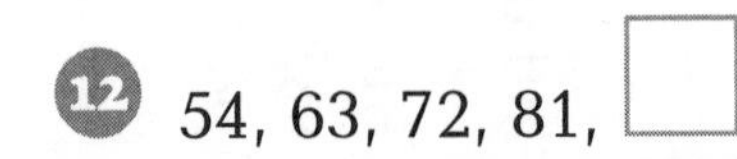

Skill 58 Grade 4

Number Patterns (for Division)

OBJECTIVE Find and complete number patterns

MATERIALS pattern blocks

Lay out a circle, a square, a triangle, and a circle with pattern blocks.

Ask: **What block should I put down next?** square **Next?** triangle

Discuss how recognizing a pattern can help us see the relationship between objects or numbers.

Next, write: 1, 2, 3, 4. Tell students that these numbers follow a pattern.

Ask: **What are the next 3 numbers?** 5, 6, 7 **How do you know?** The numbers increase by 1.

Write 10, 9, 8, 7.

Ask: **What are the next 3 numbers?** 6, 5, 4 **How do you know?** The numbers decrease by 1.

Review with students the meaning of increase get larger and decrease get smaller. In the example at the top of the page, help students count back aloud from 27 to 24, 24 to 21, and 21 to 18 to verify that the numbers decrease by 3.

Ask: **What number do you subtract each time?** 3

Relate *decrease* to *subtract*.

In the second example, have students count on 2 to verify that the numbers increase by 2.

Ask: **What number do you add each time?** 2

Relate *increase* to *add*.

TRY THESE In Exercises 1–3, students find rules to continue patterns.

- **Exercise 1** Find and apply subtract 2 rule.
- **Exercise 2** Find and apply add 6 rule.
- **Exercise 3** Find and apply subtract 9 rule.

PRACTICE ON YOUR OWN Focus on the operation in the rule. Encourage students to test the rule on the first few numbers in each sequence.

Ask: **How do you know whether to add or subtract?** When the numbers decrease, subtract. When the numbers increase, add.

CHECK Success is indicated by 3 out of 3 correct responses.

Students who successfully complete **Practice on Your Own** and **Check** are ready to move on to the next skill.

COMMON ERRORS

- Students may choose the wrong operation.

Students who made more than three errors in **Practice on Your Own**, or who were not successful in the **Check** section, may benefit from the **Alternative Teaching Strategy** on the next page.

Optional

Alternative Teaching Strategy

Number Patterns

OBJECTIVE Complete number patterns

Show students this number pattern:

3, 6, 9, 12, □, □, □

Say: **When you have a lot of numbers, one way to find a pattern is to look at the numbers two at a time.** Help students focus on the first two numbers only in the sequence.

[3, 6,] 9, 12, □, □, □

Ask: **Do the numbers increase or decrease?** increase **By how much?** 3

Then pair up 6 and 9.

3, [6, 9,] 12, □, □, □

Ask: **Do the numbers increase or decrease?** increase **By how much?** 3

Repeat for 9 and 12.

Ask: **Do you see a pattern?** yes **How do the numbers change?** They increase by 3. **What are the next 3 numbers in the pattern?** 15, 18, 21

Help students make a table of number patterns. Give them a pair of numbers that increase or decrease by 2, 3, 4, 5, 6, 7, 8, or 9. Ask students to:

- Check off increase (+) or decrease (−).
- Write the rule.
- Write the next 3 numbers.

	+	−	Rule	Next 3
27, 30	✔		+3	33, 36, 39
45, 40		✔	−5	35, 30, 25

Name ______________________ Skill ______________________

Grade 4 Skill 58

Number Patterns

Find the pattern. Write the next three numbers in the pattern.

Do the numbers increase or decrease?

27, 24, 21, 18, ■ , ■ , ■

− 3   − 3   − 3

The numbers decrease by 3. The rule is: **Subtract 3.**

Use the rule to find the next three numbers.

18 − 3   15 − 3   12 − 3

27, 24, 21, 18, **15, 12, 9**

The next three numbers are: 15, 12, 9.

Do the numbers increase or decrease?

8, 10, 12, 14, ■ , ■ , ■

+ 2   + 2   + 2

The numbers increase by 2. The rule is: **Add 2.**

Use the rule to find the next three numbers.

14 + 2   16 + 2   18 + 2

8, 10, 12, 14, **16, 18, 20**

The next three numbers are: 16, 18, 20.

Try These

Find the pattern. Write the rule. Find the next three numbers.

1. 32, 30, 28, 26, ■ , ■ , ■
   Do the numbers increase or decrease? ______ Rule: ______
   32, 30, 28, 26, ___, ___, ___

2. 18, 24, 30, 36, ■ , ■ , ■
   Do the numbers increase or decrease? ______ Rule: ______
   18, 24, 30, 36, ___, ___, ___

3. 72, 63, 54, 45, ■ , ■ , ■
   Do the numbers increase or decrease? ______ Rule: ______
   72, 63, 54, 45, ___, ___, ___

Go to the next side.

Name ____________________ Skill ____________________

Practice on Your Own

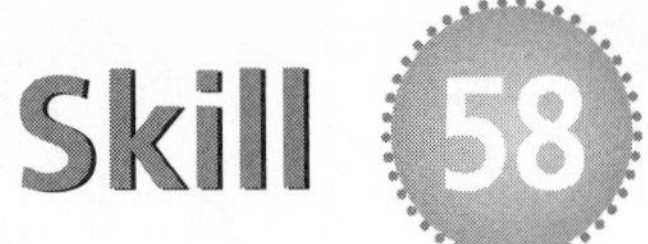

Think: Do the numbers increase or decrease? Find the rule.

Write the next three numbers in the pattern.

24, 32, 40, 48, ■, ■, ■
+ 8 + 8 + 8

The numbers increase by 8. The rule is: **Add 8.**

Use the rule to find the next three numbers.

48 + 8   56 + 8   64 + 8
24, 32, 40, 48, **56, 64, 72**

So, the next three numbers are 56, 64, 72.

Find the pattern. Write the rule. Write the next three numbers in the pattern.

1. 36, 30, 24, 18, ■, ■, ■

   Do the numbers increase or decrease? ☐

   Rule: ☐

   36, 30, 24, 18, ☐, ☐, ☐

2. 32, 40, 48, 56, ■, ■, ■

   Do the numbers increase or decrease? ☐

   Rule: ☐

   32, 40, 48, 56, ☐, ☐, ☐

3. 21, 18, 15, 12, ■, ■, ■

   Do the numbers increase or decrease? ☐

   Rule: ☐

   21, 18, 15, 12, ☐, ☐, ☐

4. 16, 14, 12, 10, ■, ■, ■

   Rule: ☐

   16, 14, 12, 10, ☐, ☐, ☐

5. 27, 36, 45, 54, ■, ■, ■

   Rule: ☐

   27, 36, 45, 54, ☐, ☐, ☐

6. 30, 35, 40, 45, ■, ■, ■

   Rule: ☐

   30, 35, 40, 45, ☐, ☐, ☐

Write the next three numbers in the pattern.

7. 12, 14, 16, 18, ☐, ☐, ☐
8. 42, 36, 30, 24, ☐, ☐, ☐
9. 24, 20, 16, 12, ☐, ☐, ☐

Check

Write the next three numbers in the pattern.

10. 49, 42, 35, 28, ☐, ☐, ☐
11. 45, 54, 63, 72, ☐, ☐, ☐
12. 48, 40, 32, 24, ☐, ☐, ☐

Skill 59 Grade 4

Read a Number Line

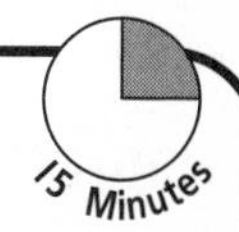

OBJECTIVE Use a number line to find, compare, add, and subtract numbers

Direct students' attention to the first number line. Begin by having them locate points with number labels, such as 2, 6, 0, and so on. Tell students that to find a point that is not numbered, like A, they should look between two that are, for example 8 and 10.

Say: **Each space on the number line represents 1 unit. You know that A is 9 since 9 is the unit between 8 and 10.**

As students look at the second example, have them put a finger on 40. Then have them move left to 25. Point out that as you move left on the number line, the numbers are less.

Ask: **In which direction did you move to go from 40 to 25?** left **Is 25 less than 40?** yes

Write 25 < 40. Point out that the small end of the symbol points to the number that is less.

Then compare 15 and 30. Have students put their finger on 15 and move to 30. Point out that as you move right, the numbers are greater.

Ask: **In which direction did you move to go from 15 to 30?** right **Is 30 greater than 15?** yes

Write: 30 > 15.

In the third example, remind students that when you add, you find more. Help them find 3 and move 4 spaces right to find 3 + 4 = 7. Have them count aloud, stopping at 4, 5, 6, and 7.

Remind students that when you subtract, you find less. Help them find 8 and move 2 spaces left to 6 to show 8 − 2 = 6.

Ask: **Do you move right or left to add?** right **Do you move right or left to subtract?** left

TRY THESE Exercises 1–3 practice finding, comparing, and adding numbers on the number line.

- **Exercise 1** Find 33.
- **Exercise 2** Compare 43 and 55.
- **Exercise 3** Add 4 + 5.

PRACTICE ON YOUR OWN Review the examples at the top of the page. Remind students to move right to add and left to subtract.

CHECK Success is determined by 4 out of 4 correct responses.

Students who successfully complete the **Practice on Your Own** and **Check** are ready to move on to the next skill.

COMMON ERRORS

- Students may move in the wrong direction on the number line.
- Students may point the greater than/less than symbol incorrectly.

Students who made more than three errors in **Practice on Your Own**, or who were not successful in the **Check** section, may benefit from the **Alternative Teaching Strategy** on the next page.

Optional

Alternative Teaching Strategy

Read a Thermometer

OBJECTIVE Read and use a thermometer representing a number line

MATERIALS copies of thermometers

Distribute the thermometers. Point out that a thermometer is like a vertical number line. Tell students they will use the thermometer to find various temperatures and changes in temperatures. Point out that, on this thermometer, each mark represents 1 unit, or 1°.

Begin the activity by having students point to temperatures on their thermometers, for example 60°, 68°, 77°, 80°, and so on.

Next, have students compare 70° and 66° by first locating 70° and moving down to find 66°. Point out that as you move down the thermometer, the temperatures are lower. So, 66° is less than 70°. Write: 66 < 70. In similar fashion, but moving up the thermometer, compare a temperature greater than another.

Help students use their thermometers to add and subtract to find changes in temperature. Discuss the fact that, if the temperature goes up a number of degrees, we add to find the new temperature. Ask students to find a temperature increase of 6° from 70°.

Say: **Yesterday the temperature was 70°. It went up 6° last night. What is it today?** 76°

Students should find 70° and count 6 spaces up. So, 70° + 6° = 76°. In similar fashion, but using subtraction, work with a temperature that decreases.

Name ______________________ Skill ______________________

Grade 4
Skill 59

Read a Number Line

Use a number line to find, compare, add, or subtract numbers.

What number does A represent?

Each tick mark on this number line represents 1 unit.

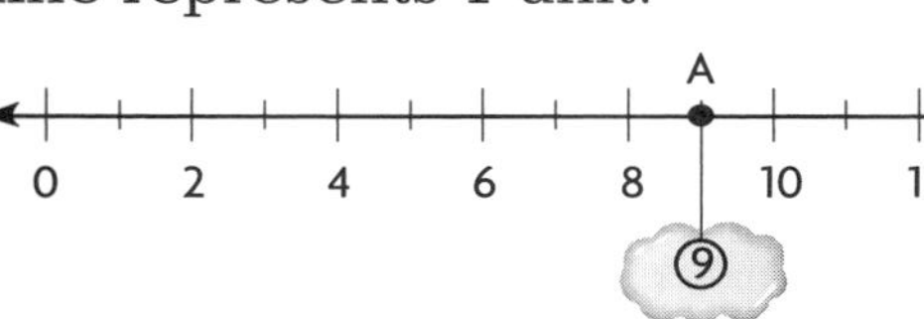

The point for A is on the mark between 8 and 10.

So, A represents 9.

Use a number line to compare 25 and 40.

As you move **left** on the number line, the numbers are **less**.

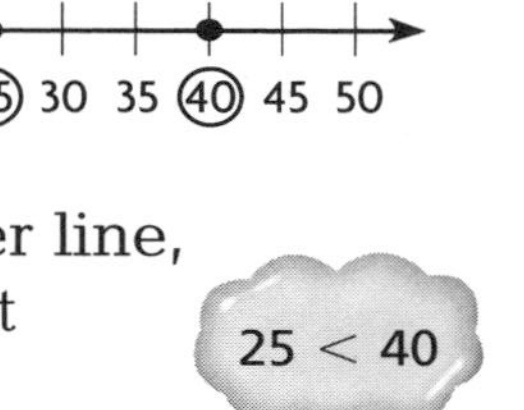

On the number line, 25 is to the left of 40.

$25 < 40$

So, 25 is less than 40.

You can use a number line to add and subtract.

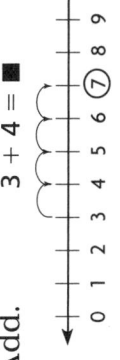

Add. $3 + 4 = ■$

0 1 2 3 4 5 6 (7) 8 9 10

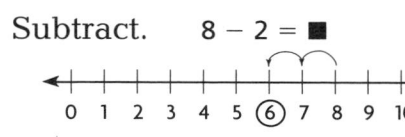

Start at 3. Count 4 spaces to the right. **So, $3 + 4 = 7$.**

Subtract. $8 - 2 = ■$

0 1 2 3 4 5 (6) 7 8 9 10

Start at 8. Count 2 spaces to the left. **So,** $8 - 2 = 6$.

Try These

1. Use the number line to find the number B represents.

   30 32 B 34 36 38 40

   B represents ☐.

2. Use the number line to compare. Write *right* or *left*. Write $>$ or $<$ for ◯.

   43 (arrow) 40 45 50 55 60 65

   43 is to the ☐ of 55.

   43 ◯ 55.

3. Use the number line to add. Write *right* or *left*. Write the sum.

   0 1 2 3 (4) 5 6 7 8 9 10

   Start at 4.

   Count 5 spaces to the ☐.

   $4 + 5 =$ ☐

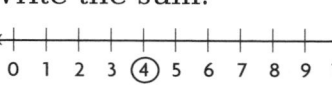

Go to the next side.

Name ______________________ Skill ______________________

Practice on Your Own

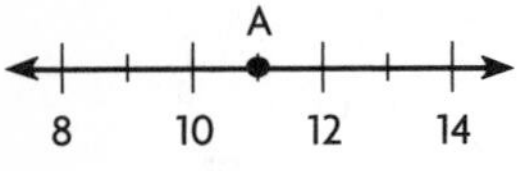

A represents 11.

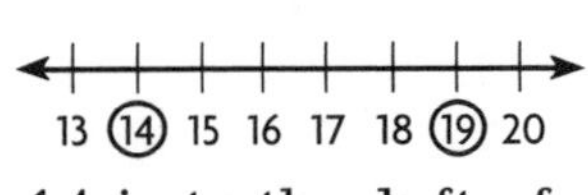

14 is to the left of 19, so, 14 < 19.

9 + 3 = 12

Find the number each letter represents.

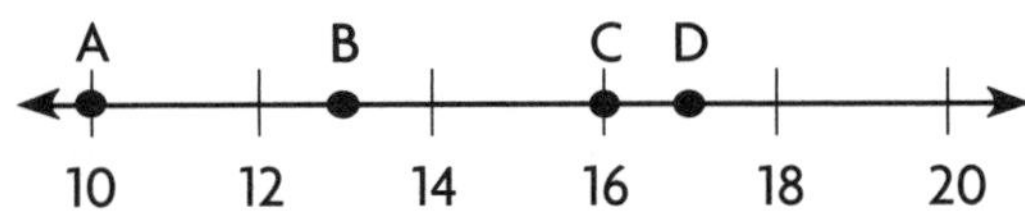

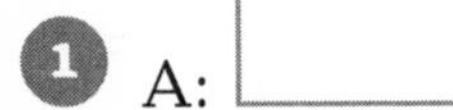
1 A: ☐

2 B: ☐

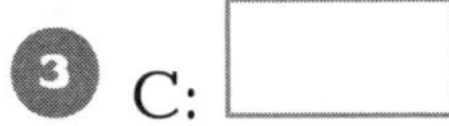
3 C: ☐

4 D: ☐

Use the number line to compare. Write >, < or = for each ◯.

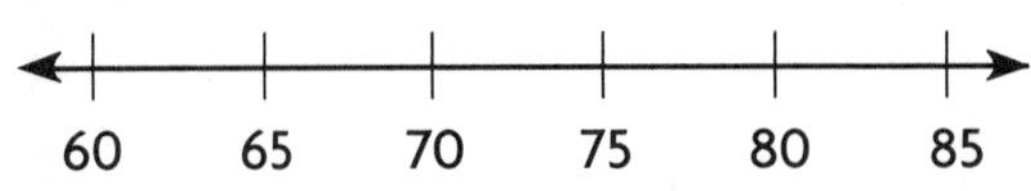

5 65 ◯ 70

6 75 ◯ 65

7 71 ◯ 78

8 82 ◯ 70

Use the number line to add and subtract.

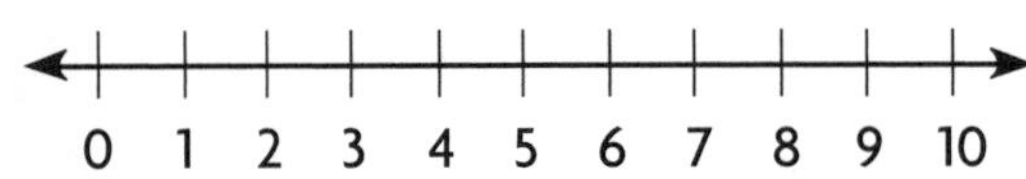

9 1 + 8 = ☐

10 4 + 3 = ☐

11 9 − 6 = ☐

12 10 − 8 = ☐

Check

Use the number line.

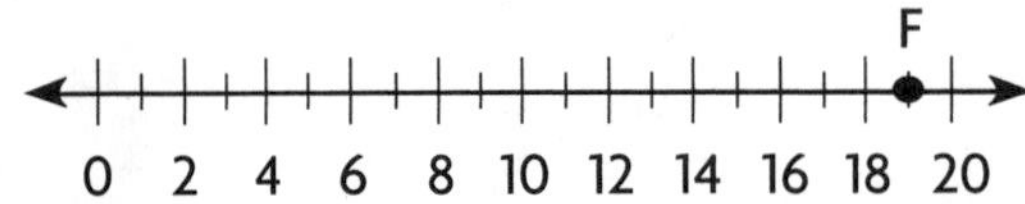

13 F: ☐

14 18 ◯ 11

15 10 − 7 = ☐

16 3 + 5 = ☐

Locate Points on a Coordinate Grid (First Quadrant Only)

15 Minutes

OBJECTIVE Locate points on a coordinate grid, first quadrant only

Read together the description of the miner's map at the top of the page. Examine the grid.

Ask: **What do the numbers across the bottom of the grid and up the side remind you of?** two number lines, one horizontal and the other vertical **At which point do the lines meet?** 0

Focus on the term *ordered pair*. Examine the grid closely. Point out that all the places marked with symbols can be found by naming ordered pairs. This might be a good time to look at the legend and read the names and symbols of all the places that can be located.

Help students trace the path to the miner's cabin. For point (3, 2), show them how to first move their finger 3 spaces to the right. Then they move 2 spaces up.

Ask: **Where do I start?** at 0 **What does the first number tell you?** Move 3 spaces right. **The second number?** Move 2 spaces up.

Emphasize the importance of the order of the numbers in the ordered pair. Ask students to find (2, 3). Contrast the location of this point to that of (3, 2).

TRY THESE Exercises 1–3 require students to identify places located at various points on the grid.

- **Exercise 1** Move right 1, up 1.
- **Exercise 2** Move right 5, up 4.
- **Exercise 3** Move right 2, up 4.
- **Exercise 4** Move right 4, up 5.

PRACTICE ON YOUR OWN Before students begin the exercises, examine together the map of the state park. Study the legend. Review the term *ordered pair*. Stress care in moving right first, then up. Remind students to always start at point 0.

CHECK Determine if students know how to use ordered pairs to locate points on a grid. Success is determined by 3 out of 3 correct responses.

Students who successfully complete the **Practice on Your Own** and **Check** are ready to move on to the next skill.

COMMON ERRORS

- Students may move up first, then right.
- Students may forget to start at point 0.

Students who made more than two errors in **Practice on Your Own**, or who were not successful in the **Check** section, may benefit from the **Alternative Teaching Strategy** on the next page.

Optional

Alternative Teaching Strategy

Locate Points on a Coordinate Grid

20 Minutes

OBJECTIVE Locate points on a coordinate grid

MATERIALS grids, marked as shown below

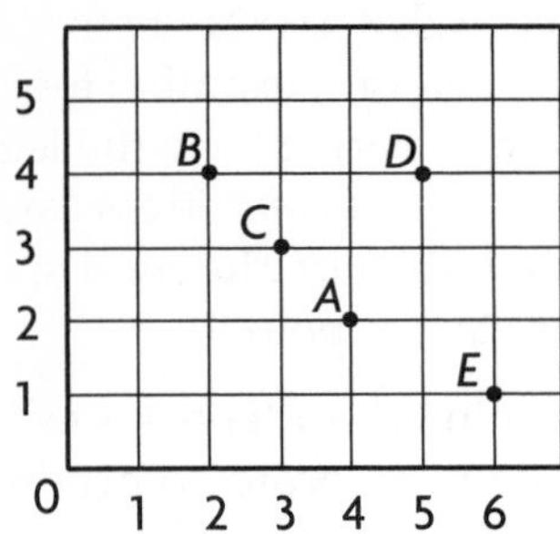

Distribute grids. Tell students they will use the ordered pair (4, 2) to locate a point on the grid.

Ask: **Where should you start?** point 0 **In which direction should you move first?** right **How many spaces?** 4

Observe students as they move across 4 spaces, using their finger as a guide. You may have them draw an arrow as shown on the miner's map. Have them stop and check that they are at the right point so far.

Ask: **In which direction should you move next?** up **How many spaces?** 2

Observe as they move up 2 spaces, again using a finger or an arrow to guide them. Have them name the letter, A, at point (4, 2).

Give students the ordered pairs for the rest of the points on the grid. Students can work in pairs, checking each other as they name the letters at the points.

Working in pairs, one student can mark another point, for example F, given the ordered pair (1,1) and ask the partner to name the letter at that point (1, 1).

Name ______________________ Skill ______________________

Grade 4 **Skill 60**

Locate Points on a Coordinate Grid

The horizontal and vertical lines on the map are called a **grid**. An **ordered pair** of numbers like **(3,2)** names a point on a grid. You can use ordered pairs to locate points on a grid.

Look at the map. Name the place at point (3,2).

Step 1 Start at point 0 on grid.

Step 2 The first number in the ordered pair tells you how many spaces to move to the right. Move 3 spaces to the right.

Step 3 The second number tells you how many spaces to move up. Move up 2 spaces.

So, the miner's cabin is located at point (3,2).

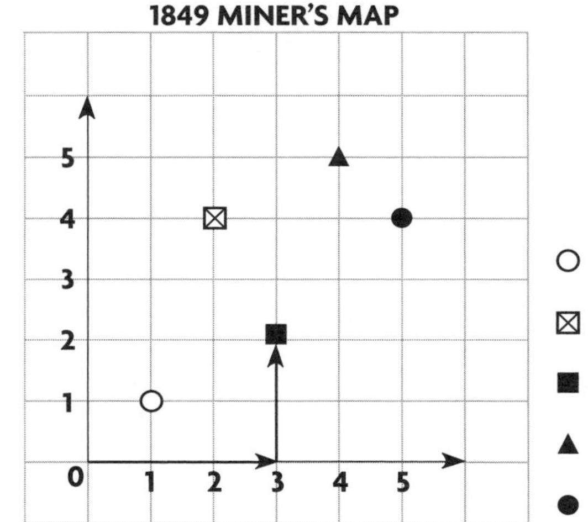

Try These

Use the map above. Name the place you find at each point.

1. (1,1)
Start at point 0. Move ☐ spaces to the right.
Move ☐ spaces up.
The place at point (1,1) is ________.

2. (5,4)
Start at point 0. Move ☐ spaces to the right.
Move ☐ spaces up.
The place at point (5,4) is ________.

3. (2,4)
Start at point 0. Move ☐ spaces to the right.
Move ☐ spaces up.
The place at point (2,4) is ________.

4. (4,5)
Start at point 0. Move ☐ spaces to the right.
Move ☐ spaces up.
The place at point (4,5) is ________.

Go to the next side.

Name ______________________ Skill ______________________

Practice on Your Own

Skill 60

Think:
The ordered pair (4,3) means move **4** spaces **to the right** and then move **3** spaces **up**.
Remember: Start at 0.
Look at the map of the state park. Name the place at point (4,3)
The place at point (4,3) is **Campsite A.**

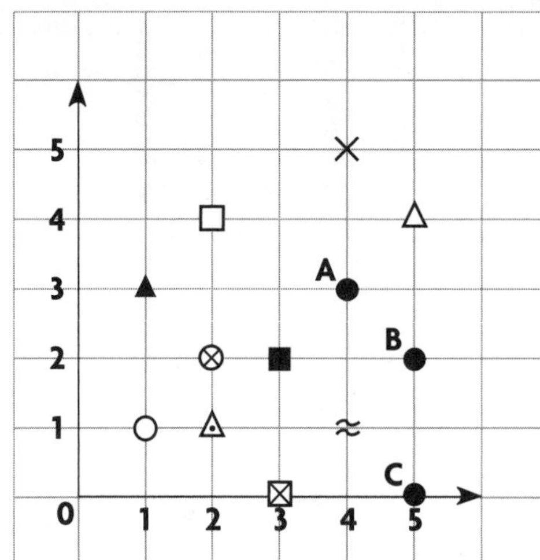

○ Bicycle Rack
■ Bridge
△ Lookout Tower
✕ Eagle's Nest
● Campsite A, B, & C
⊠ Nature Center
□ Water Fall
▲ Ranger's Cabin
⊗ Picnic Tables
≈ Swimming Hole
⚠ First Aid Station

Use the map above. Name the place you find at each point.

1 (1, 1)
Start at point 0. Move ☐ spaces to the right.
Move ☐ spaces up.
Place: ☐

2 (1, 3)
Start at point 0. Move ☐ spaces to the right.
Move ☐ spaces up.
Place: ☐

3 (3, 2)
Start at point 0. Move ☐ spaces to the right.
Move ☐ spaces up.
Place: ☐

4 (2, 4)
Move ☐ spaces to the right.
Move ☐ spaces up.
The place at point (2,4): ☐

5 (4,5)
Move ☐ spaces to the right.
Move ☐ spaces up.
The place at point (4,5): ☐

6 (5,2)
Move ☐ spaces to the right.
Move ☐ spaces up.
The place at point (5,2): ☐

Check

Use the map above. Name the place you find at each point.

7 (4,1)
Place: ☐

8 (2,1)
Place: ☐

9 (5,0)
Place: ☐

Skill 61 Grade 4

Use Function Tables

OBJECTIVE Complete a function table when the rule and input are given

MATERIALS Picture of a function machine

Begin the lesson by using the drawing of a function machine and reviewing the parts of the machine. Call students' attention to the part of the machine labeled "Rule." Explain that the rule for a function machine will use one of the computational operations: addition, subtraction, multiplication, or division.

Draw students' attention to Example A.

Say: **You can see the first input number and the first output number. What was done to the input number to get the output number?** Added 4

What was done to the second input number? Added 4

Continue until you reach input 10. Ask students what they should do to 10. Add 4

Use similar questions for Example B.

Ask: **In Example B, what is the rule?** Subtract 3

What should you do to 10 to find the output? Subtract 3

TRY THESE In Exercises 1–2, students complete number sentences using the given rule and find two output numbers.

- **Exercise 1** Rule: Add 2.
- **Exercise 2** Rule: Subtract 1.

PRACTICE ON YOUR OWN Review the examples at the top of the page. Have students complete each function table.

In Exercises 1–2, students complete number sentences using the given rule and find two output numbers.

CHECK Determine if students can complete the function tables.

Success is determined by 2 out of 3 correct responses.

Students who successfully complete the **Practice on Your Own** and **Check** are ready to move on to the next skill.

COMMON ERRORS

- Students may have forgotten the basic facts.
- Students do not apply the given rule to each input number, or may use the wrong operation.

Students who made more than three errors in the **Practice on Your Own**, or who were not successful in the **Check** section, may benefit from the **Alternative Teaching Strategy** on the next page.

Optional

Alternative Teaching Strategy

Use a Function Machine

OBJECTIVE Use a function machine to recognize a pattern by its rule

MATERIALS Copies of teacher-created black line master of a function machine

Demonstrate the use of the function machine on a copy made from the black line master. Write "add 2" on the Rule line. As you provide students with input numbers 2, 3, 4, and 5, have students tell you the output numbers. Record output answers. 4, 5, 6, 7

Recall that the rule does not change each time a new number is given.

Students work as partners in this activity. Distribute papers with a picture of a function machine.

Ask one student in each pair to begin by writing "add 3" on the Rule line. Have the same student write "2" on the input line.

Ask: **How are you going to find the output number?** Use the rule on the machine

What is the output number? 5

Have the partner record the output number.

Continue this activity using input numbers 6, 7, 8, and 9.

Repeat the activity, changing the rules to include subtraction, multiplication, and division.

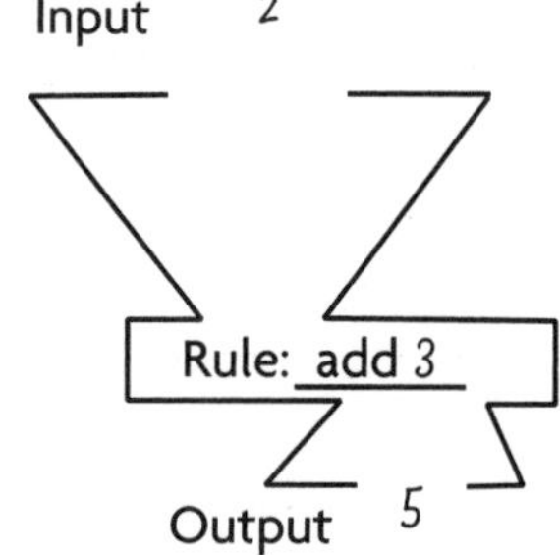

Name ______________________ Skill ______________________

Grade 4 **Skill 61**

Use Function Tables

You can think of a **function table** as a machine. You put in a number, called the input. The machine uses a rule on the number, and puts out another number, called the output. To use a rule for a function table, you use addition, subtraction, multiplication, or division.

Example A Addition

Rule: Add 4.

Use the input number and the rule to find the output number.

INPUT	OUTPUT
3	7
5	9
6	10
10	■

INPUT	RULE	=	OUTPUT
3	+ 4	=	7
5	+ 4	=	9
6	+ 4	=	10
10	+ 4	=	**14**

Example B Subtraction

Rule: Subtract 3.

Use the input number and the rule to find the output number.

INPUT	OUTPUT
3	0
6	3
8	5
10	■

INPUT	RULE	=	OUTPUT
3	− 3	=	0
6	− 3	=	3
8	− 3	=	5
10	− 3	=	**7**

Try These

Use the rule to complete each table.

1. Rule: Add 2.

INPUT	OUTPUT
3	5
5	7
6	■
10	■

INPUT	RULE	=	OUTPUT
3	+ 2	=	5
5	+ 2	=	7
6	+ 2	=	
10	+ 2	=	

2. Rule: Subtract 1.

INPUT	OUTPUT
4	3
5	4
7	■
9	■

INPUT	RULE	=	OUTPUT
4	− 1	=	3
5	− 1	=	4
7	− 1	=	
9	− 1	=	

Go to the next side.

Name ______________________ Skill ______________________

Practice on Your Own

Use the rule on the input number to find the output number.

Multiplication

Rule: Multiply by 5.

INPUT	OUTPUT
1	5
3	15
4	20
6	■

INPUT	RULE =	OUTPUT
1	× 5 =	5
3	× 5 =	15
4	× 5 =	20
6	× 5 =	**30**

Division

Rule: Divide by 2.

INPUT	OUTPUT
6	3
8	4
12	6
16	■

INPUT	RULE =	OUTPUT
6	÷ 2 =	3
8	÷ 2 =	4
12	÷ 2 =	6
16	÷ 2 =	**8**

Use the rule to complete each table.

1. Rule: Multiply by 2.

INPUT	OUTPUT
2	4
3	6
4	■
9	■

INPUT	RULE =	OUTPUT
2	× 2 =	4
3	× 2 =	6
4	× 2 =	☐
9	× 2 =	☐

2. Rule: Divide by 3.

INPUT	OUTPUT
3	1
6	2
12	■
24	■

INPUT	RULE =	OUTPUT
3	÷ 3 =	1
6	÷ 3 =	2
12	÷ 3 =	☐
24	÷ 3 =	☐

Check

Use the rule to complete each table.

3. Rule: Multiply by 2.

INPUT	OUTPUT
5	10
7	☐
10	☐
12	☐

4. Rule: Add 3.

INPUT	OUTPUT
2	5
4	☐
7	☐
10	☐

5. Rule: Subtract 2.

INPUT	OUTPUT
2	0
5	☐
6	☐
12	☐

Answer Card

Algebra and Functions

Grade 4

Skill 53

TRY THESE

1. 4
2. 9, 9
3. 8, 8
4. 9, 9

PRACTICE

1. 7
2. 11, 11
3. 9, 9
4. 13 ,13
5. 9, 9
6. 17, 17
7. 11, 11
8. 6
9. 7
10. 6
11. 4

CHECK

12. 4
13. 12
14. 3
15. 14

Skill 54

TRY THESE

1. + 5 or add 5; 30, 5, 35, 35
2. + 3 or add 3; 13, 3, 16, 16
3. + 2 or add 2; 30, 2, 32, 32

PRACTICE

1. + 3 or add 3; 11, 3, 14, 14
2. + 6 or add 6; 21, 6, 27, 27
3. + 8 or add 8; 29, 8, 37, 37
4. + 5 or add 5; 24
5. + 8 or add 8; 33
6. + 4 or add 4; 31
7. 50
8. 44
9. 53

CHECK

10. 55
11. 32
12. 53

Skill 55

TRY THESE

1. − 5 or subtract 5; 5 − 5 = 0, 0
2. − 3 or subtract 3; 6 − 3 = 3, 3
3. − 6 or subtract 6; 8 − 6 = 2, 2

PRACTICE

1. − 3 or subtract 3; 5 − 3 = 2, 2
2. − 2 or subtract 2; 17 − 2 = 15, 15
3. − 6 or subtract 6; 9 − 6 = 3, 3
4. − 4 or subtract 4; 3
5. − 5 or subtract 5; 11
6. − 7 or subtract 7; 14
7. 9
8. 14
9. 8

CHECK

10. 22
11. 8
12. 17

Skill 56

TRY THESE

1. 3, 9, 3, 3
2. 7, 42, 6, 6
3. 8, 24, 3, 3

PRACTICE

1. 2, 8, 4, 4
2. 5, 30, 6, 6
3. 9, 54, 6, 6
4. 5
5. 5
6. 9
7. 3
8. 4
9. 6

CHECK

10. 8
11. 9
12. 7

Skill 57

TRY THESE

1. 24
2. add 5; 90, 5, 95, 95
3. add 10; 80, 10, 90, 90

PRACTICE

1. add 2; 14, 2, 16, 16
2. add 5; 30, 5, 35, 35
3. add 10; 50, 10, 60, 60
4. add 3; 24
5. add 4; 48
6. add 6; 54
7. 42
8. 72
9. 72

CHECK

10. 45
11. 36
12. 90

Answer Card

Algebra and Functions

Grade 4

SKILL 58

TRY THESE

1. decrease, – 2 or subtract 2; 24, 22, 20
2. increase, + 6 or add 6; 42, 48, 54
3. decrease, – 9 or subtract 9; 36, 27, 18

PRACTICE

1. decrease, – 6 or subtract 6; 12, 6, 0
2. increase, + 8 or add 8; 64, 72, 80
3. decrease, – 3 or subtract 3; 9, 6, 3
4. – 2 or subtract 2; 8, 6, 4
5. + 9 or add 9; 63, 72, 81
6. + 5 or add 5; 50, 55, 60
7. 20, 22, 24
8. 18, 12, 6
9. 8, 4, 0

CHECK

10. 21, 14, 7
11. 81, 90, 99
12. 16, 8, 0

SKILL 59

TRY THESE

1. 33
2. left, <
3. right, 9

PRACTICE

1. 10
2. 13
3. 16
4. 17
5. <
6. >
7. <
8. >
9. 9
10. 7
11. 3
12. 2

CHECK

13. 19
14. >
15. 3
16. 8

SKILL 60

TRY THESE

1. 1, 1, Old Oak Tree
2. 5, 4, Bear's den
3. 2, 4, Fresh water well
4. 4, 5, Entrance to mine

PRACTICE

1. 1, 1, Bicycle Rack
2. 1, 3, Ranger's Cabin
3. 3, 2, Bridge
4. 2, 4, Water Fall
5. 4, 5, Eagle's Nest
6. 5, 2, Campsite B

CHECK

7. Swimming Hole
8. First Aid Station
9. Campsite C

SKILL 61

TRY THESE

1. 8, 12
2. 6, 8

PRACTICE

1. 8, 18
2. 4, 8

CHECK

3. 14, 20, 24
4. 7, 10, 13
5. 3, 4, 10

Measurement and Geometry

Measurement

Skill 62 Grade 4

Time to the Half Hour and Quarter Hour

OBJECTIVE Tell time to the half and quarter hour

MATERIALS analog and digital clocks

You may wish to begin the lesson by reviewing the parts of an analog clock.

Direct students' attention to Example A.

Say: **You know that at 9:00 the minute hand is on the 12. When the minute hand moves from 12 to 6, you can count by fives to find the number of minutes after 9.**

Ask: **What time is it when the hour hand is on 9 and the minute hand is on 6?** 30 minutes after 9 **How did you decide the time?** When I counted by fives from 12 to 6, I got 30 minutes.

Draw the students' attention to the position of the hour hand between 9 and 10. Explain that when the minute hand moves from 12 to 6, the hour hand moves *between* 9 and 10.

Relate 30 minutes to "halfway around the clock." Explain that when the clock shows 9:30, the time can be expressed in more than one way. Have students read the different ways in the Example.

Continue to ask similar questions as you work through Example B with the students. For Example C, explain that when the minute hand passes the 6, students can tell time *before* the hour. Have students count by fives from 9 to 12 on the clock.

Ask: **How many minutes is it before 5 o'clock?** 15 minutes before 5 o'clock

Relate 15 minutes to a "quarter of the way around the clock." Then have students read the time as shown in Example C.

TRY THESE Exercises 1-4 model time to the half hour and quarter hour.

- **Exercise 1** Half hour after the hour.
- **Exercise 2** Quarter hour after the hour.
- **Exercise 3** Quarter hour before the hour.
- **Exercise 4** Quarter hour after the hour.

PRACTICE ON YOUR OWN Review the example at the top of the page. Have students note that the digital clock is one way of showing time, 1:45.

CHECK Determine if students know how to read and write time to the half hour and quarter hour and express it in more than one way.

Success is determined by 3 out of 3 correct responses.

Students who successfully complete the **Practice on Your Own** and **Check** are ready to move on to the next skill.

COMMON ERRORS

- Some students may confuse the hour and minute hands, mistaking a time such as 12:30 for 6:00.
- Some students may not understand the difference between time before and time after the hour.

Students who made more than three errors in the **Practice on Your Own**, or who were not successful in the **Check** section, may benefit from the **Alternative Teaching Strategy** on the next page.

Optional

Alternative Teaching Strategy

Model Time on a Clock Face

20 Minutes

OBJECTIVE Use model clocks to tell time to the half and quarter hour

MATERIALS analog clock face, paper circles with clock faces but no hands

Remind students that there are 60 minutes in an hour, and it takes 5 minutes for the minute hand to go from one number to the next.

Distribute the paper clocks. On the analog clock demonstrate time to the hour, for example, 4:00. Then move the minute hand to show 4:15. Have the students count by fives from 12 to tell how many minutes after 4 the clock shows. Have students tell you that the time is now 15 minutes after 4 or 4:15.

Ask the students to draw the hands on their clock face to show 4:15. Then have the students fold the circles in half, and then in half again so that there is a fold from 12 to 6 and from 9 to 3. Have students confirm that each part represents one fourth or one quarter of the clock.

Point out that the minute hand moved one quarter of the way around the clock as it went from 4:00 to 4:15. Recall that 4:15 is also quarter past 4.

Repeat the activity for 4:30 and 4:45. Have students write each time three ways.

Then recall that once the minute hand passes the six, students can also tell time before the hour. Have students count by fives from 9 to 12. Point out that 4:45 can also be expressed as 15 minutes before 5 or quarter to 5.

Repeat the activity until the students demonstrate understanding. Then have them use only the analog clock to express time to the quarter and half hour.

15 minutes after 4
quarter after 4
four fifteen or 4:15

30 minutes after 4
half past 4
four thirty or 4:30

15 minutes before 5
quarter to 5
four forty-five or 4:45

Name ______________________ Skill ______________________

Grade 4 Skill 62

Time to the Half Hour and Quarter Hour

Read the time. Count by fives.

Example A

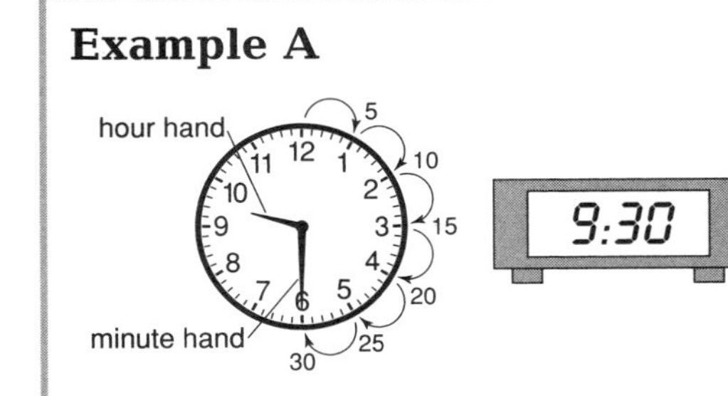

A half hour is 30 minutes.

The time is: 30 minutes after 9
half past 9
nine thirty or 9:30

Example B

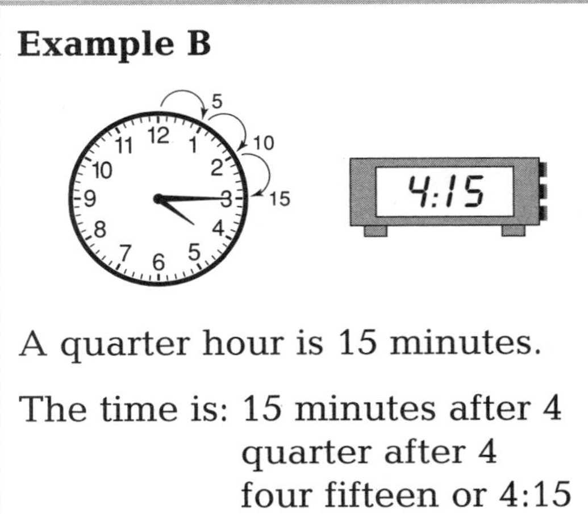

A quarter hour is 15 minutes.

The time is: 15 minutes after 4
quarter after 4
four fifteen or 4:15

Example C

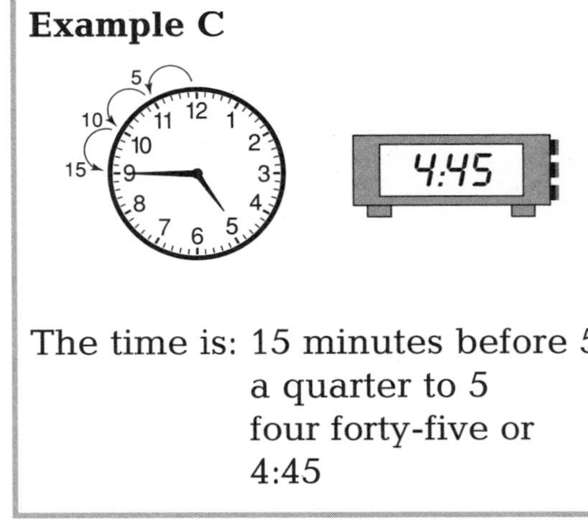

The time is: 15 minutes before 5
a quarter to 5
four forty-five or
4:45

Try These

Count by fives.

1. 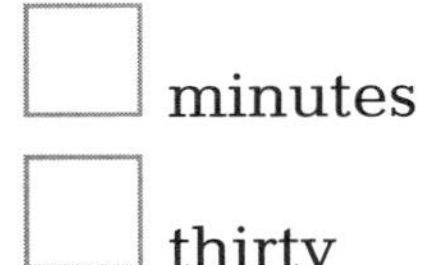

☐ minutes after ☐

☐ thirty

☐ : ☐

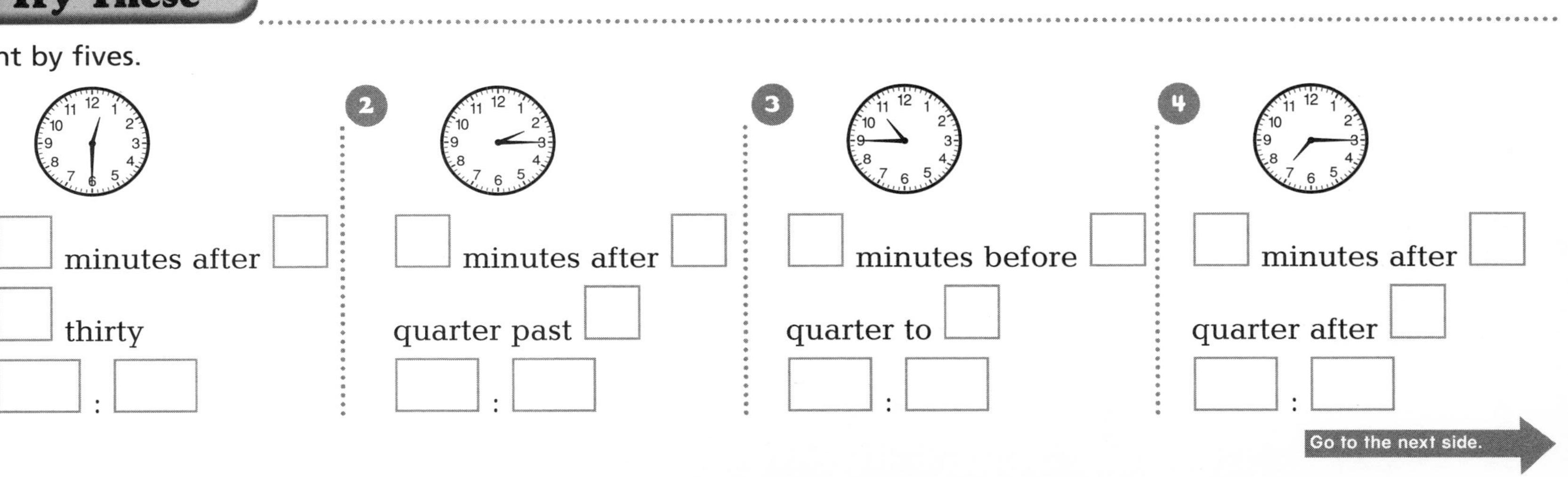

2. ☐ minutes after ☐

quarter past ☐

☐ : ☐

3. ☐ minutes before ☐

quarter to ☐

☐ : ☐

4. ☐ minutes after ☐

quarter after ☐

☐ : ☐

Go to the next side.

Name ______________________ Skill ______________________

Practice on Your Own

Skill 62

The time is:

15 minutes before 2

quarter to 2

1:45

Write the time.

☐ minutes after ☐

half past ☐

☐ : ☐

2

☐ minutes after ☐

quarter after ☐

☐ : ☐

3

quarter to ☐

☐ minutes before ☐

☐ : ☐

4

☐ thirty

5

quarter after ☐

6

quarter to ☐

Write the time as it would look on a digital clock.

7 three thirty

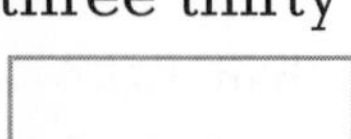

8 eight fifteen

9 six forty-five

Check

Write the time.

11

12

Skill 63 Grade 4 Time to the Minute

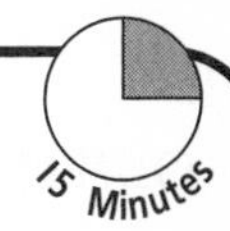

OBJECTIVE Tell time to the minute by counting minutes after and before the hour

MATERIALS large analog clock

You may wish to use a large clock to model the lesson. Point out the hour and minute hands.

Direct students' attention to the clock in Example A. Point out that the clock shows that it is after 10 o'clock.

Say: **To read the actual time, what can you do?** **Count the minutes from 12 to the minute hand.**

Explain that since it takes five minutes for the minute hand to move from number to number, they can count by fives to the 4, and then count by ones to the minute hand. After the students have counted aloud, say: **You have counted 23 minutes. What is the time?** **23 minutes after 10 or 10:23**

Direct students' attention to the clock in Example B. Explain that when the minute hand passes the 6, it may be easier to tell time *before* the hour. Point out that the clock shows that it is not yet 7 o'clock.

Say: **To read the time, how can you count the minutes before the hour?** **Start at the 12 and count back by fives to 10 and then count one more to the minute hand.** **You counted 11 minutes before the hour. What time is it?** **11 minutes before 7**

Have students count the minutes after the hour and note that 11 minutes before 7 is also 6:49. Explain that digital clocks show time in minutes after the hour.

TRY THESE Exercises 1–4 model writing the time before and after the hour in this order:

- **Exercise 1** Time before the hour.
- **Exercises 2–4** Time after the hour.

PRACTICE ON YOUR OWN Review the two examples at the top of the page. Ask students to explain the two ways to tell time that are shown. Ask which time would a digital clock show.

CHECK Determine if students know how to count the number of minutes before and after the hour. Success is determined by 3 out of 3 correct responses.

Students who successfully complete the **Practice on Your Own** and **Check** are ready to move on to the next skill.

COMMON ERRORS

- Students may be unable to distinguish between the hour and minute hands.
- Students may count too many or too few "fives."
- Students may confuse time before the hour and time after the hour.

Students who made more than three errors in the **Practice on Your Own**, or who were not successful in the **Check** section, may benefit from the **Alternative Teaching Strategy** on the next page.

Optional

Alternative Teaching Strategy

Model Telling Time to the Minute

15 Minutes

OBJECTIVE Use a clock to tell time before and after the hour

MATERIALS analog clocks, paper, pencil

If necessary, familiarize the students with the clock. Point out the hour and minute hands, and the minute marks that are between the numbers.

Explain to the students that they will be counting the minutes to tell the time.

Begin by having students use the clocks to model telling time to the hour, half hour, and quarter hour. Then show 9:18 on the clock, pointing out that since the time is after 9 o'clock, they count the minutes from the 12 to the minute hand.

Students may recall that since there are 5 minutes between each number, they can count by fives to the 3. Then they can count by ones to the minute hand. Have them record the time as 18 minutes after nine, or 9:18. Have students recall that on a digital clock, the time is displayed as 9:18.

Repeat the activity for 8:42. Have students model the time by counting minutes after the hour. Then recall that when the minute hand passes the six, they can also tell the time by counting the minutes before the next hour. Have them count back by fives from the 12 to the 9 and then by ones to the minute hand. Then have them record the time three ways:

42 minutes after 8
18 minutes before 9
8:42

Ask which time is displayed on a digital clock.

Continue with other examples of time before and after the hour. Pay attention to students who have difficulty counting by fives and then ones, or get confused when switching from time after the hour to time before the hour.

When students show an understanding of how to tell time to the minute, have them stop modeling the time and tell time from your clock only.

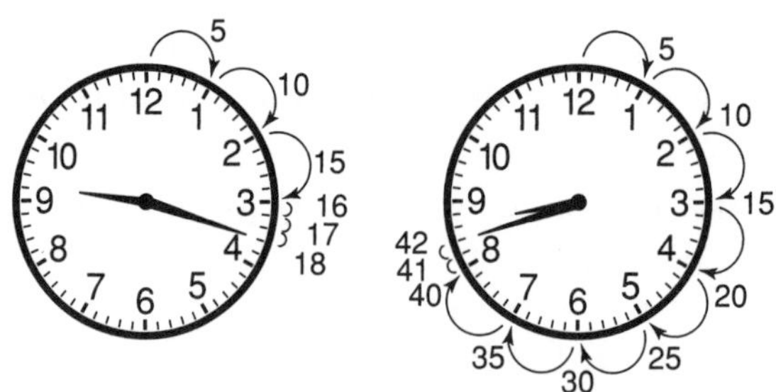

18 minutes after 9
9:18

42 minutes after 8
8:42

18 minutes before 9
8:42

Name ______________________ Skill ______________________

Time to the Minute

Grade 4 Skill 63

Read the time.

Example A

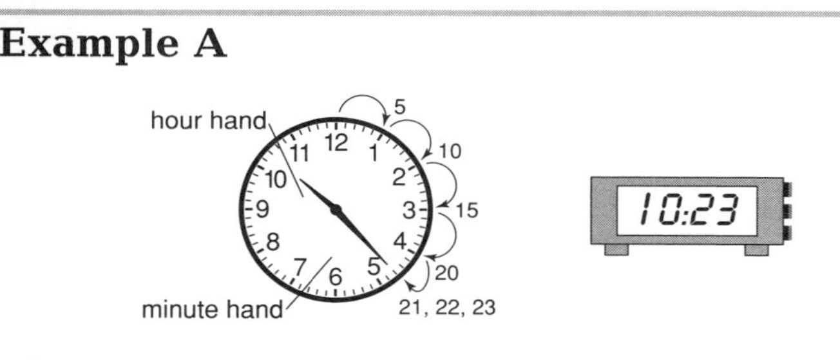

Count by fives and ones: 5, 10, 15, 20, 21, 22, 23.
The time is 23 minutes after 10 or 10:23.

Example B

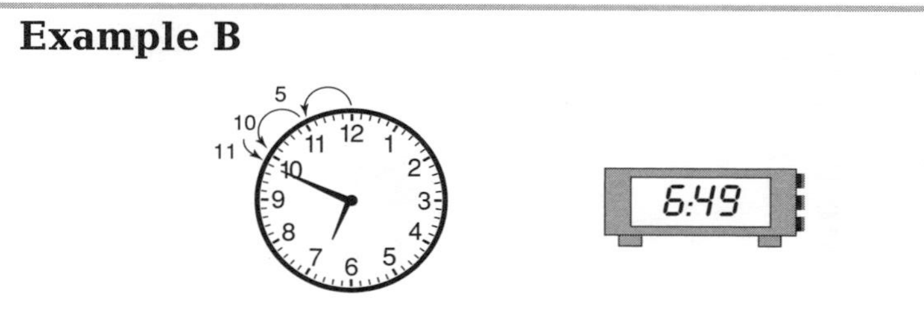

Count back from 12 by fives and ones: 5, 10, 11.
The time is 11 minutes before 7 or 6:49.

Try These

Count by fives and ones. Write the time.

1. 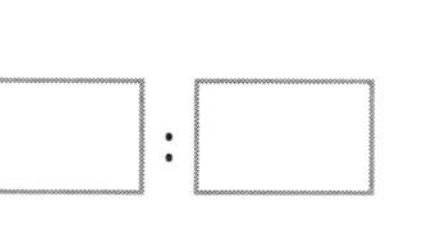 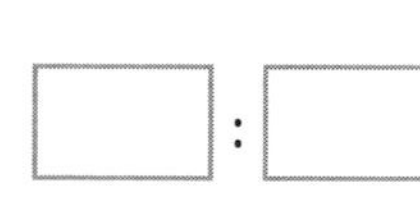 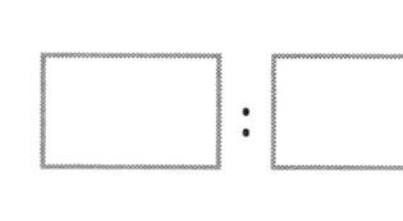 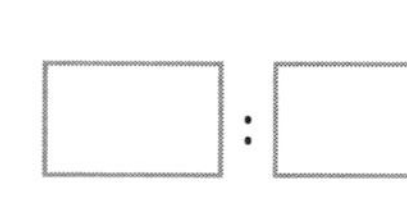
   ____ : ____

2.  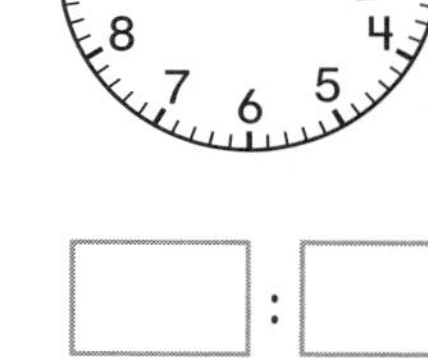 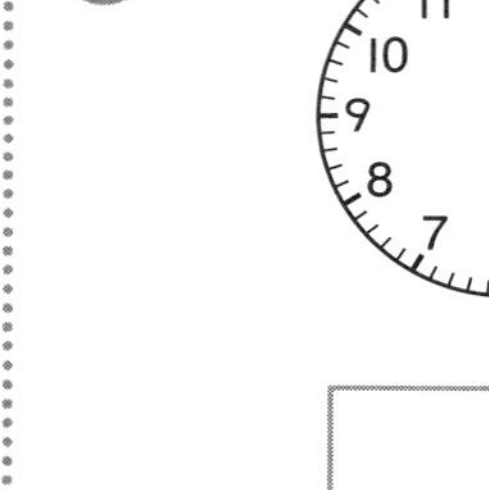 
   ____ : ____

3. 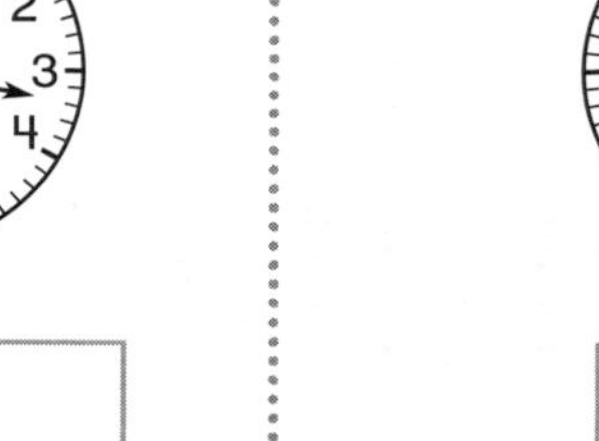
   ____ : ____

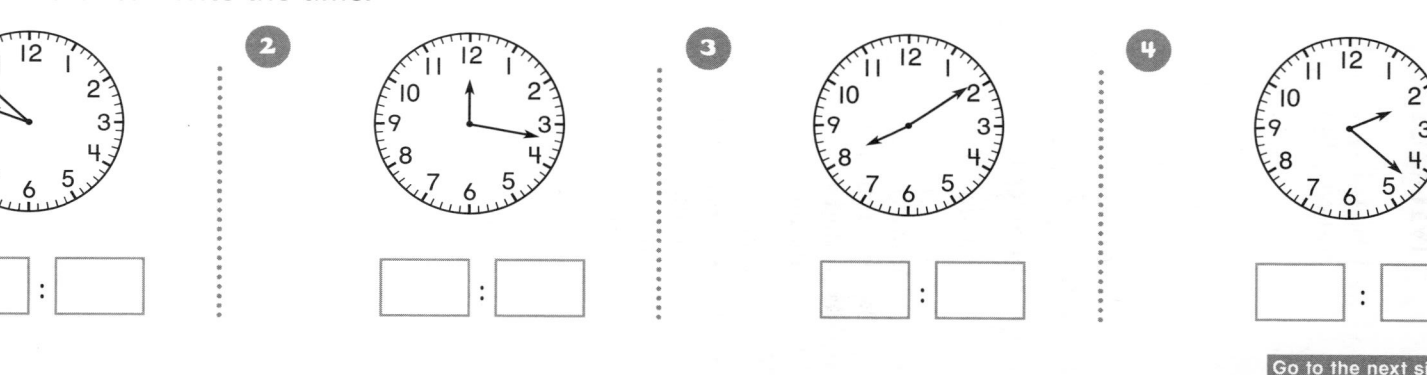

4. ____ : ____

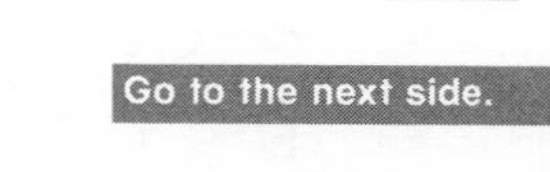

Name ______________________ Skill ______________________

Practice on Your Own

Skill 63

Time after the hour

Count 30 minutes from 12. Then count by fives and ones: 30, 35, 36. The time is 36 minutes after 2 or 2:36.

Time before the hour

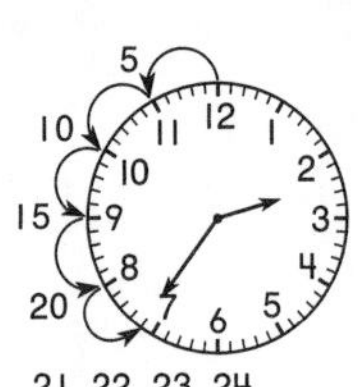

Count by fives and ones back from 12: 5, 10, 15, 20, 21, 22, 23, 24. The time is 24 minutes before 3 or 2:36.

Write the time two ways.

1. ☐ minutes after ☐
☐ : ☐

2. ☐ minutes before ☐
☐ : ☐

3. ☐ minutes after ☐
☐ : ☐

4. ☐ minutes after ☐
☐ : ☐

5. ☐ minutes after ☐
☐ : ☐

6. ☐ minutes before ☐
☐ : ☐

Write the time as it would look on a digital clock.

7. 11 minutes before two
☐ : ☐

8. 18 minutes after ten
☐ : ☐

9. 18 minutes before seven
☐ : ☐

Check

Write the time.

10. ☐

11. ☐

12. ☐

Skill 64 Grade 4

Use a Calendar

OBJECTIVE Use a calendar

Begin the skill by reviewing the meaning of ordinal numbers and practicing some examples. Generate the following table from students' responses to questions like: **How do you say the ordinal number for 3?** third **How do you write it in a date?** 3rd

	Say	Write
1	first	1st
2	second	2nd
3	third	3rd
4	fourth	4th
5	fifth	5th
6	sixth	6th
7	seventh	7th
8	eighth	8th
9	ninth	9th
10	tenth	10th

As students read through the skill, ask them to point to each date on the calendar. Then have them use ordinal numbers to describe other days of the month. Ask them to find other information such as: **How many Mondays are there in April?** 5 **If swim meets are on Mondays and Fridays, how many meets will there be in April?** 9

Students need to understand that the spaces in a calendar represent days belonging to other months. So, there are only 4 Tuesdays in April, not 5.

TRY THESE Exercises 1–3 give students practice in getting information from a calendar.

- **Exercises 1 and 2** Use ordinal numbers to describe days of the month.
- **Exercises 3 and 4** Use the calendar to find information about days of the month.

PRACTICE ON YOUR OWN Review the material at the top of the page. Alert students to the fact that they will need to write dates using ordinal numbers. For example, for the first Saturday of December, write December 5th, not December 5.

CHECK Success is determined by 3 out of 3 correct responses.

Students who successfully complete the **Practice on Your Own** and **Check** are ready to move on to the next skill.

COMMON ERRORS

- Students may count spaces representing days from other months as belonging to the month they are looking at. For example, there are 5 Wednesdays in December but only 4 Mondays.

Students who made more than two errors in the **Practice on Your Own**, or who were not successful in the **Check** section, may benefit from the **Alternative Teaching Strategy** on the next page.

Optional

Alternative Teaching Strategy

Hands-On: Use a Calendar

OBJECTIVE Use a calendar

MATERIALS 9 × 12-inch copies of April calendar shown in Skill 64

Give out copies of the April calendar. Have students cover or fold the calendar page so that only the first week of April is visible.

Have students use ordinal numbers to describe the days of the first week only. For example, ask them to name the first day of the week, the second, and so on. Then repeat, this time mixing up the order.

Next, have students describe the dates of the first week. They should be able to tell you that April 5th is a Thursday, April 2nd a Monday, and so on.

Tell students to next look at the complete calendar for April. Ask them a variety of questions about the rest of the month, such as:

What is the fourth Friday of April? April 27th **Which day of the week is April 11th?** Wednesday **How many Wednesdays are in the month?** 4

You might also ask students to circle all the Tuesdays, put a square around every other Monday, a triangle around the first and last Fridays, or to cross out all the Saturdays.

APRIL

Sunday	Monday	Tuesday	Wednesday	Thursday	Friday	Saturday
1	2	3	4	5	6	7
8	9	10	11	12	13	14
15	16	17	18	19	20	21
22	23	24	25	26	27	28
29	30					

Conclude the activity by having students write in dates of after–school or weekend activities, appointments, or events of their own. Have them describe the dates of their activities using ordinal numbers.

Name ______________________ Skill ______________________

Grade 4 Skill 64

Use a Calendar

You can use **ordinal** numbers to describe days on a calendar.
Ordinal numbers tell the order or position of things.

The table shows some ordinal numbers.

SAY	WRITE
first	1st
second	2nd
third	3rd
fourth	4th
fifth	5th
sixth	6th
seventh	7th
eighth	8th
ninth	9th
tenth	10th

APRIL

Sun	Mon	Tue	Wed	Thu	Fri	Sat
1	2	3	4	5	6	7
8	9	10	11	12	13	14
15	16	17	18	19	20	21
22	23	24	25	26	27	28
29	30					

Use ordinal numbers to describe dates on the calendar.

The *first* Saturday is April 7th.
The *fourth* Monday is April 23rd.

April *13th* is a Friday.
April *25th* is a Wednesday.

Here is some other information you can find on the calendar.

- There are four Fridays in April.
- If baseball practices are on Tuesdays and Thursdays, there will be 8 practices in April.

Try These

Use the April calendar above. Use ordinal numbers where you can. Complete.

1. April 17th is the ______ Tuesday of the month.
2. The ______ Sunday is April 1st.
3. Which day of the week is April 12th? ______
4. How many Tuesdays are in the month? ______

Go to the next side.

Name ______________________ Skill ______________________

Skill 64

Practice on Your Own

DECEMBER						
Sun	Mon	Tue	Wed	Thu	Fri	Sat
		1	2	3	4	5
6	7	8	9	10	11	12
13	14	15	16	17	18	19
20	21	22	23	24	25	26
27	28	29	30	31		

December 3 is the *first* Thursday of the month.

October is the *tenth* month.

MONTHS OF THE YEAR			
January	31 days	July	31 days
February	28 days	August	31 days
March	31 days	September	30 days
April	30 days	October	31 days
May	31 days	November	30 days
June	30 days	December	31 days

Use the December calendar. Write the date.

1. first Saturday ________
2. third Wednesday ________
3. fourth Monday ________
4. fifth Wednesday ________

Answer the questions.

5. What day of the week is December 21st? ________
6. How many Tuesdays are in the month of December? ________
7. If you circled all the Mondays and Wednesdays in December, how many dates would be circled? ________
8. Write an ordinal number for the last day of December. ________
9. Name the fifth month of the year. ________
10. What is the date of the third Tuesday of December? ________
11. What is the month after January? ________
12. Which month comes before September? ________
13. How many months are in one year? ________

Check

14. What day of the week is December 12th? ________
15. What date is the fourth Saturday of December? ________
16. Name the eleventh month of the year. ________

Measure to the Nearest Inch, Half-Inch

OBJECTIVE Measure objects to the nearest inch and half-inch

MATERIALS inch rulers

Begin by pointing out to students that one inch is about the length from the tip of the thumb to the first joint.

Have students look at the inch ruler in the first picture. Point out the inch marks and numbers.

Say: **Notice how one end of the pencil is lined up with the zero mark on the ruler.**

Have students trace the dashed line from the pencil point down to the ruler.

Ask: **Between which two inch marks on the ruler is the other end of the pencil?** 2 in. and 3 in. **Which inch mark is closer, the 2-inch mark or the 3-inch mark?** 3-in. **So, what is the length of the pencil to the nearest inch?** 3 inches

Direct students' attention to the next picture. Explain that the marks between the inch marks are half-inch marks. Have students find the $\frac{1}{2}$-inch mark and the $1\frac{1}{2}$-inch mark.

Have students trace the dotted line from the pencil point down to the ruler.

Ask: **Between which marks on the ruler is the other end of the pencil?** between the $\frac{1}{2}$-inch and the 1-inch marks **Is the end of the pencil closer to the $\frac{1}{2}$-inch mark or the 1-inch mark?** closer to the 1-inch mark

TRY THESE Exercises 1–2 have inch rulers in place. Dashed lines help students measure each pencil to the nearest inch or half inch.

- **Exercise 1** Length to the nearest inch.
- **Exercise 2** Length to the nearest half-inch.

PRACTICE ON YOUR OWN Review the example at the top of the page. Suggest to students that they use the dashed line to see where the pencil point would be on the ruler.

Ask: **Between which two marks does the dashed line fall on the ruler?** between the $2\frac{1}{2}$-inch mark and the 3-inch mark

CHECK Determine if students can use a ruler to measure to the nearest inch or half-inch. Success is indicated by 2 out of 2 correct responses.

Students who successfully complete the **Practice on Your Own** and **Check** are ready to move on to the next skill.

COMMON ERRORS

- Students may confuse inch and half-inch marks.
- Students may record the next whole inch when they write a length to the nearest half-inch.

Students who made more than two errors in the **Practice on Your Own**, or who were not successful in the **Check** section, may benefit from the **Alternative Teaching Strategy** on the next page.

Optional

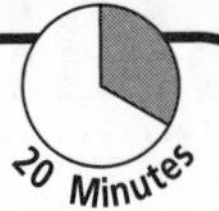

Alternative Teaching Strategy
Use Inch Rulers to Measure

OBJECTIVE Use inch rulers to measure objects to the nearest inch and half-inch

MATERIALS inch rulers with inch and half-inch marks, paper, small objects

You may wish to have students work in pairs. Distribute rulers and have students draw a line segment that is three inches long.

Emphasize starting the line at the end of the ruler and ending the line at the 3-inch mark. Encourage students to be as accurate as possible. Have students exchange papers to check each other's work.

Now have each student use the ruler to draw a line segment whose length falls between any two inch marks. Have partners exchange papers and measure each other's line segment.

To guide students, say: **This line segment does not end precisely on an inch mark. You can measure the line segment to the nearest inch.**

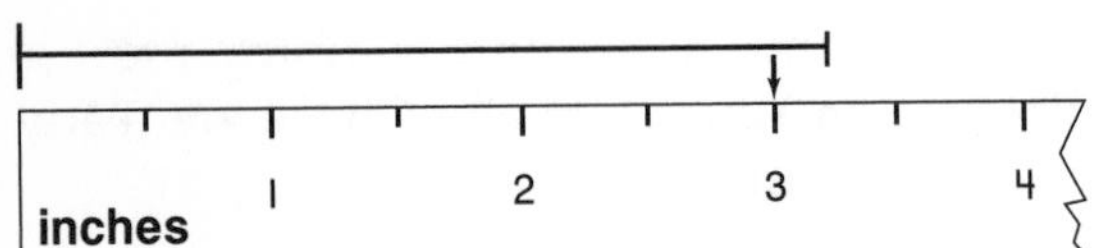

This line is 3 inches long, to the nearest inch.

Then ask: **To which inch mark on the ruler is the end of the line segment nearest?**

Have students record the nearest inch and then compare that measurement with their partner's.

Point out the half-inch intervals on the ruler. Suggest that the students draw a line segment that is three and one-half inches long, and have partners check each other's work.

Then have students draw a line segment whose length falls between 2 and $2\frac{1}{2}$ inches.

Say: **This line segment does not end exactly on an inch mark or on a half-inch mark. Which inch mark or half-inch mark is closest to the end of the line segment?**

Repeat the activity several times until students demonstrate competence with measuring to the nearest inch and half inch. Then let student pairs measure small objects in the classroom. Partners can check each other's results.

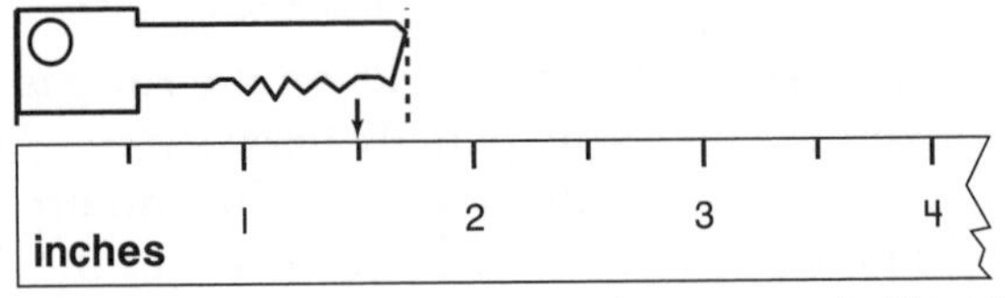

This key is $1\frac{1}{2}$ inches long, to the nearest half inch.

Name ______________________ Skill ______________________

Grade 4 **Skill 65**

Measure to the Nearest Inch, Half-Inch

The inch (in.) is a customary unit of length. Use the inch to measure small objects.

Find the length of the pencil to the nearest inch.

inches 1 2 3

The length is closer to 3 inches than to 2 inches.

So, to the nearest inch, the pencil is 3 in. long.

Think: Line up one end of the object with the end of the ruler. Then measure.

Find the length of the pencil to the nearest half-inch.

inches 1 2

The length is closer to 1 inch than $\frac{1}{2}$ inch. So, to the nearest half-inch, the pencil is 1 in. long.

Try These

1. Measure to the nearest inch.

☐ inches

2. Measure to the nearest half-inch.

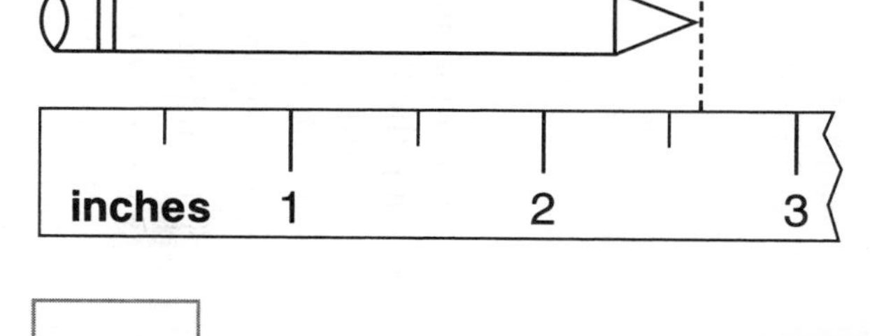

☐ inches

Go to the next side.

Name ______________________ Skill ______________________

Practice on Your Own

Skill 65

Find the length of the pencil to the nearest half-inch.

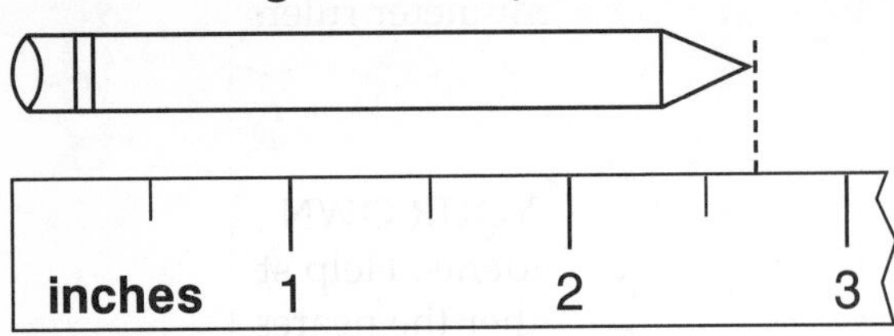

Think:
Line up one end of the object with the end of the ruler.
Then measure.

To the nearest half-inch, the pencil is $2\frac{1}{2}$ in.

Measure to the nearest inch.

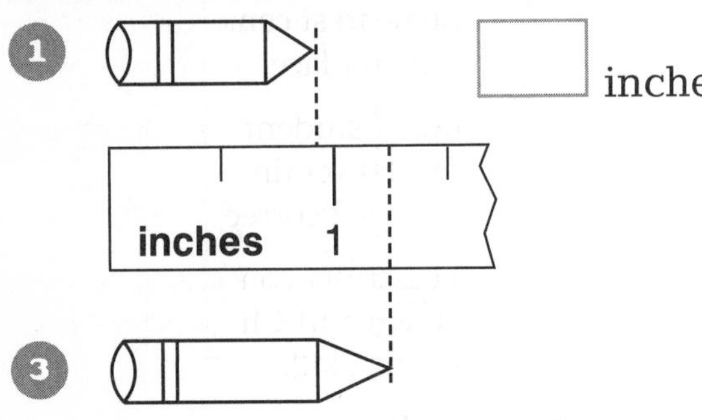

1. ☐ inches

3. ☐ inches

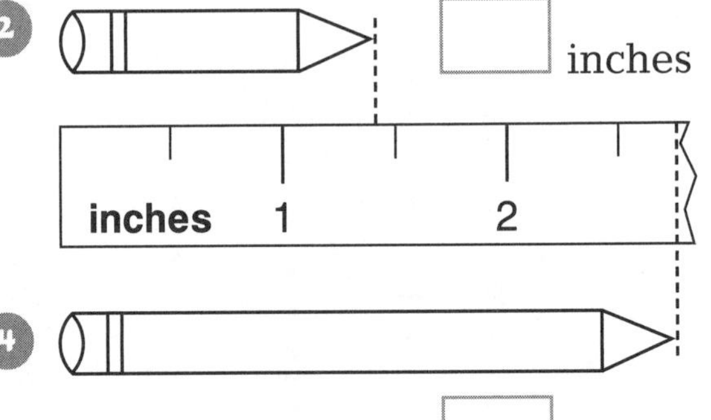

2. ☐ inches

4. ☐ inches

Measure to the nearest half-inch.

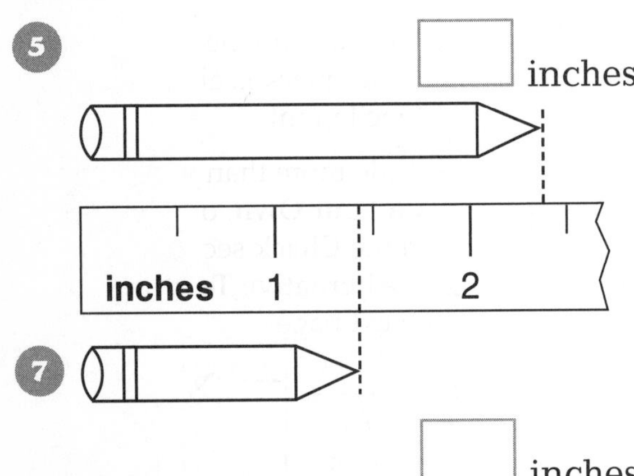

5. ☐ inches

7. ☐ inches

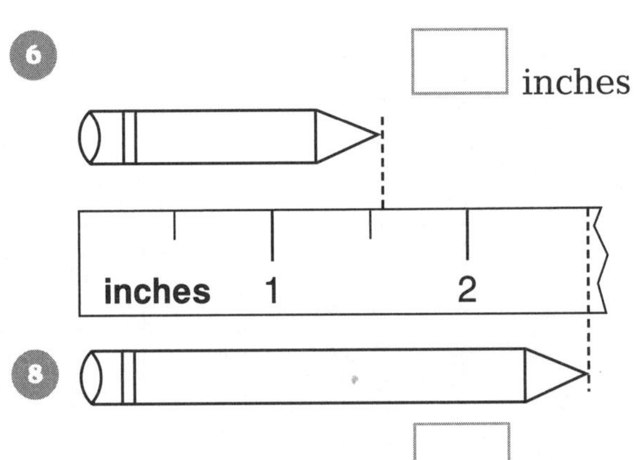

6. ☐ inches

8. ☐ inches

Check

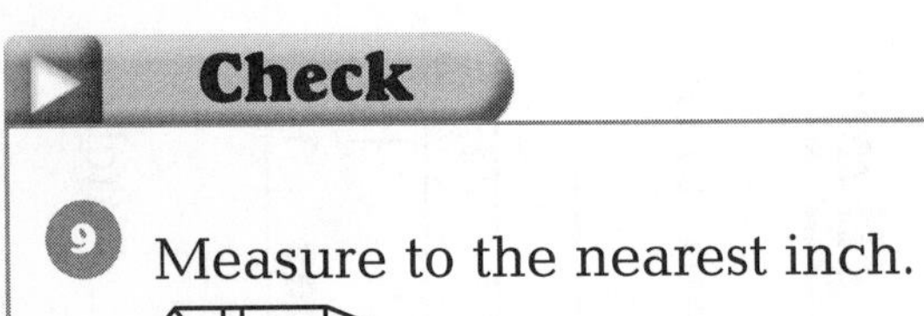

9. Measure to the nearest inch.

☐ inches

10. Measure to the nearest half-inch.

☐ inches

Measure to the Nearest Centimeter

OBJECTIVE Measure to the nearest centimeter

MATERIALS centimeter rulers

10 Minutes

Begin by recalling that a centimeter (cm) is a metric unit of length used to measure small objects. Have students look closely at the metric ruler on the page. Point out where the ruler begins on the left.

Direct students' attention to the first example.

Emphasize how the ruler is lined up with the left end of the pencil.

Point out that the dashed line comes down from the very tip of the pencil to the ruler at a point between 9 cm and 10 cm.

Ask: **What is the length of the pencil to the nearest centimeter?** **9 cm** **How did you decide?** **Possible response: The dashed line marks a point on the ruler that is closer to the 9 cm mark than to the 10 cm mark.**

Confirm for students that, in this case, the nearest centimeter is to the left of the point marked by the dashed line.

Ask similar questions as students measure the second pencil.

TRY THESE In Exercises 1–3, students measure the lengths of pencils to the nearest centimeter. Metric rulers are in place on the page and dashed lines are provided to help students measure.

- **Exercises 1–2** Measure to the nearest centimeter. The nearest cm is to the right of the dashed line.
- **Exercise 3** Nearest cm is to the left of the dashed line.

PRACTICE ON YOUR OWN Review the example with students. Help students focus on deciding whether the nearest centimeter is to the right or to the left of the pencil point. **to the left, 7 cm**

Exercises 1–4 Students measure lengths to the nearest centimeter with cues in place.

CHECK Determine if students can measure an object to the nearest centimeter. Success is indicated by 2 out of 2 correct responses.

Students who successfully complete the **Practice on Your Own** and **Check** are ready to move on to the next skill.

COMMON ERRORS

- Students may not align the ruler precisely at the left end and thus measure inaccurately.
- Students may mistakenly believe that the nearest centimeter mark is always to the right of the pencil point.

Students who made more than two errors in the **Practice on Your Own**, or who were not successful in the **Check** section, may benefit from the **Alternative Teaching Strategy** on the next page.

Optional

Alternative Teaching Strategy

20 Minutes

Measure Objects to the Nearest Centimeter

OBJECTIVE Measure objects to the nearest centimeter

MATERIALS centimeter ruler, paper clips (3.2 cm long), small objects, and paper

Distribute materials and instruct students to look closely at the metric ruler. Point out where the ruler begins. **Some metric rulers begin at the very edge of the ruler; others have a small space and then a mark to label the beginning. Tell students that this is the *zero mark*, even if it is not labeled 0.**

Discuss any questions students might have about the centimeter ruler and measuring.

Then call students' attention to the length of one centimeter. Ask students to point to the 1 cm mark on the ruler. Have them measure the width of one of their fingers; ask them to see which finger fits most easily between the beginning of the ruler and the one centimeter mark.

Now, have students measure the length of a paper clip to its nearest centimeter.

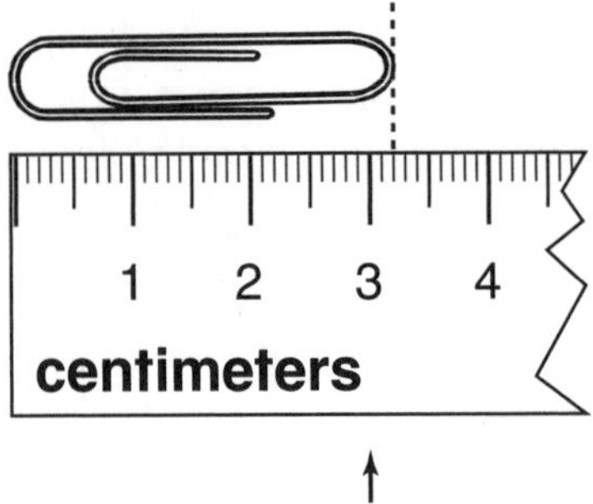

Say: **Line up one end of end of the paper clip with the zero mark on the ruler. Look at the other end. Between which two centimeter marks is that end?** 3 cm and 4 cm **Is it closer to the 3 or to the 4 on the ruler?** closer to the 3

Repeat this activity using several small objects such as a large paper clip, eraser, key, or crayon. Monitor students as they work and have them compare their results.

When students have demonstrated sufficient competence, have them work independently to measure small objects around the classroom.

Name ______________________ Skill ______________________

Grade 4 **Skill 66**

Measure to the Nearest Centimeter

A centimeter (cm) is a metric unit of length. Use the centimeter to measure small objects.

Find the length of the pencil to the nearest centimeter.

1 2 3 4 5 6 7 8 9 10
centimeters

So, to the nearest centimeter, the pencil is 9 cm.

Find the length of the pencil to the nearest centimeter.

1 2 3 4 5 6 7 8
centimeters

So, to the nearest centimeter, the pencil is 7 cm.

Think: Line up one end of the object with the end of the ruler. Then measure.

Try These

Measure each pencil to the nearest centimeter.

1. 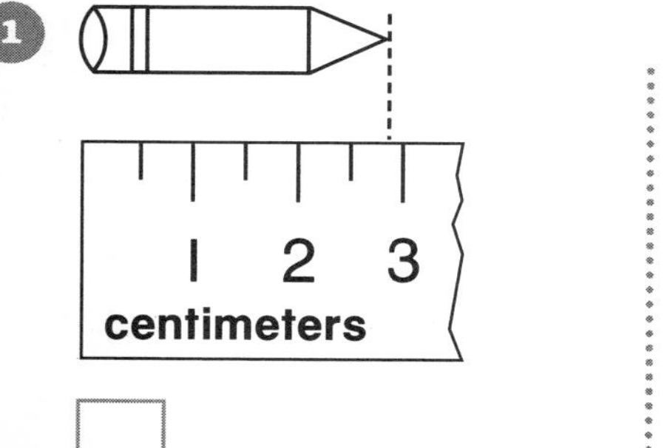

☐ centimeters

2. 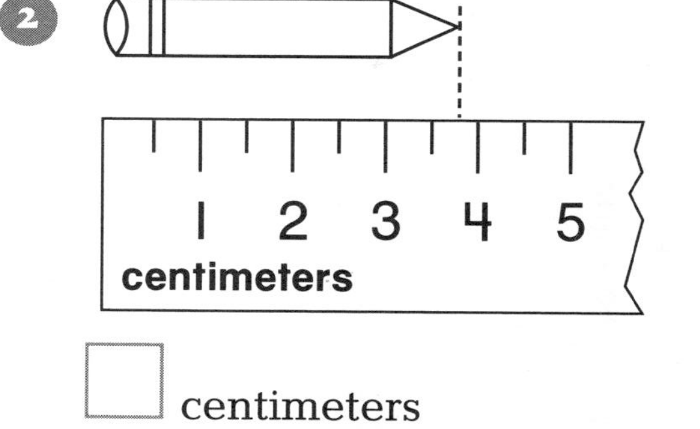

☐ centimeters

3. 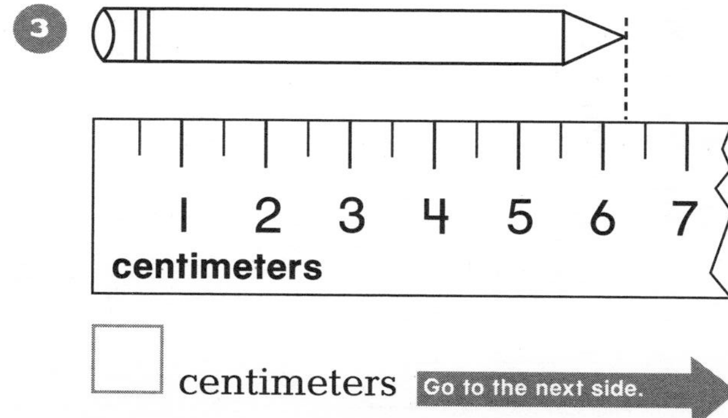

☐ centimeters

Go to the next side.

Name ______________________ Skill ______________________

Practice on Your Own

Skill 66

Find the length of the pencil to the nearest centimeter.

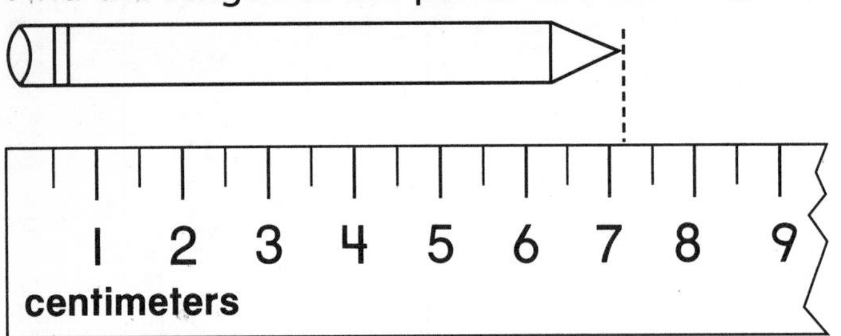

Think:
Line up one end of the object with the end of the ruler. Then measure.

To the nearest centimeter, the pencil is 7 centimeters.

Measure to the nearest centimeter.

1. 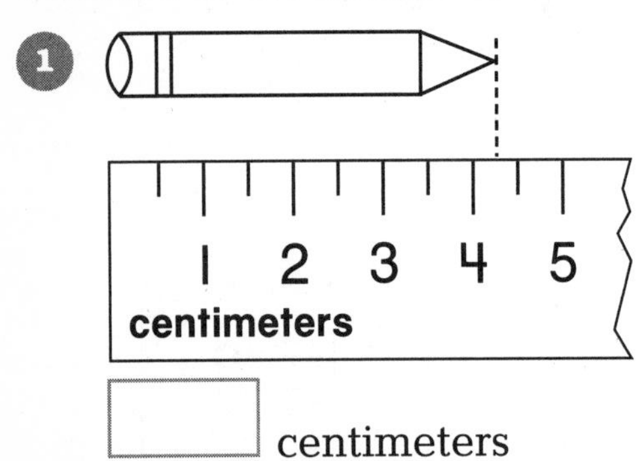

☐ centimeters

2. 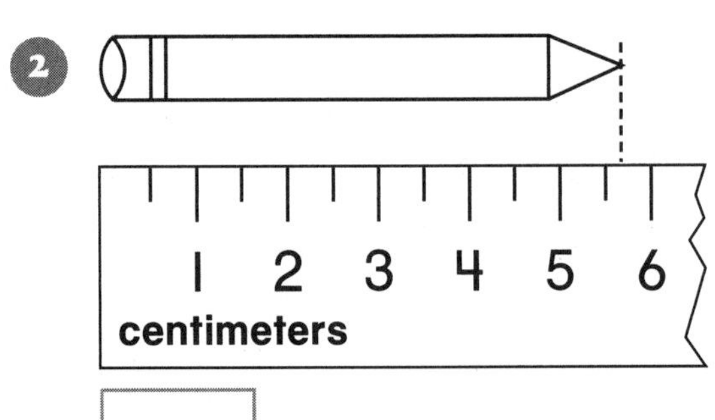

☐ centimeters

3. 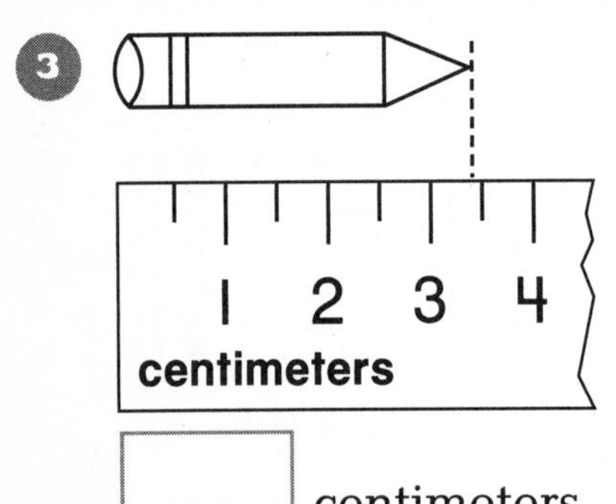

☐ centimeters

4. 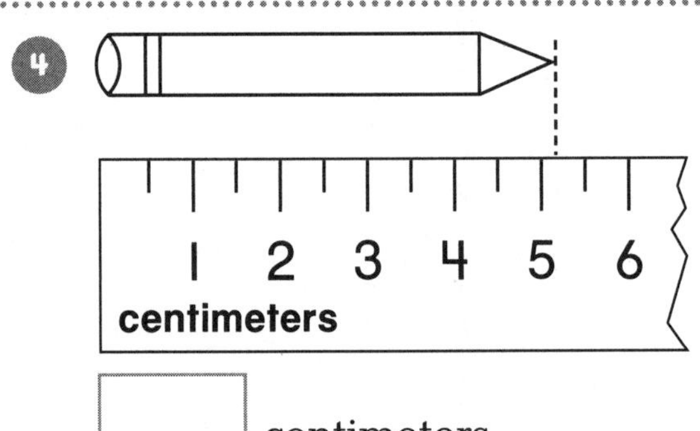

☐ centimeters

Check

Measure to the nearest centimeter.

5. 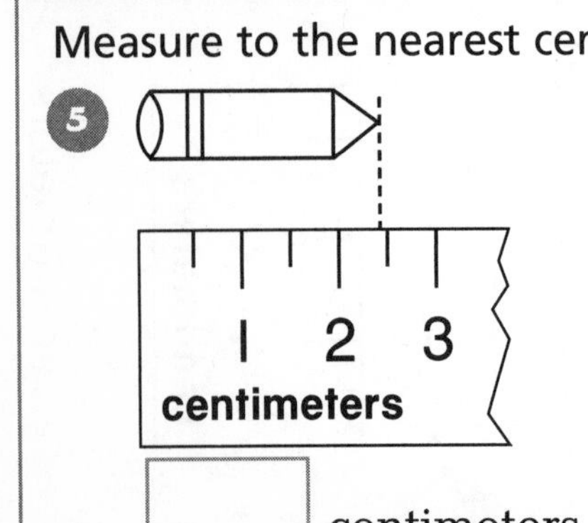

☐ centimeters

6. 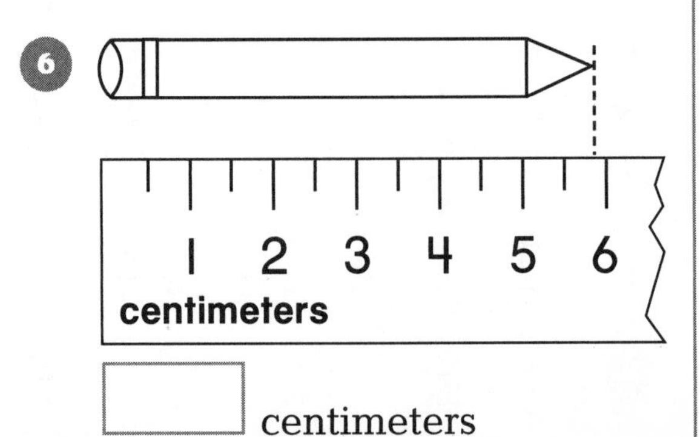

☐ centimeters

Answer Card

Measurement

Grade 4

SKILL 62

TRY THESE

1. 30, 12, 12, 12:30
2. 15, 2, 2, 2:15
3. 15, 11, 11, 10:45
4. 15, 7, 7, 7:15

PRACTICE

1. 30, 7, 7, 7:30
2. 15, 5, 5, 5:15
3. 3, 15, 3, 2:45
4. 11
5. 12
6. 4
7. 3:30
8. 8:15
9. 6:45

CHECK

Possible answers are given.

10. 9:30 or half past 9 or 30 minutes after 9 or nine-thirty
11. 6:15 or 15 minutes after 6 or quarter after 6
12. 5:45 or 15 minutes before 6 or quarter to 6

SKILL 63

TRY THESE

1. 9:52
2. 12:17
3. 8:09
4. 2:22

PRACTICE

1. 28, 3, 3:28
2. 8, 10, 9:52
3. 17, 12, 12:17
4. 9, 8, 8:09
5. 29, 4, 4:29
6. 22, 12, 11:38
7. 1:49
8. 10:18
9. 6:42

CHECK

10. 7:57 or 3 minutes before 8
11. 2:22 or 22 minutes after 2
12. 5:43 or 17 minutes before 6

SKILL 64

TRY THESE

1. 3rd
2. 1st
3. Thursday
4. 4

PRACTICE

1. December 5th
2. December 16th
3. December 28th
4. December 30th
5. Monday
6. 5
7. 9
8. 31st
9. May
10. 15th
11. February
12. August
13. 12

CHECK

14. Saturday
15. December 26th
16. November

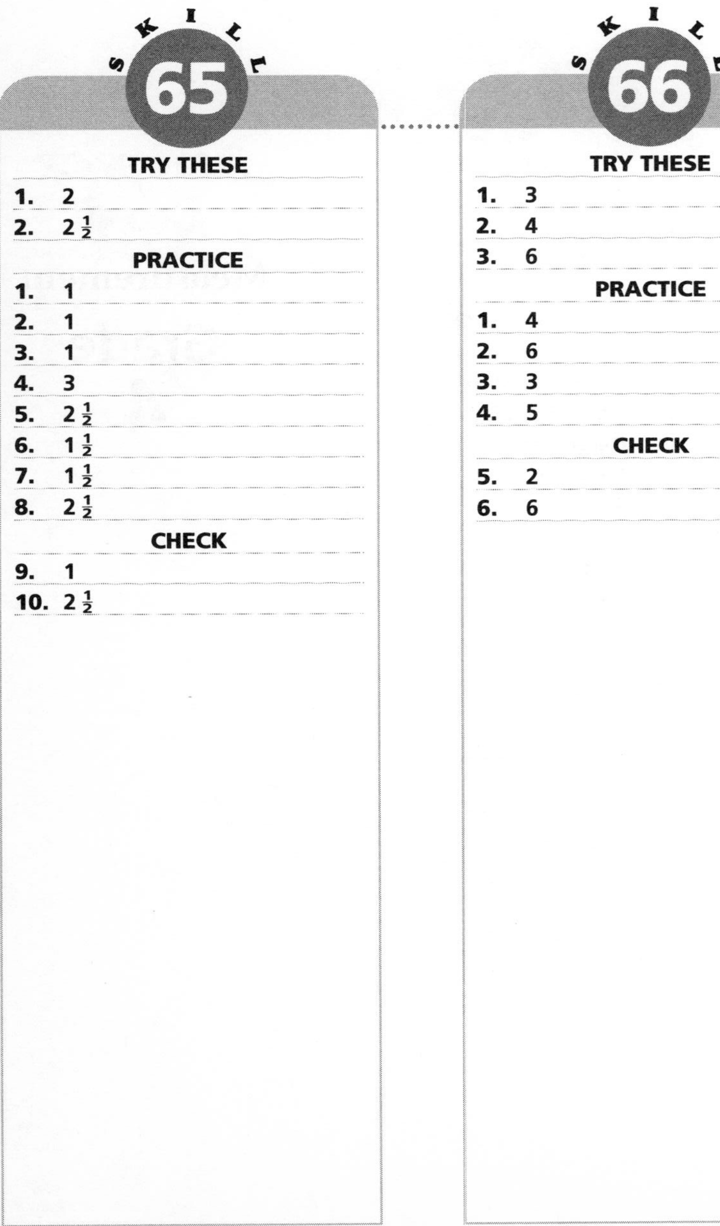

Answer Card

Measurement

Grade 4

Skill 65

TRY THESE

1. 2
2. $2\frac{1}{2}$

PRACTICE

1. 1
2. 1
3. 1
4. 3
5. $2\frac{1}{2}$
6. $1\frac{1}{2}$
7. $1\frac{1}{2}$
8. $2\frac{1}{2}$

CHECK

9. 1
10. $2\frac{1}{2}$

Skill 66

TRY THESE

1. 3
2. 4
3. 6

PRACTICE

1. 4
2. 6
3. 3
4. 5

CHECK

5. 2
6. 6

Measurement and Geometry

Geometry

Skill 67 Grade 4 Identify Angles

OBJECTIVE Identify a right angle, and an angle that is greater than and less than a right angle

MATERIALS index cards

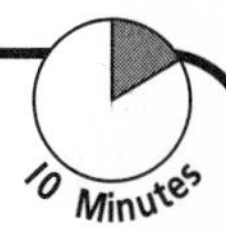

Review how an angle is formed when two rays meet at an endpoint.

You may wish to use index cards to model a square corner.

Say: **Model A shows an angle that is the same as a square corner. Can you think of other things that have square corners or right angles?** math book, note paper, table top

Students may note that the index card in Model A has 4 square corners. Then explain that the angle in Model B is less than a right angle.

Ask: **How can you tell the angle is less than a right angle?** The "opening" between rays is less than the opening between rays for a right angle.

Continue with similar questioning as you discuss Model C.

TRY THESE Exercises 1–3 prompt the students as they identify each of the three angles.

- **Exercise 1** Less than a right angle.
- **Exercise 2** Greater than a right angle.
- **Exercise 3** Right angle.

PRACTICE ON YOUR OWN Review the examples with students. Ask students to describe each angle in their own words.

CHECK Determine if students know how to identify a right angle, and using a right angle as a guide, identify angles that are greater than and less than a right angle.

Success is indicated by 3 out of 3 correct responses.

Students who successfully complete the **Practice on Your Own** and **Check** are ready to move on to the next skill.

COMMON ERRORS

- Students may confuse the terms "greater than" and "less than" when identifying angles.
- Students may not recognize the angles when shown in different positions.

Students who made more than 3 errors in the **Practice on Your Own**, or who were not successful in the **Check** section, may benefit from the **Alternative Teaching Strategy** on the next page.

Optional

Alternative Teaching Strategy

20 Minutes

Use Tagboard Angle Models

OBJECTIVE Use an angle model to represent right angles and angles greater than and less than a right angle

MATERIALS for each student: two strips of tagboard, one fastener, and one index card, drawing paper, markers

Have students create their own angle model using the two strips of tagboard. Join the strips at the ends with the paper fastener.

Use the index card to illustrate a square corner. Have students position the card inside the opening of the angle model to verify that a square corner is formed. Remind students that an angle that is the same as a square corner is called a right angle.

Then have students use the model to show angles greater than and less than a right angle.

Index card

Ask students to trace the model to draw a right angle on drawing paper, and use a marker to color in the space between the rays.

Have students draw an angle greater than and less than a right angle, coloring in the space between the rays. Ask students to compare the size of the colored areas to the that of the right angle.

Repeat the activity with other angles. When the students understand how the angles relate to one another, have them draw angles without using the model and identify them as right angles, less than right angles, or greater than right angles.

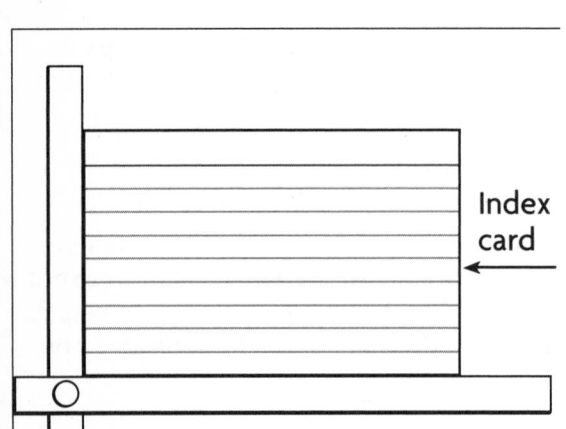

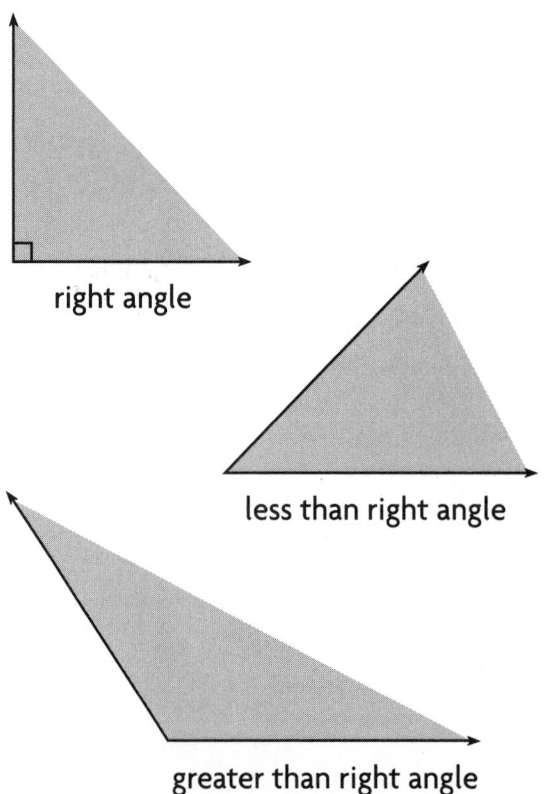

Name ______________________ Skill ______________________

Grade 4
Skill 67

Identify Angles

An angle is formed when two rays meet at an endpoint.
Use the corner of an index card, or a sheet of paper, to find a right angle.

Model A

A **right angle** is a square corner.

Model B

This angle is **less than** a right angle.

Model C

This angle is **greater than** a right angle.

Try These

Answer the questions about each angle. Write *yes* or *no*.

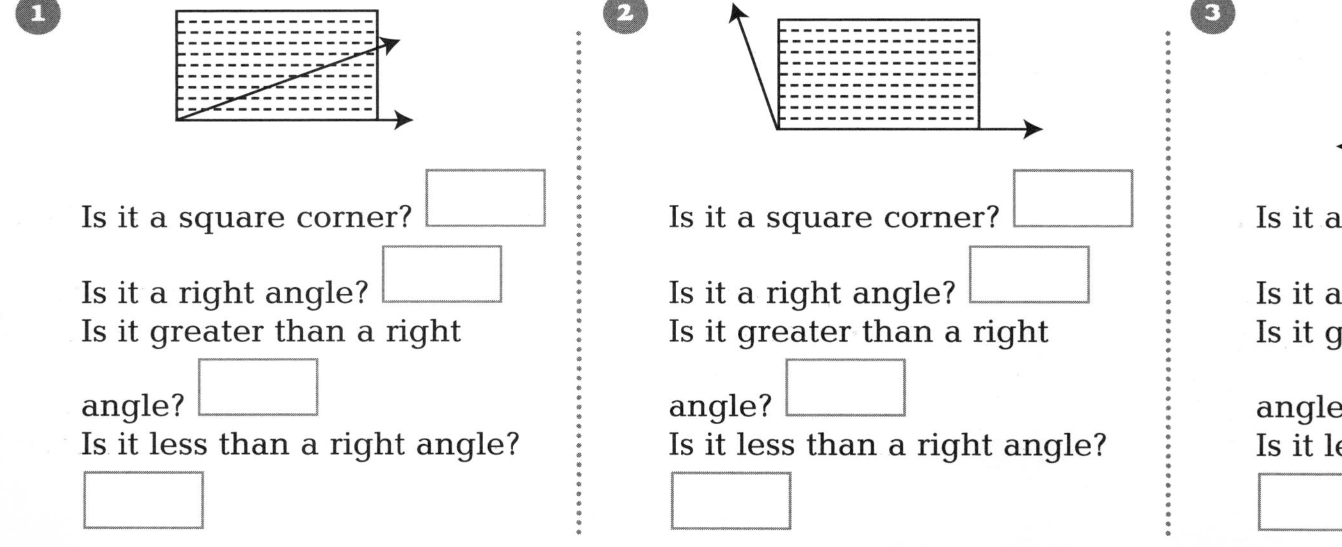

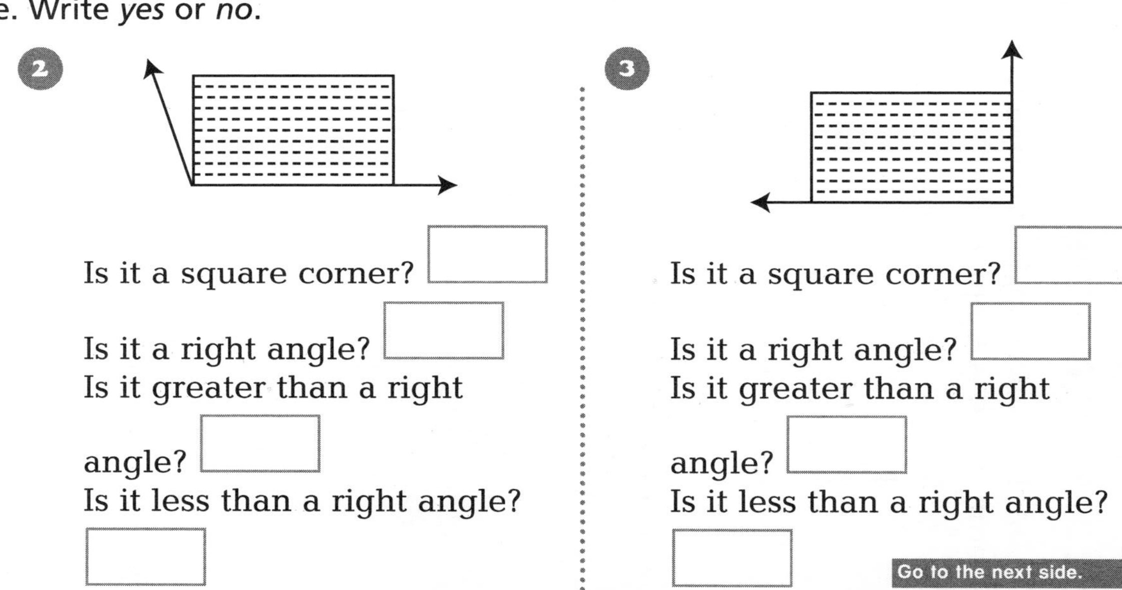

1

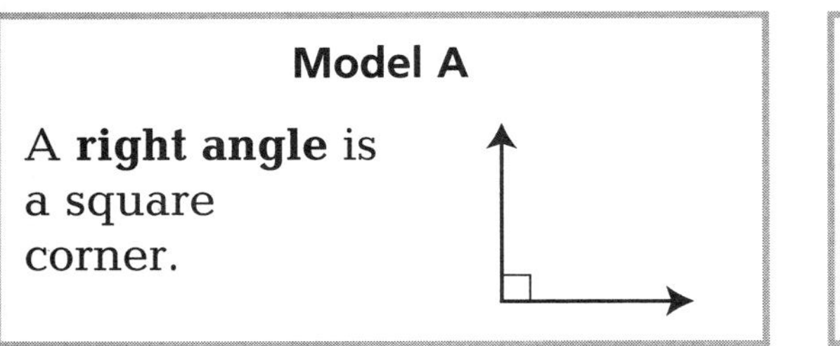

Is it a square corner? ______

Is it a right angle? ______

Is it greater than a right angle? ______

Is it less than a right angle? ______

2

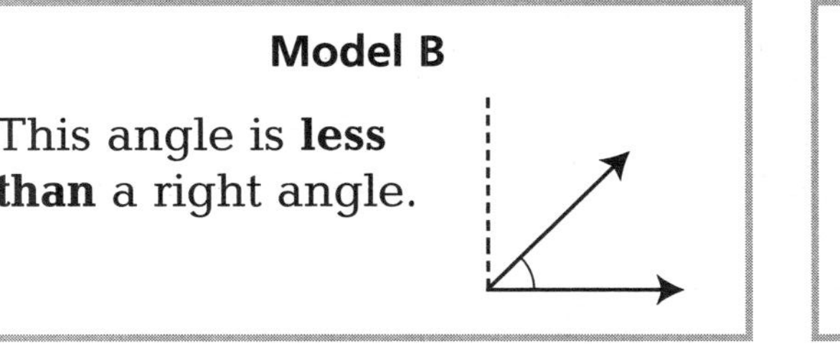

Is it a square corner? ______

Is it a right angle? ______

Is it greater than a right angle? ______

Is it less than a right angle? ______

3

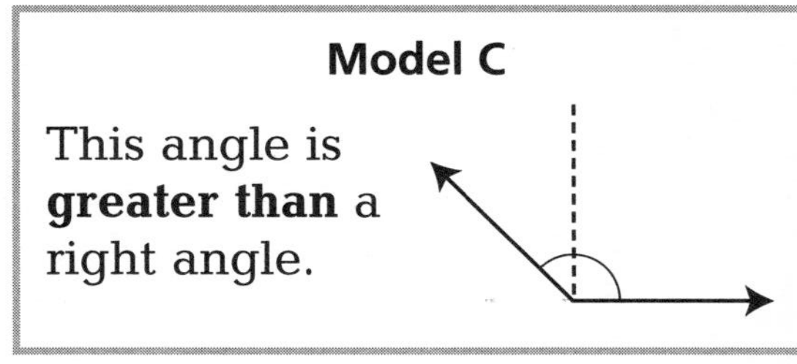

Is it a square corner? ______

Is it a right angle? ______

Is it greater than a right angle? ______

Is it less than a right angle? ______

Go to the next side.

Name ______________________ Skill ______________________

Practice on Your Own

Skill 67

An angle is formed when two rays meet at an endpoint.
A right angle is a square corner.

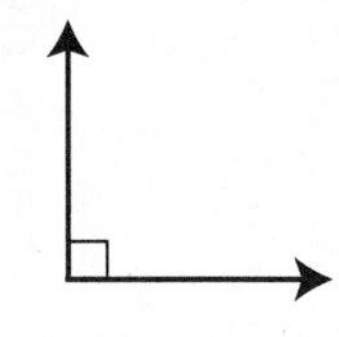

right angle

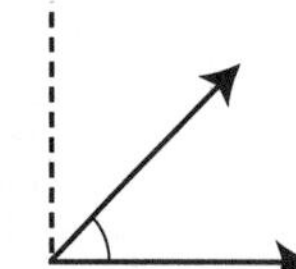

less than a right angle

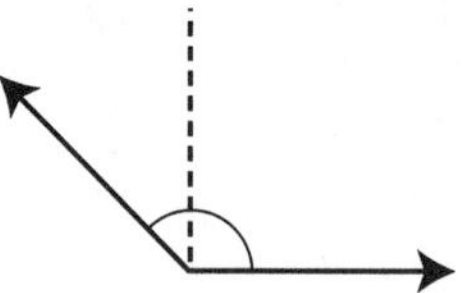

greater than a right angle

Circle the answer that correctly describes the angle.

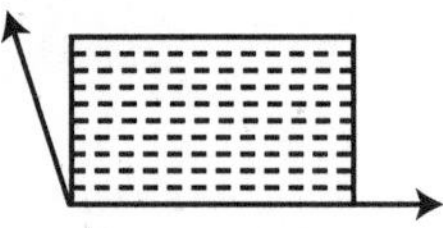

right angle
less than a right angle
greater than a right angle

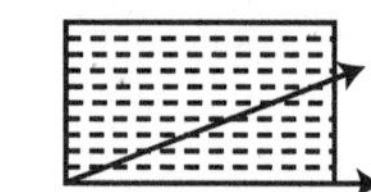

right angle
less than a right angle
greater than a right angle

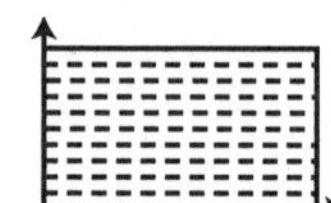

right angle
less than a right angle
greater than a right angle

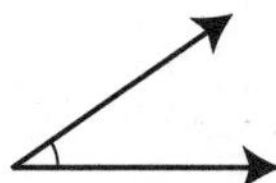

right angle
less than a right angle
greater than a right angle

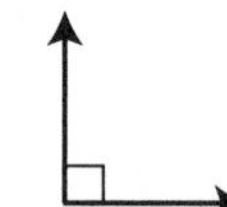

right angle
less than a right angle
greater than a right angle

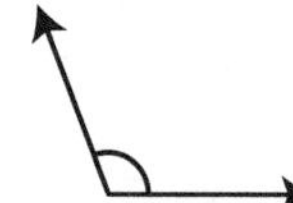

right angle
less than a right angle
greater than a right angle

Tell if each angle is a *right* angle, *less than* a right angle, or *greater than* a right angle.

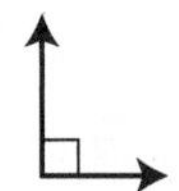

____

____

9 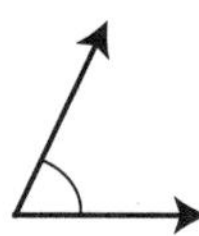

____

Check

Tell if each angle is a *right* angle, *less than* a right angle, or *greater than* a right angle.

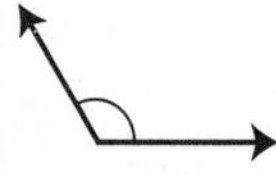

____

11 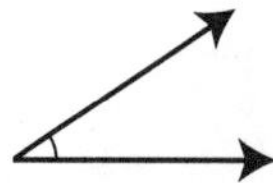

____

12 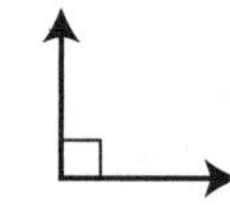

____

Skill 68 Grade 4

Compare Figures

15 Minutes

OBJECTIVE Identify congruent figures

MATERIALS 2 congruent triangles, 2 congruent rectangles

Begin by reminding students that congruent figures have the same size and shape.

As students look at the first model, say: **Look at the size and shape of the figures. Are these figures the same size?** Yes **Are they the same shape?** Yes

Explain that because they are the same size and the same shape, they are congruent.

Have students look at the second model.

Say: **Look at the figures. Are they the same shape?** No **How do you know?** The figures are not the same shape. One is a rectangle and one is a triangle.

For the third model, say: **The figures are the same shape. Are they congruent?** No, the triangles are the same shape but they are not the same size.

Display the congruent triangles, laying one upon the other to show that they are the same size and shape. Rotate the triangle.

Ask: **Are the triangles still congruent?** Yes, they are still the same size and shape.

Show the congruent rectangles in different positions and ask the same question. Remind students that congruent figures are the same size and shape, but are in different positions.

TRY THESE In Exercises 1–4 students look at size, shape, and position to test for congruence.

- **Exercise 1** Different shape.
- **Exercises 2 and 4** Different position.
- **Exercise 3** Different size.

PRACTICE ON YOUR OWN Review the examples at the top of the page. Have students explain why the figures in the first model are congruent, and those in the second are not congruent.

CHECK Determine if students can select congruent figures, and that they understand position does not affect congruency. Success is determined by 3 out of 3 correct responses.

Students who successfully complete the **Practice on Your Own** and **Check** are ready to move on to the next skill.

COMMON ERRORS

- Students may neglect to check both size and shape when determining congruency.
- Students may have difficulty determining if figures are congruent when they are in different positions.

Students who made more than three errors in the **Practice on Your Own**, or who were not successful in the **Check** section, may benefit from the **Alternative Teaching Strategy** on the next page.

Optional

Alternative Teaching Strategy
Model Congruent Figures

15 Minutes

OBJECTIVE Use models to show congruent figures

MATERIALS a template of various sized squares, a template of various sized rectangles, a template of various sized triangles, tracing paper

Display the templates for each plane figure and have students name the plane figure. Emphasize that the figures on each template are the same shape but different sizes.

Distribute the templates and tracing paper. Direct students to trace a square. Then have students trace a triangle.

Ask: **Are your traced figures the same shape?** No

Explain that the traced figures cannot be congruent because congruent figures have the same shape and size.

Then, have each student choose a figure from a template and trace it. When tracings are complete, ask: **Are your traced figure and the figure on the template the same size and shape?** Yes

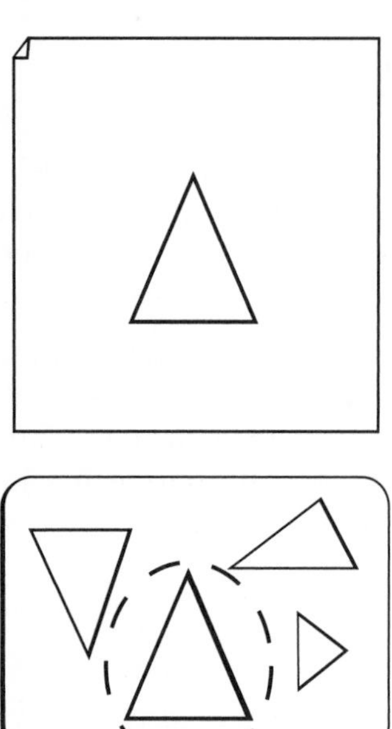

congruent

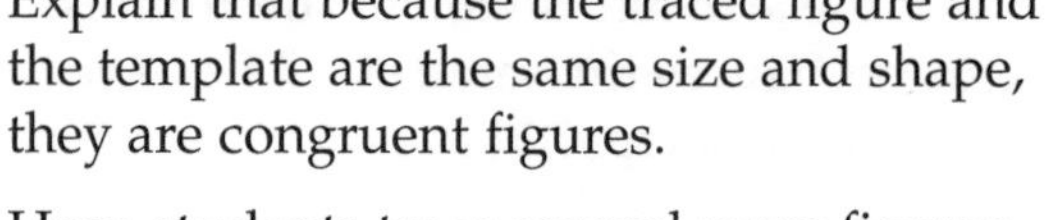

Explain that because the traced figure and the template are the same size and shape, they are congruent figures.

Have students trace several more figures. Choose a tracing and a figure on the template with the same shape but different size.

Ask: **These figures have the same shape. Are they congruent?** No **How do you know?** They are different sizes.

As students demonstrate an understanding of congruence, have them pick a template and a tracing and tell whether they are congruent. Have students explain why they are or are not congruent.

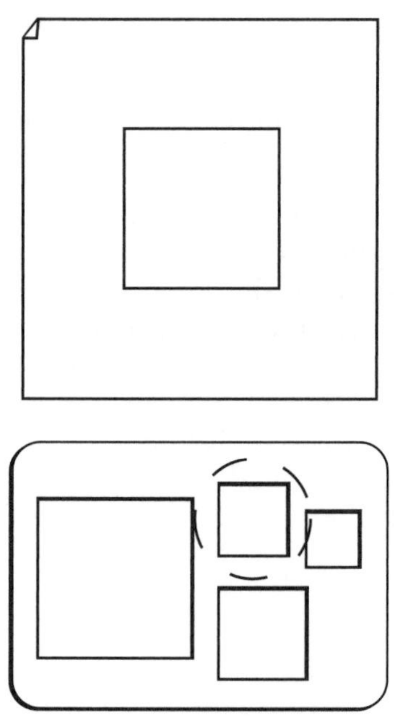

not congruent

Name ______________ Skill ______________

Compare Figures

Grade 4 Skill 68

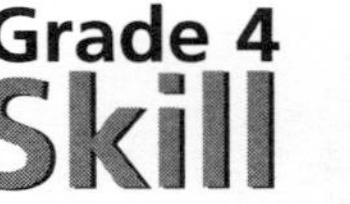

Figures that are the same shape and the same size are **congruent**.

Congruent

These squares are the same *size* and the same *shape*.

squares

They are congruent.

Not Congruent

These figures are *not* the same shape.

rectangle triangle

They are **not** congruent.

Not Congruent

These triangles are the same shape but *not* the same size.

triangles

They are **not** congruent.

Try These

Answer the questions about the figures. Write *yes* or *no*.

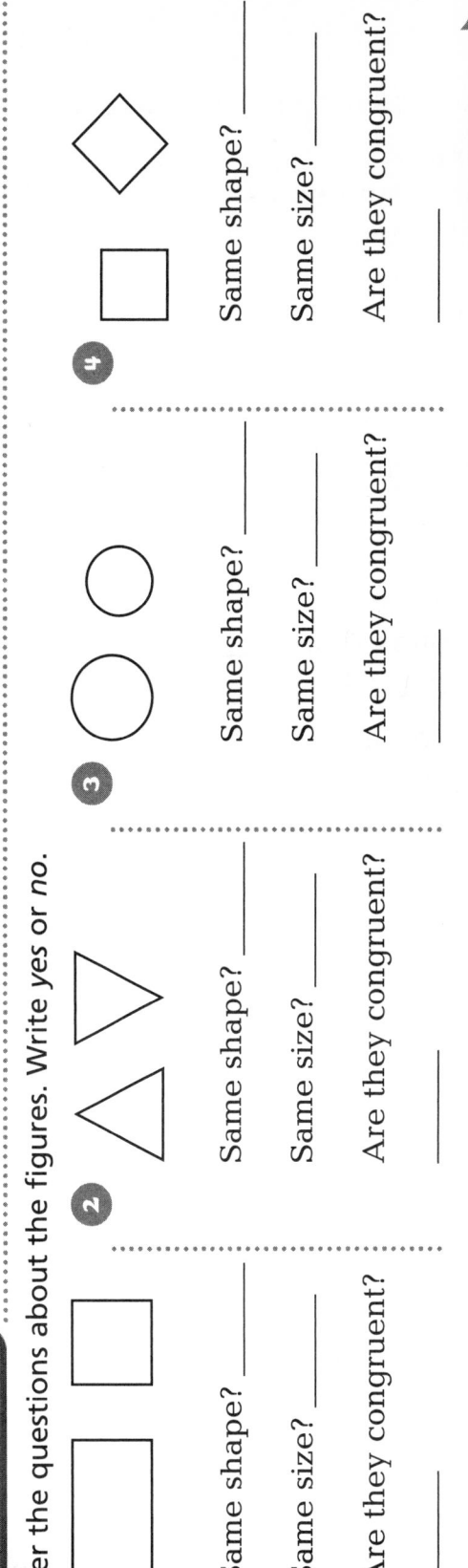

1. Same shape? ______
   Same size? ______
   Are they congruent? ______

2. Same shape? ______
   Same size? ______
   Are they congruent? ______

3. Same shape? ______
   Same size? ______
   Are they congruent? ______

4. Same shape? ______
   Same size? ______
   Are they congruent? ______

Go to the next side.

Name ______________________ Skill ______________________

Practice on Your Own

Skill 68

Think:
Congruent figures are the same *shape* and the same *size*.

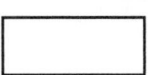

Remember:
Figures do not have to be in the same position to be congruent.

They are congruent. They are **not** congruent.

Answer the questions about the figures. Write *yes* or *no*.

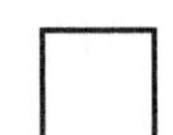

1. Same shape? _____
   Same size? ______
   Are they congruent? ______

2. Same shape? _____
   Same size? ______
   Are they congruent?______

3. Same shape? _____
   Same size? ______
   Are they congruent? ______

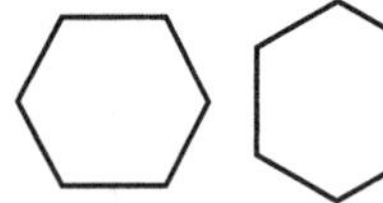

4. Are they congruent? _______

5. Are they congruent? _______

6. Are they congruent? _______

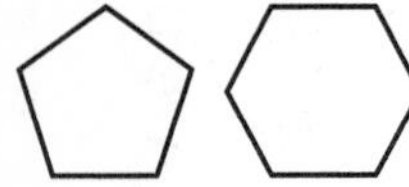
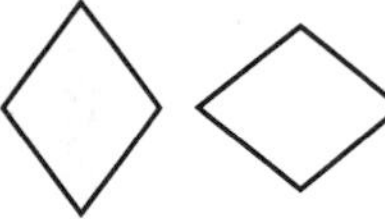

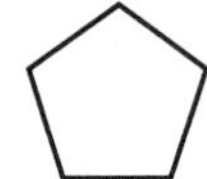
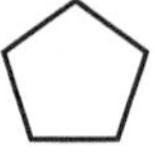

7. Congruent? _______

8. Congruent? _______

9. Congruent? _______

Check

Write *yes* or *no*.

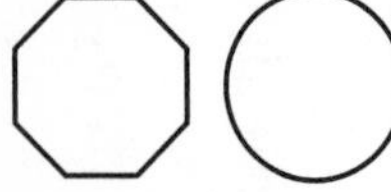

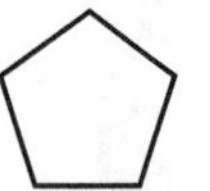
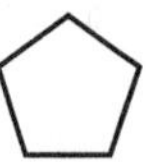

10. Congruent? _______

11. Congruent? _______

12. Congruent? _______

Skill 69 Grade 4 Identify Symmetric Figures

OBJECTIVE Identify lines of symmetry

Begin the skill by reminding students that if they fold a figure in half along a line so that both parts match, the fold is a line of symmetry.

Call the students' attention to Model A. Explain that a figure can have more than one line of symmetry.

Ask: **How many lines of symmetry does the rectangle have?** 2 **How do you know?** I can fold the figure in half two ways and the parts will match.

Ask similar questions for Model B.

For Model C, ask: **Do the two parts match?** No

Help students understand that matching parts should have the same size and shape. So the line shown on the circle is not a line of symmetry.

TRY THESE Exercises 1–3 prompt students to match parts to determine if the line is a line of symmetry.

- **Exercise 1** Line of symmetry.
- **Exercise 2** Not a line of symmetry.
- **Exercise 3** Line of symmetry.

PRACTICE ON YOUR OWN Work through the models at the top of the page. Have students identify the line of symmetry in each figure and tell why it is a line of symmetry.

CHECK Determine if students are able to identify a line of symmetry in a figure. Success is indicated by 3 out of 3 correct responses.

Students who successfully complete the **Practice on Your Own** and **Check** are ready to move on to the next skill.

COMMON ERRORS

- Students may identify a line that does not divide the figure into two matching parts as a line of symmetry.
- Students may not recognize that a figure has more than one line of symmetry.

Students who made more than two errors in the **Practice on Your Own**, or who were not successful in the **Check** section, may benefit from the **Alternative Teaching Strategy** on the next page.

Optional

Alternative Teaching Strategy

15 Minutes

Use Models to Identify Lines of Symmetry

OBJECTIVE Fold figures to identify one or more lines of symmetry

MATERIALS centimeter square dot paper (Teacher Resource 40), scissors

Remind students when they fold a figure along a line of symmetry, both parts match.

Provide students with the centimeter square dot paper. Have them outline and cut out a 5 by 5 square.

Say: **Begin by folding the square in half from top to bottom. Do both parts match?** yes **Does the square have at least one line of symmetry?** yes

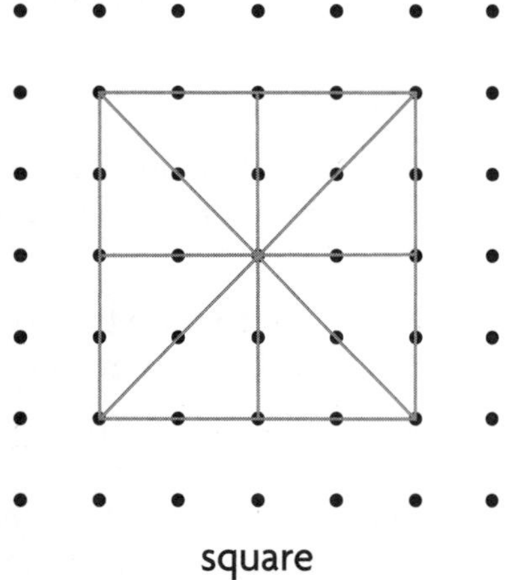

square

Have students trace over the lines of symmetry on the dot paper as they find them.

Continue: **Suppose you fold the square in half from one corner to the opposite corner. Is the fold line a line of symmetry?** yes

Guide students as they discover the two diagonal lines of symmetry for the square.

Repeat this activity for a 5 by 7 rectangle. When students show an understanding of lines of symmetry in a square and rectangle, have them look for lines of symmetry in an equilateral triangle 3, a trapezoid 1, and a parallelogram 0. Ask students each time to explain how they know if a figure has a line of symmetry.

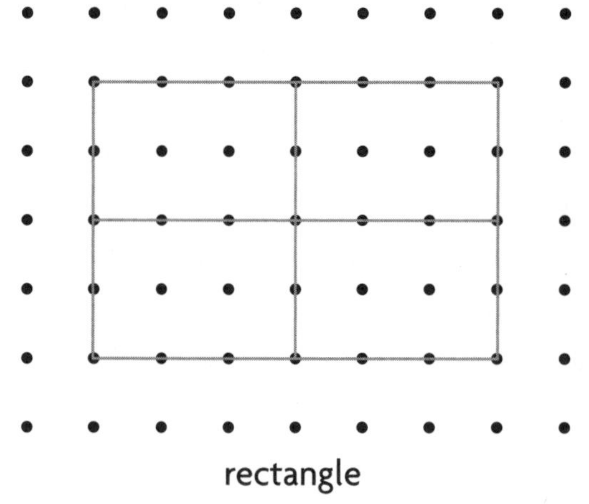

rectangle

Name ______________________ Skill ______________________

Grade 4 Skill 69

Identify Symmetric Figures

A line of symmetry is an imaginary line that divides a figure in half.
If you fold a figure along a line of symmetry, both parts match. A figure can have more than one line of symmetry.

Model A

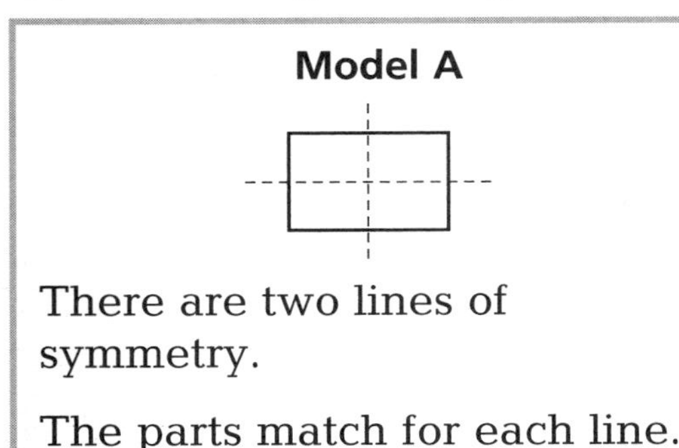

There are two lines of symmetry.

The parts match for each line.

Model B

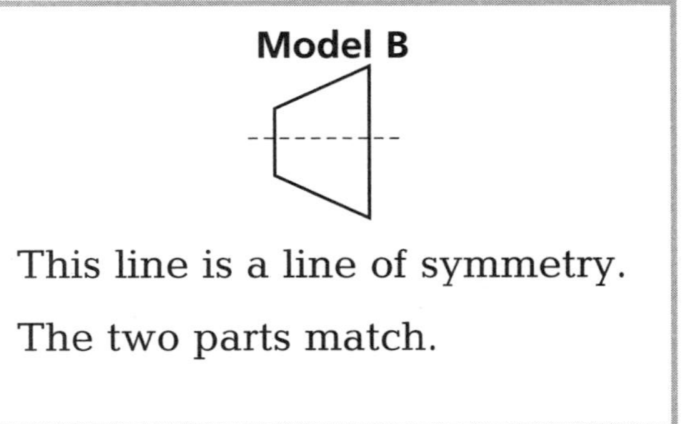

This line is a line of symmetry.

The two parts match.

Model C

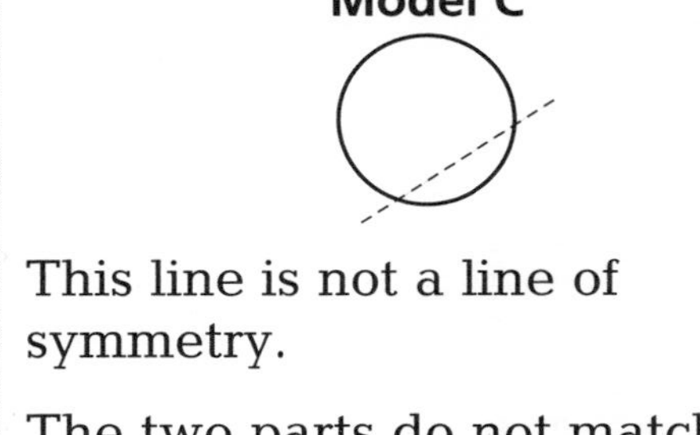

This line is not a line of symmetry.

The two parts do not match.

Try These

Complete.

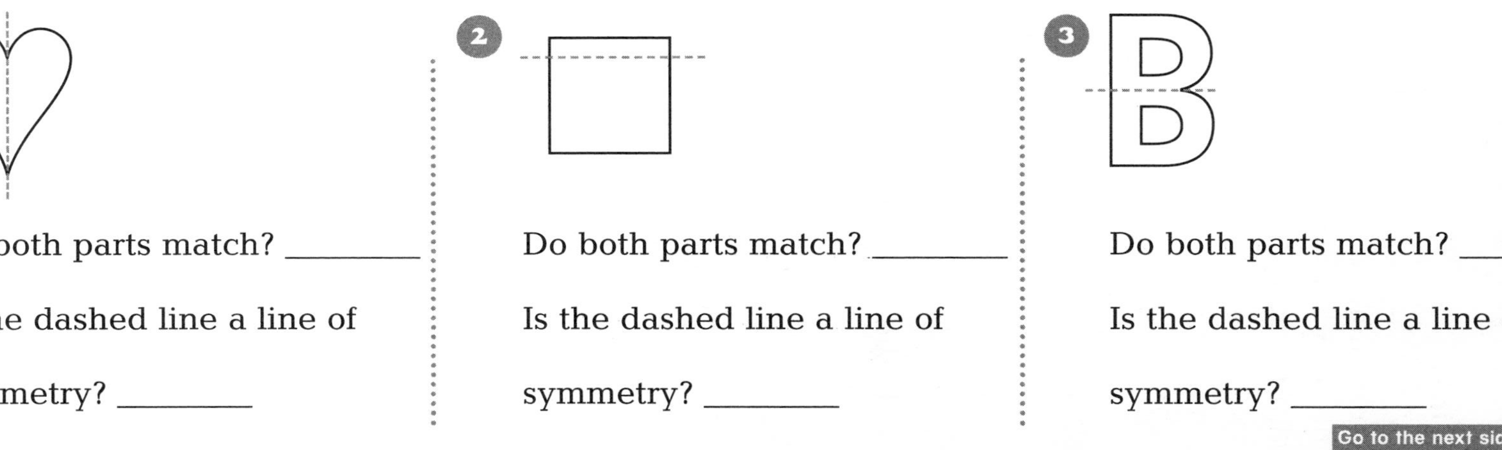

1. Do both parts match? ________

   Is the dashed line a line of symmetry? ________

2. Do both parts match? ________

   Is the dashed line a line of symmetry? ________

3. Do both parts match? ________

   Is the dashed line a line of symmetry? ________

Go to the next side.

Name ______________________ Skill ______________________

Practice on Your Own

Think:
If you fold a figure along a line of symmetry, both parts match.

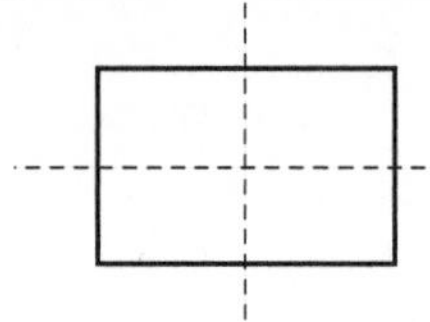

lines of symmetry

A figure can have more than one line of symmetry.

not a line of symmetry

Complete.

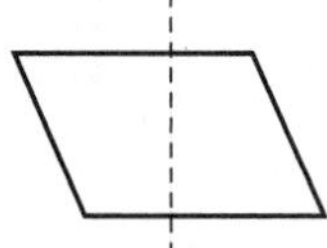

Do both parts match? ______

Is the dashed line a line of symmetry? ______

2

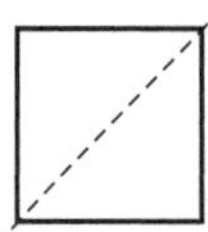

Do both parts match? ______

Is the dashed line a line of symmetry? ______

3

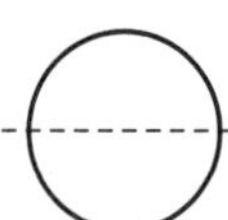

Do both parts match? ______

Is the dashed line a line of symmetry? ______

4

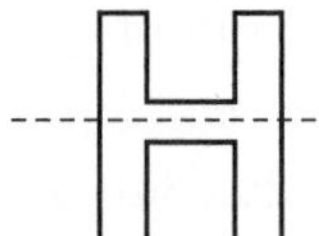

Is the dashed line a line of symmetry? ______

5

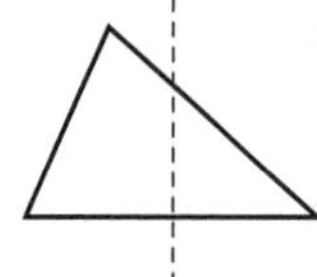

Is the dashed line a line of symmetry? ______

6

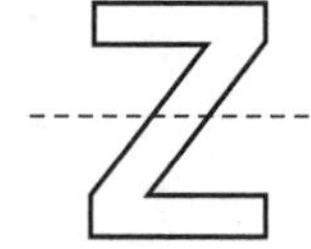

Is the dashed line a line of symmetry? ______

Decide if you can draw at least one line of symmetry. Write *yes* or *no*.

7

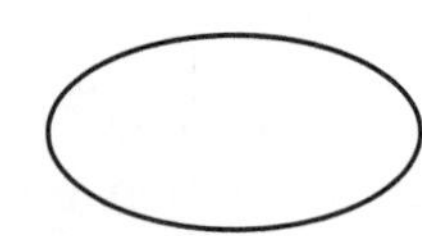

______

8 ______

9 ______

Check

Can you draw at least one line of symmetry? Write *yes* or *no*. Then draw the line, if possible.

10

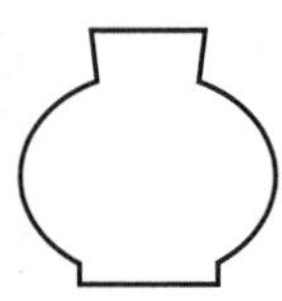

______

11

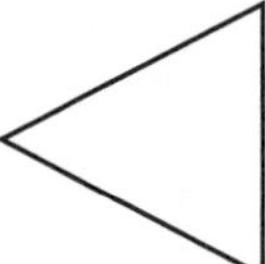

______

12

______

Skill 70 Grade 4

Sort Polygons

OBJECTIVE Count sides and angles to name polygons

Begin by reviewing the definition of *polygon*. To explain the meaning of a *closed figure*, tell students that if they drew one of the plane figures without lifting the pencil from the paper, they would end at the same point at which it started. Then compare an open figure to a closed figure by displaying an example of both.

Review the definitions of sides and angles. Explain that they may recognize some of the polygons, and may not recognize some of the others. Remind students that any closed figure with straight sides is a polygon, even if the sides are not the same length.

Ask: **How many sides does a triangle have?** 3 **How many angles?** 3

Repeat the questions as the students examine each plane figure.

Ask: **What do you notice about the sides and angles for each figure?** **Each figure has the same number of sides as it has angles.**

Explain that polygons are named by the number of sides and angles they have. You may wish to point out how the prefixes in the names suggest the number of sides and angles.

Prefix	Meaning
tri	three
quadri	four
penta	five
hexa	six
octa	eight

TRY THESE Exercises 1–3 ask the students to count the sides and angles before naming polygon.

- **Exercise 1** A square.
- **Exercise 2** A triangle.
- **Exercise 3** A hexagon.

PRACTICE ON YOUR OWN Review the five polygons with students. Remind them that counting the sides and angles can help them name the polygon.

CHECK Determine if students know how to count the number of sides and angles correctly, and know the names of the five polygons.

Success is indicated by 3 out of 3 correct responses.

Students who successfully complete the **Practice on Your Own** and **Check** are ready to move on to the next skill.

COMMON ERRORS

- Students may count the number of sides or angles incorrectly, by counting a side or angle twice or by skipping a side or angle.
- Students may count the sides and angles correctly, but may not remember the correct name of the polygon.

Students who made more than 3 errors in the **Practice on Your Own**, or who were not successful in the **Check** section, may benefit from the **Alternative Teaching Strategy** on the next page.

Optional

Alternative Teaching Strategy

15 Minutes

Sort Polygons

OBJECTIVE Use models to identify the number of sides and angles in a polygon

MATERIALS several examples of each type of polygon, regular and not regular, markers and paper

Recall the definition of a polygon as a closed figure with straight sides. Display a closed and an open figure. Then begin the activity by having students trace a regular or equilateral triangle.

Suggest that they use markers to draw each side of the triangle in a different color. Then have students count and record the number of angles in the triangle.

Ask: **How many sides does a triangle have?** 3 **How many angles does a triangle have?** 3 **Does this polygon have the same number of sides as angles?** Yes.

Have students choose another regular polygon to trace and to repeat the counting process. Continue the activity with quadrilaterals, pentagons, hexagons, and octagons.

When students show understanding, introduce some polygons that are not regular. Have students use their results to verify that polygons are named by the number of sides, and that they have the same number of sides as they have angles.

Name ______________________ Skill ______________________

Sort Polygons

Grade 4 Skill 70

A **polygon** is a closed plane figure with *straight* sides. A *closed figure* begins and ends at the same point. A polygon is named by the number of sides and angles it has.

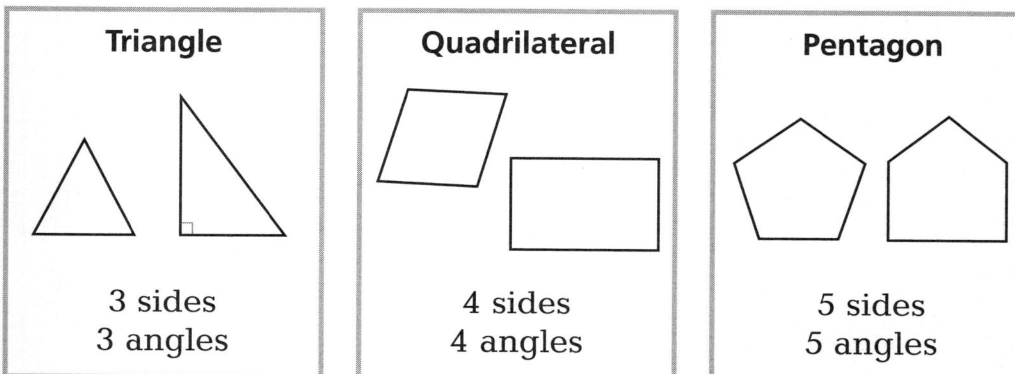

Triangle

3 sides
3 angles

Quadrilateral

4 sides
4 angles

Pentagon

5 sides
5 angles

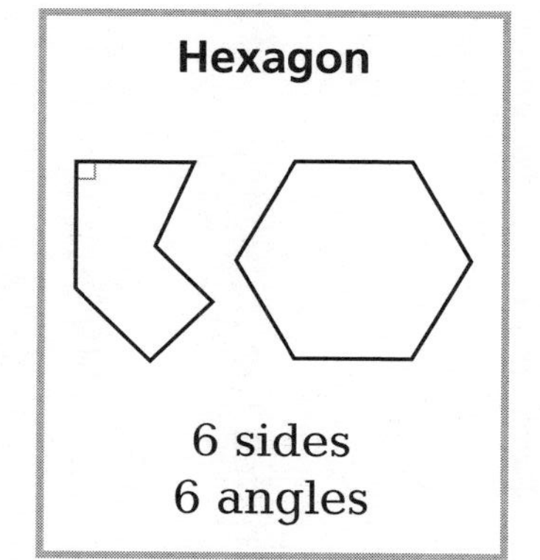

Hexagon

6 sides
6 angles

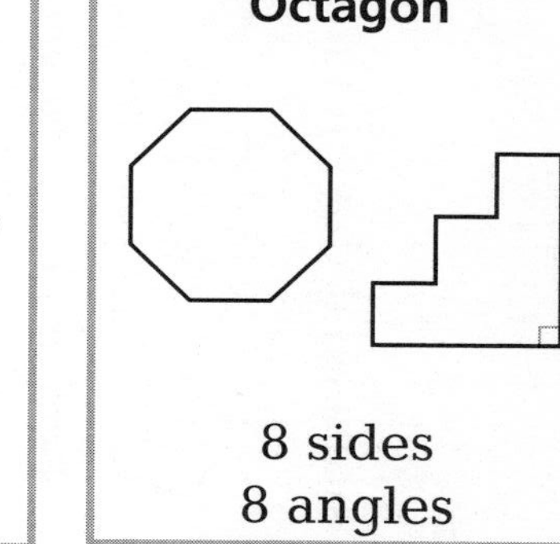

Octagon

8 sides
8 angles

A polygon has the same number of sides and angles.

Try These

Tell how many sides and angles. Then name the polygon.

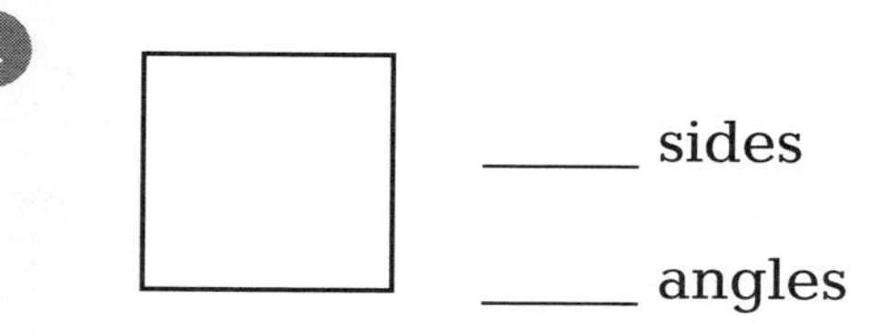

1. ______ sides
   ______ angles
   ____

2. ______ sides
   ______ angles
   ____

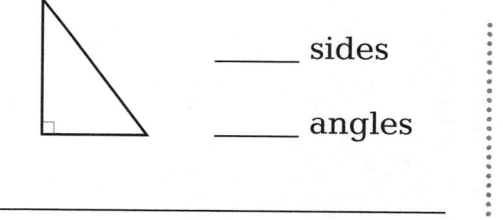

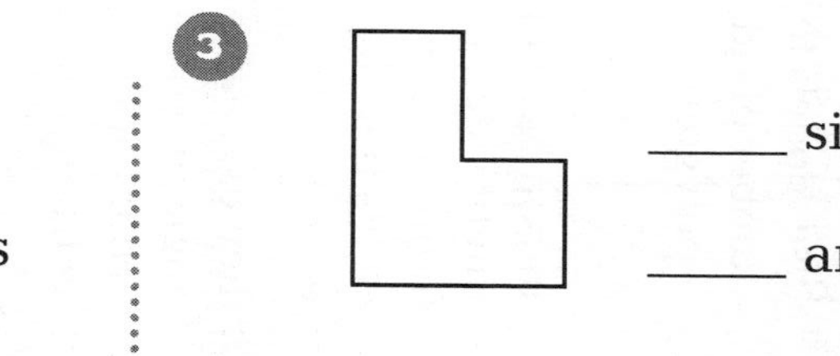

3. ______ sides
   ______ angles
   ____

Go to the next side.

Name ______________________ Skill ______________________

Practice on Your Own

Skill 70

A polygon has the same number of sides as it does angles.

Polygons

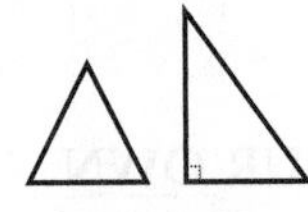
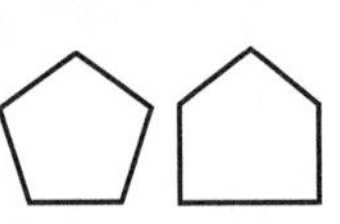
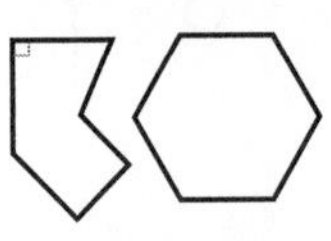
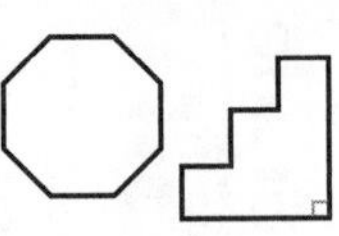

	Triangle	Quadrilateral	Pentagon	Hexagon	Octagon
Sides	3	4	5	6	8
Angles	3	4	5	6	8

Tell how many sides and angles. Then name the polygon.

1 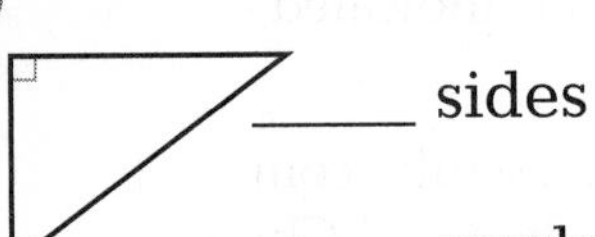 _____ sides _____ angles ______________

2  _____ sides _____ angles ______________

3  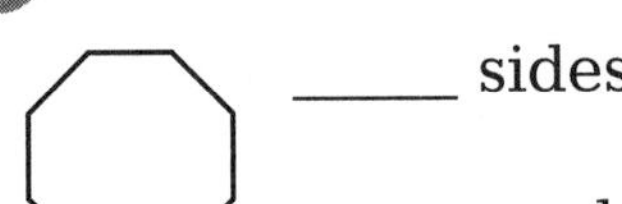 _____ sides _____ angles ______________

Name the polygon.

4 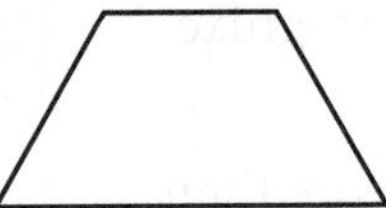 ______________

5 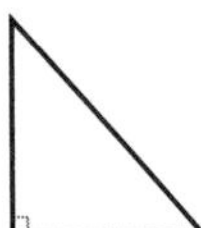 ______________

6 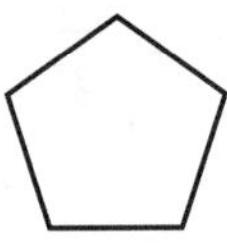 ______________

7 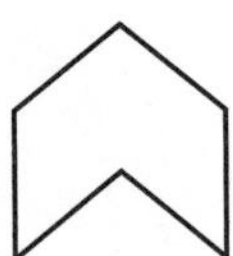 ______________

8 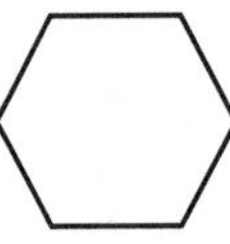 ______________

9 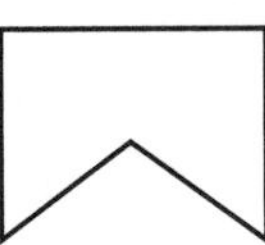 ______________

Check

Name the polygon.

10 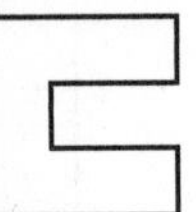 ______________

11 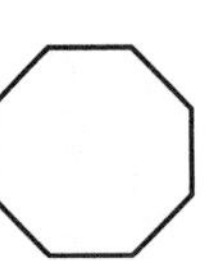 ______________

12 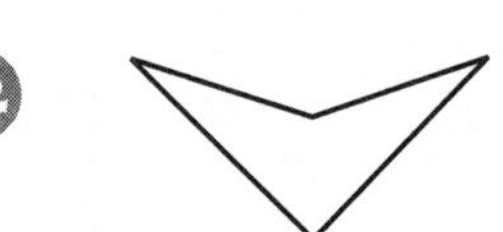 ______________

Skill 71 Grade 4

Identify Plane Figures

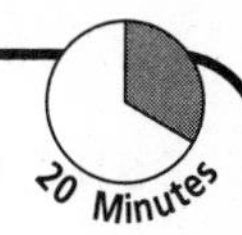

OBJECTIVE Identify plane figures: circles, squares, rectangles, and triangles

Begin by discussing the definition of a plane figure at the top of the page. Remind students that a closed figure begins and ends at the same point.

Have students name any figures they know on the card. Then ask them to name some objects that have the same shapes as the figures on the card, for example: coins, windows, pieces of paper, and traffic signs.

Ask a student to read the definition of each figure.

Then ask: **Which plane figure is formed by a curved line?** the circle **Which plane figures have four sides and four angles?** the squares and rectangles **How is a square different from a rectangle?** Only squares have four equal sides. **How are the three figures in the last example alike?** They have three sides and three angles. **How are they different?** Different shapes

TRY THESE In Exercises 1–4, students choose the figure that matches the name.

- **Exercise 1** square
- **Exercise 2** circle
- **Exercise 3** rectangle
- **Exercise 4** triangle

PRACTICE ON YOUR OWN Review the plane figures at the top of the page. Before students complete the exercises, remind them to look carefully to make sure the figure they choose is a closed figure, or has the correct number of sides and angles.

CHECK Determine if students know how to identify the plane figures based on their attributes. Success is indicated by 4 out of 4 correct responses.

Students who successfully complete **Practice on Your Own** and **Check** are ready to move on to the next skill.

COMMON ERRORS

- Students may confuse square figures with rectangles that are not squares.
- Students may not recognize a plane figure that is rotated.

Students who made more than three errors in the **Practice on Your Own**, or who were not successful in the **Check** section, may benefit from the **Alternative Teaching Strategy** on the next page.

Optional

Alternative Teaching Strategy

Use Models to Identify Plane Figures

OBJECTIVE Identify plane figures: circles, squares, rectangles, and triangles

MATERIALS Pre-cut squares and rectangles, construction paper, scissors, pencils

Have students look at the squares and rectangles. Point out the square and tell students that all four sides are equal.

Have students fold the squares diagonally, corner to corner, and show that the edges of the square match exactly.

Then have students fold the rectangles diagonally, corner to corner, and show that the edges do not match. Tell students that this is because all sides of the rectangle are not equal. Have them fold the figure in half to show that opposite sides are equal.

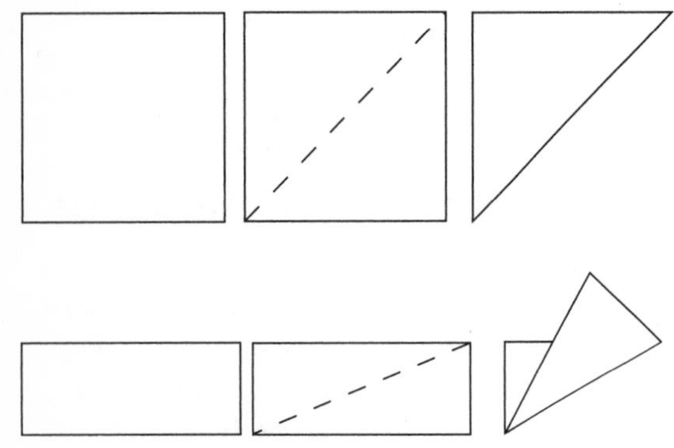

Give students construction paper and have them draw a large triangle and cut it out. Choose two students' triangles to compare and point out that although they may look different, they are triangles if they have three sides and three angles. Hold one model in various positions and ask students if it is still a triangle.

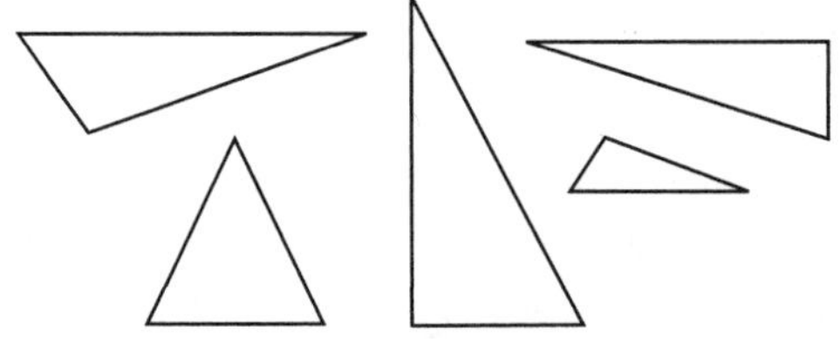

Continue with other plane figures. The goal is to have students identify the plane figures by recognizing their attributes.

Conclude by having each student draw a figure and explain its attributes to the group.

Name ______________________ Skill ______________________

Grade 4 **Skill 71**

Identify Plane Figures

A **plane figure** is a closed figure in a plane.
A plane figure is formed by lines that are curved, straight, or both.

A *closed figure* begins and ends at the same point.

A *circle* is made up of points that are the same distance from the center point.

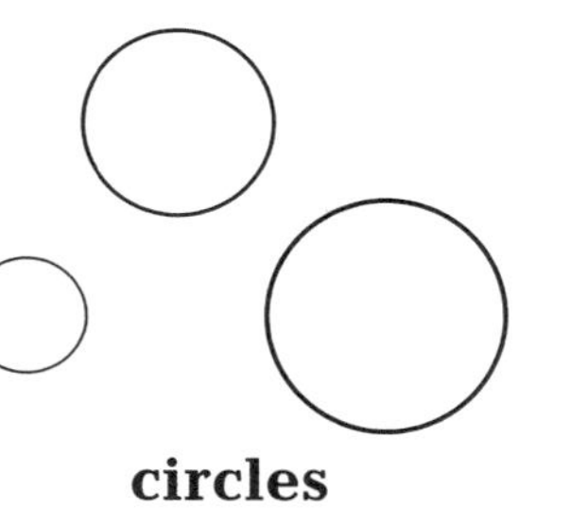

circles

A *square* is a plane figure with 4 equal sides and 4 right angles.

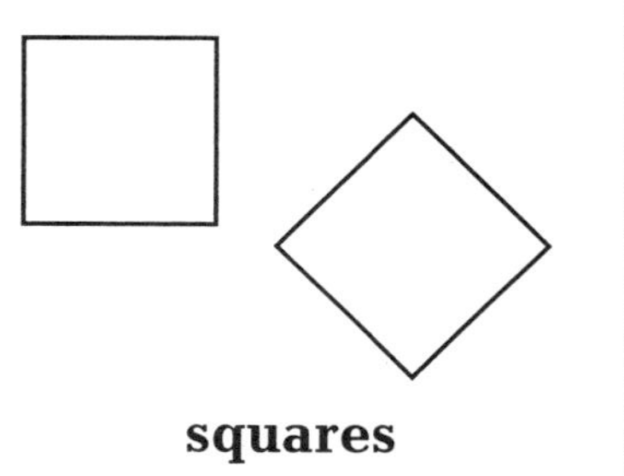

squares

A *rectangle* is a plane figure with opposite sides that are equal and 4 right angles.

rectangles

A *triangle* is a plane figure with 3 sides and 3 angles.

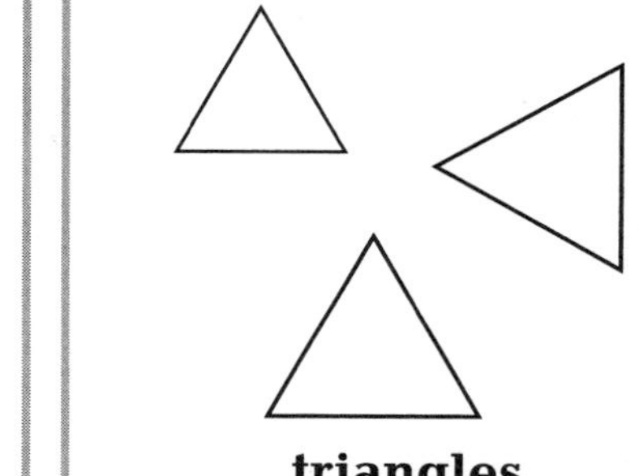

triangles

Try These

Circle the figure that matches the name.

1. **square**

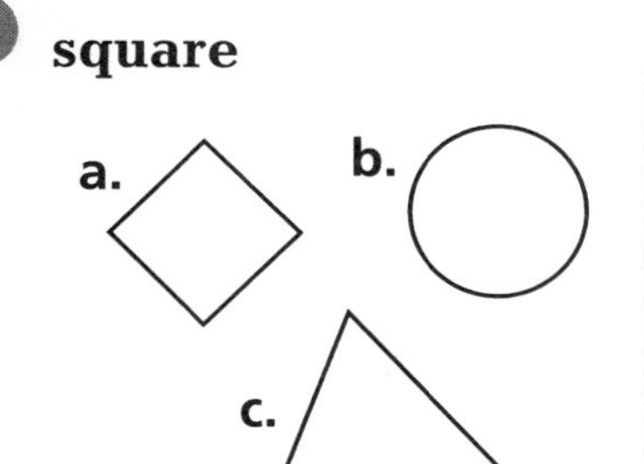

2. **circle**

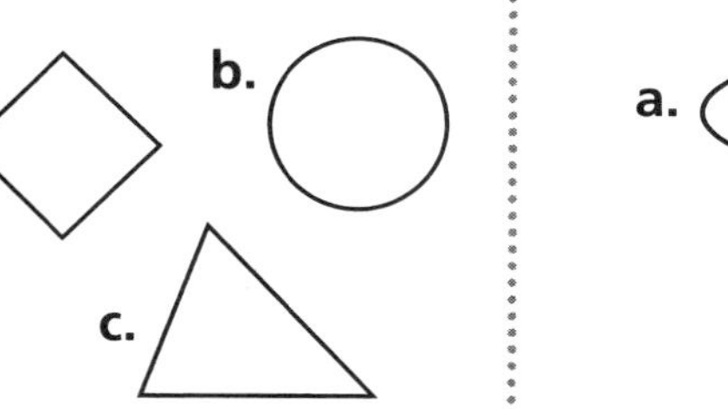

a. b. c.

3. **rectangle**

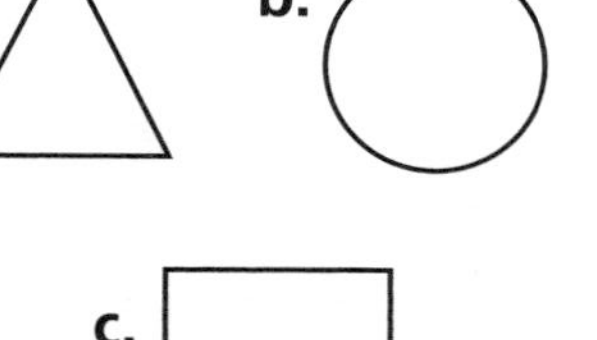

a. b. c.

4. **triangle**

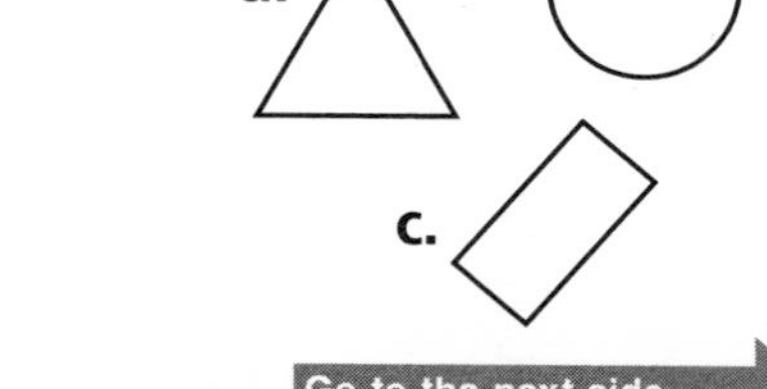

a. b. c.

Go to the next side.

Name ____________________ Skill ____________________

Practice on Your Own

Skill 71

Study the figures.

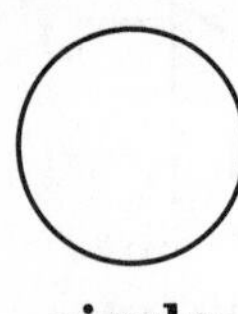
circle

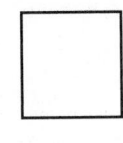

square

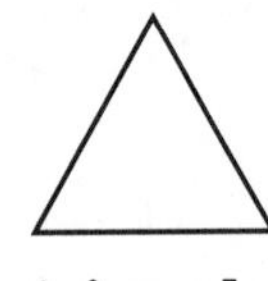
triangle

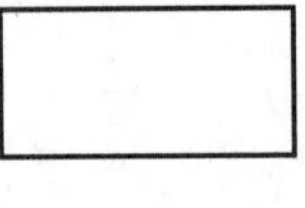
rectangle

Circle the figure that matches the name.

1. square

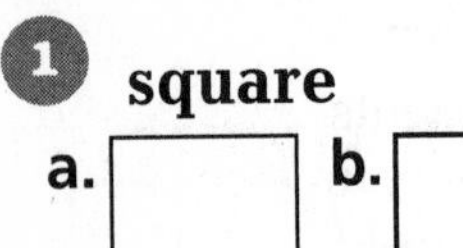

a. b. c.

2. circle

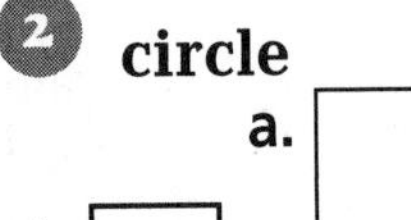
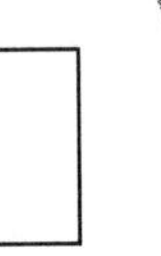

a. b. c.

3. rectangle

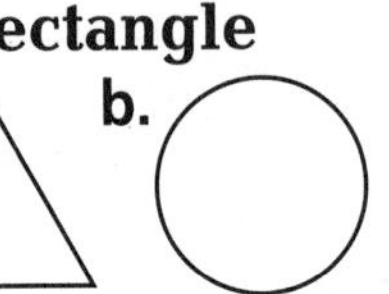

a. b. c.

4. triangle

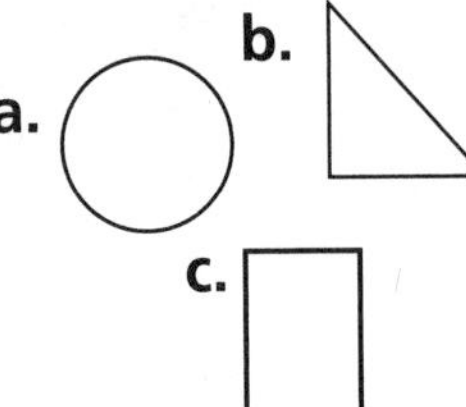

a. b. c.

5. square

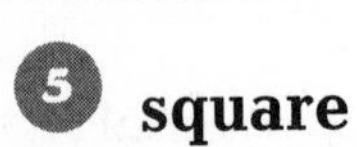
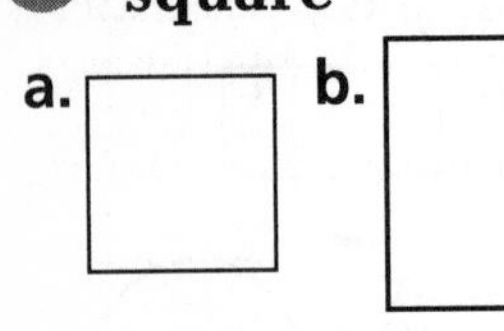

a. b. c.

6. circle

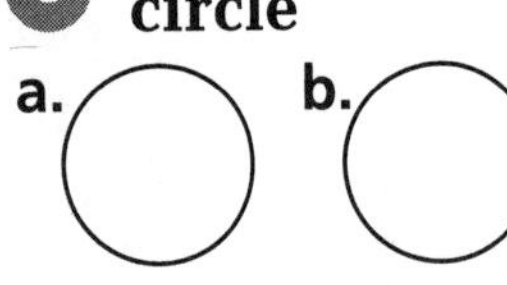
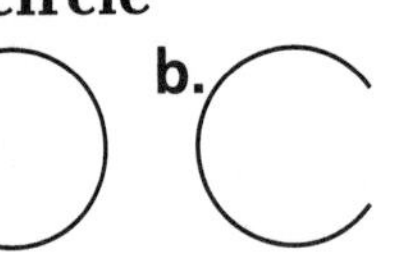

a. b. c.

7. rectangle

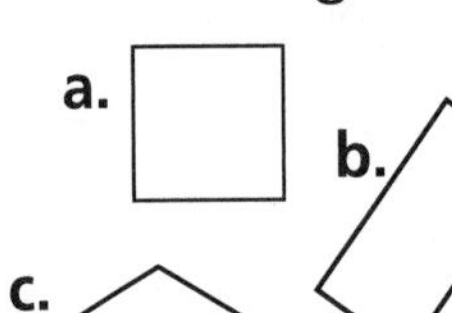

a. b. c.

8. triangle

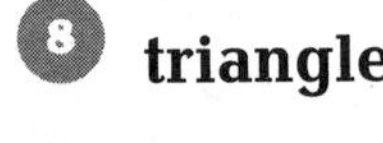
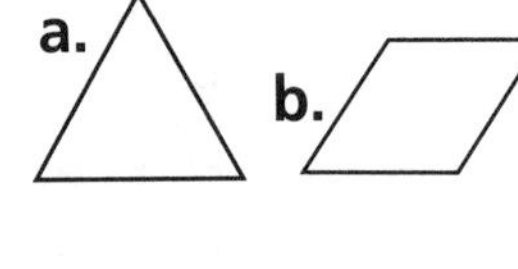

a. b. c.

Write the name of each figure.

9.  ____________

10. 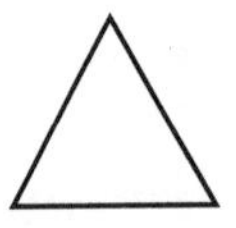 ____________

11. 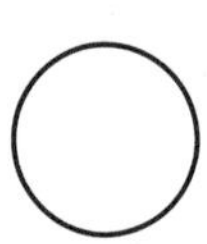 ____________

12.  ____________

Check

Write the name of each figure.

13. 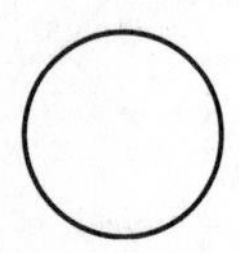 ____________

14. 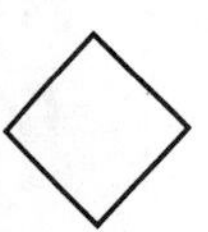 ____________

15. 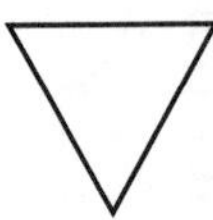 ____________

16. 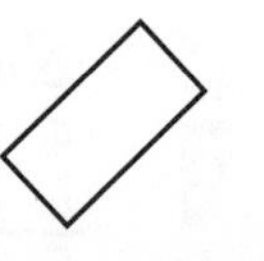 ____________

Skill 72 Grade 4

Find Perimeter (by Counting)

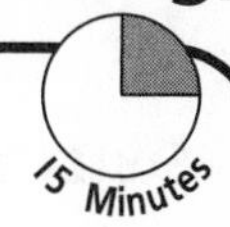

OBJECTIVE Find the perimeter of a figure

Begin by reviewing the definition of perimeter as the distance around a figure. Explain that the number of units around a figure can be counted to find the perimeter.

As students look at the first figure, say: **This figure has 4 sides, each side is 1 unit long. Count the units to find the perimeter.** 4 units

For the second figure ask: **How many sides does the figure have?** 4 **Count the units. What is the perimeter of the figure?** 12 units

Then ask: **How can you find the perimeter without counting each unit?** Count the units on each side of the figure and find the sum.

Discuss how this method is easier and more accurate, especially when the perimeters are greater or the figures more complex.

TRY THESE Exercises 1–3 model the types of exercises students will find on the **Practice on Your Own** page.

- **Exercise 1** Figure with 4 sides.
- **Exercise 2** Figure with 4 sides.
- **Exercise 3** Figure with 6 sides.

PRACTICE ON YOUR OWN Review the example at the top of the page. Ask students how many sides the figure has, and how many addends are in the addition sentence. Remind students to check the number of sides and the number of addends as they work through the page.

CHECK Determine if students understand that to find perimeter they count the units on each side of the figure and find the sum. Success is indicated by 3 out of 3 correct responses.

Students who successfully complete the **Practice on Your Own** and **Check** are ready to move on to the next skill.

COMMON ERRORS

- Students may not count the number of units correctly.
- Students may not include all the sides in the calculation.
- Students may not add correctly.

Students who made more than two errors in the **Practice on Your Own**, or who were not successful in the **Check** section, may benefit from the **Alternative Teaching Strategy** on the next page.

Optional

15 Minutes

Alternative Teaching Strategy
Use Models to Find Perimeter

OBJECTIVE Find the perimeter of a figure using grid paper

MATERIALS tiles, centimeter grid paper

Students may benefit from working in pairs or small groups. Explain that the activity is about finding perimeter. Explain that perimeter is the distance around a figure.

Using tiles, form a 3 × 5 unit rectangle. Then point out that to find the perimeter of the tile figure, you count the number of units around the outside of the figure. Demonstrate how the side of a tile represents a unit, and count each side of a corner unit as one unit each. Have students count the units as you point to each unit.

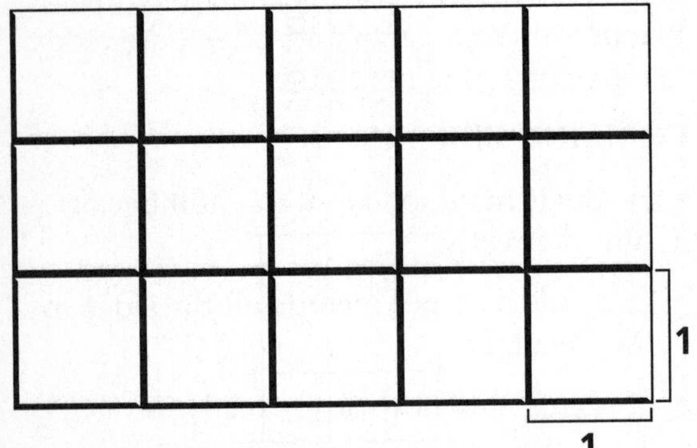

Distribute the grid paper. Have students draw a 5 × 4 unit rectangle. Have students count the units to find the perimeter.

What is the perimeter? 18 units

Then suggest that they can count the units on each side and add them to find the perimeter.

How many units are on each side? 5, 4, 5, 4

What addition sentence can you write?
5 + 4 + 5 + 4 = 18 units

Repeat the activity with similar examples.

Caution students to check that they have counted the units correctly, included all the sides in their addition, and added correctly.

When the students show understanding of how to find the perimeter, have each student outline a figure on the grid paper and exchange figures with another student. After they have found the perimeter, have students describe their methods.

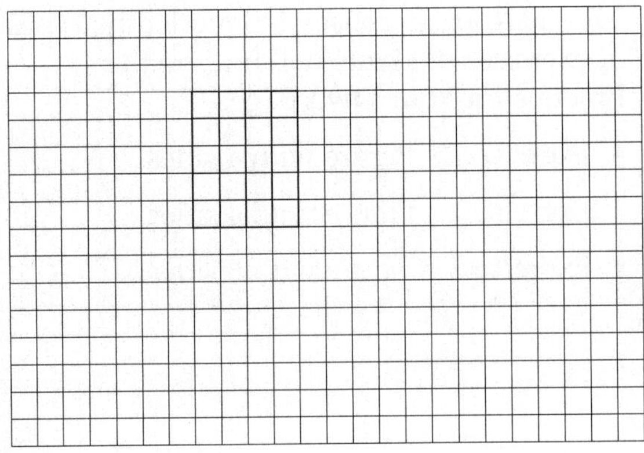

Name ______________________ Skill ______________________

Find Perimeter (by Counting)

Grade 4 **Skill 72**

The **perimeter** is the distance around a figure. To find the perimeter, count the number of units around the figure.

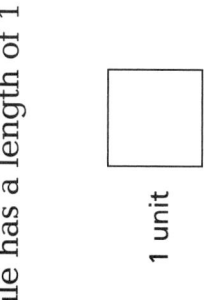

Each side of the square tile has a length of 1 unit.

1 unit

1 unit

The perimeter is 4 units.

Count the units on each side.

Side A: 4 units
Side B: 2 units
Side C: 4 units
Side D: 2 units

Think:
4 + 2 + 4 + 2 = 12

The perimeter is 12 units.

The figure has many sides. Check that you have counted all the sides.

Side A: 3 units
Side B: 1 unit
Side C: 1 unit
Side D: 1 unit
Side E: 2 units
Side F: 2 units

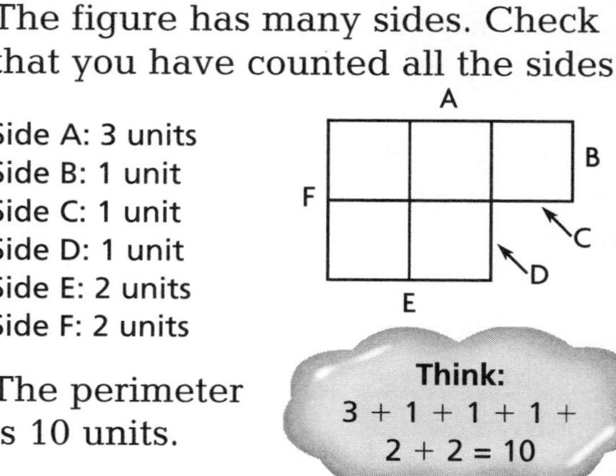
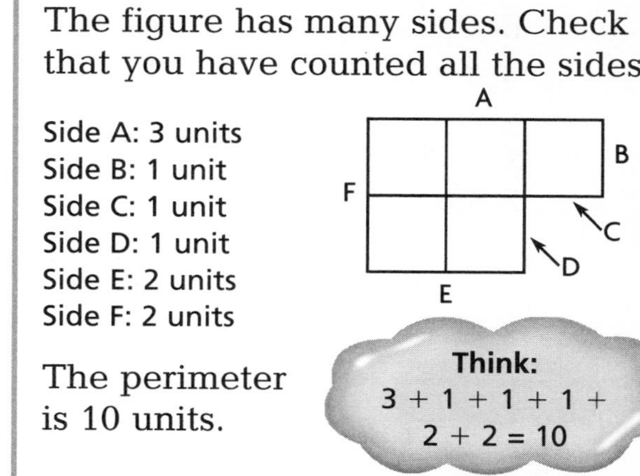

The perimeter is 10 units.

Try These

Complete.

1. Side A: 3 units
Side B: 2 units
Side C: 3 units
Side D: 2 units

The perimeter is 10 units.

2. 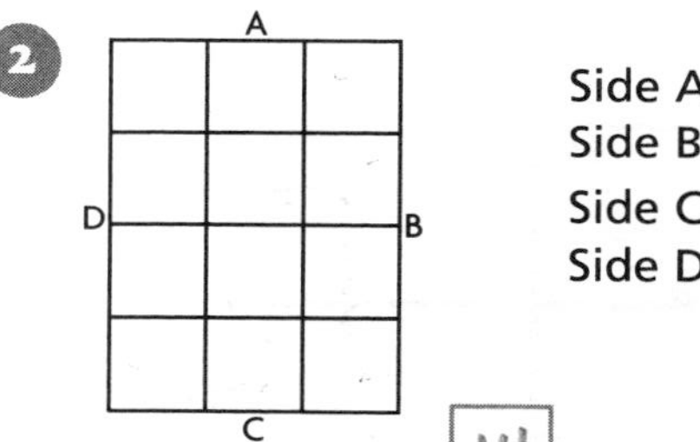

Side A: 3 units
Side B: 4 units
Side C: 3 units
Side D: 4 units

The perimeter is 14 units.

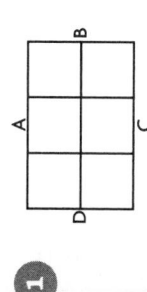

3. 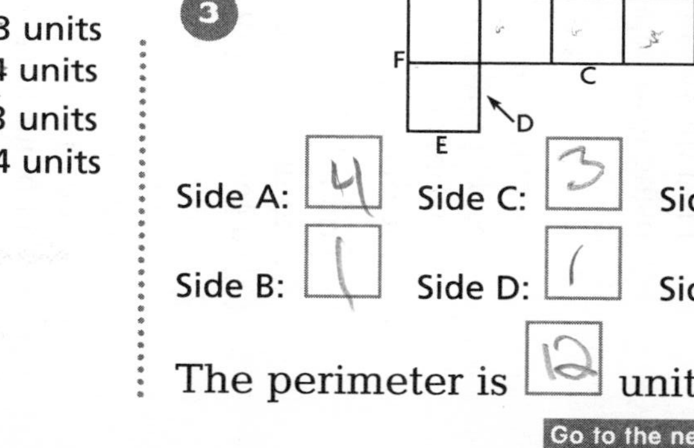

Side A: 4 Side C: 3 Side E: 1

Side B: 1 Side D: 1 Side F: 2

The perimeter is 12 units.

Go to the next side.

Name ________________________ Skill ________________________

Practice on Your Own

Think:
Count the units on each side.
3 + 3 + 3 + 3 = 12

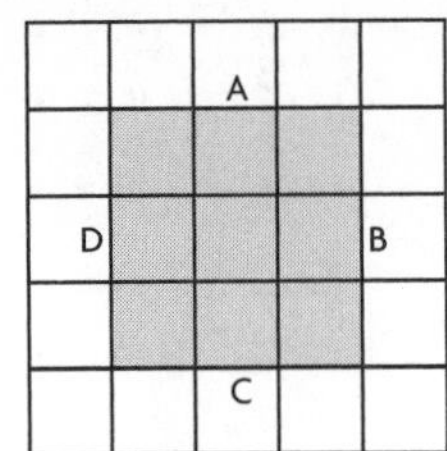

Side A: 3 units
Side B: 3 units
Side C: 3 units
Side D: 3 units

The perimeter is 12 units.

Find the perimeter.

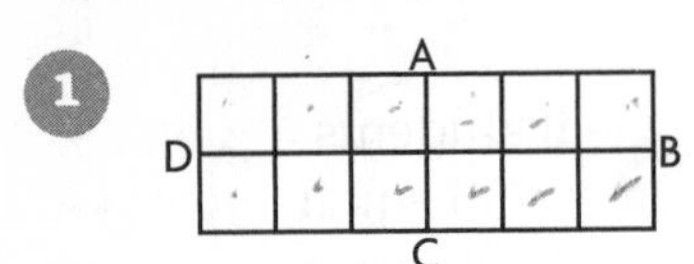

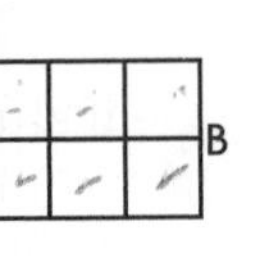

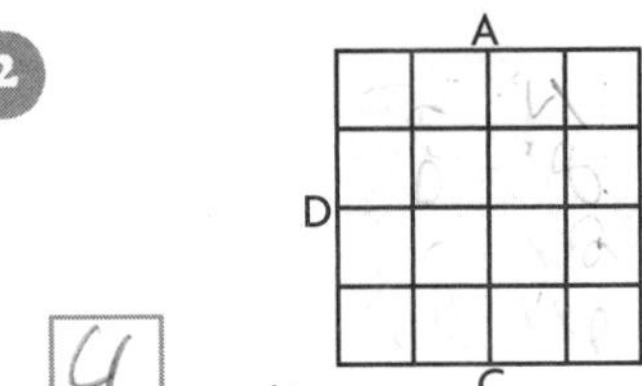

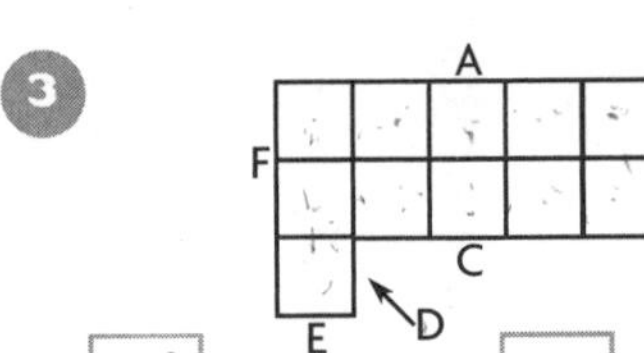

1
A:  6 units
B: 6 units
C: 6 units
D: 6 units
Perimeter: 24 units

2
A: 4 units
B: 4 units
C: 4 units
D: 4 units
Perimeter: 16 units

3
A: 5 units E: 3 units
B: 2 units F: 2 units
C: 4 units
D: 1 units
Perimeter: 17 units

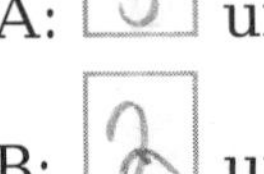

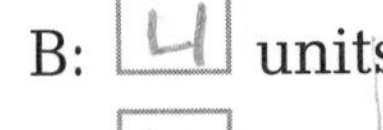
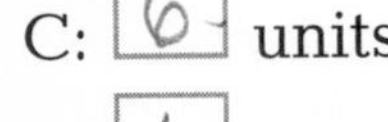
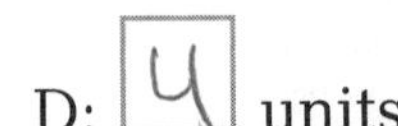

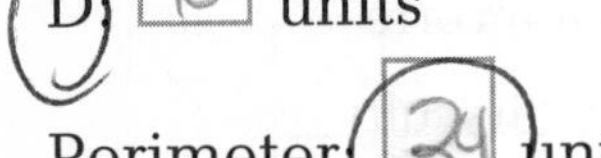
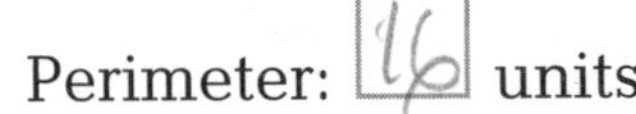
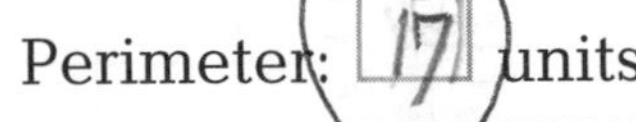

4

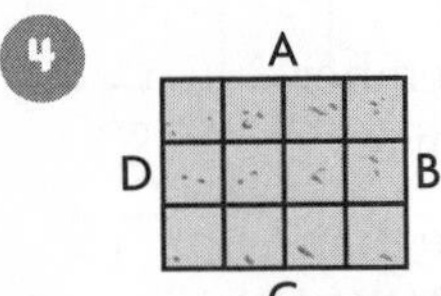

Perimeter: 12 units

5

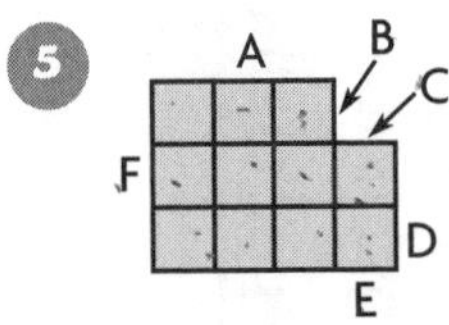

Perimeter: 15 units

6

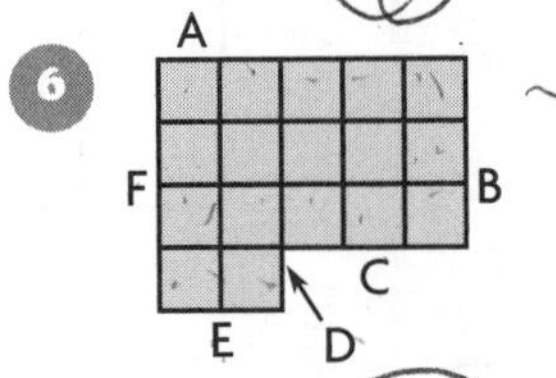

Perimeter: 17 units

Check

Find the perimeter.

7

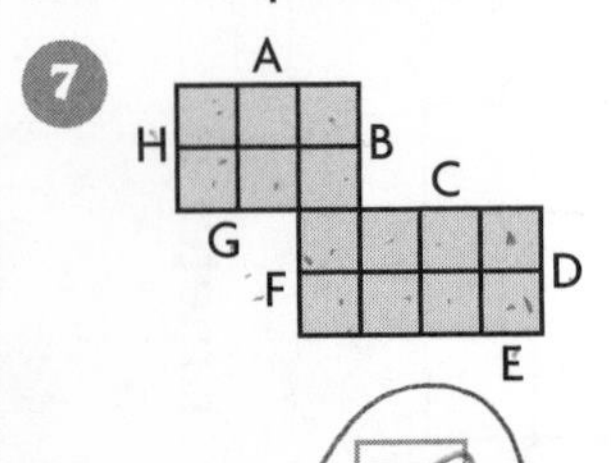

Perimeter: 22 units

8

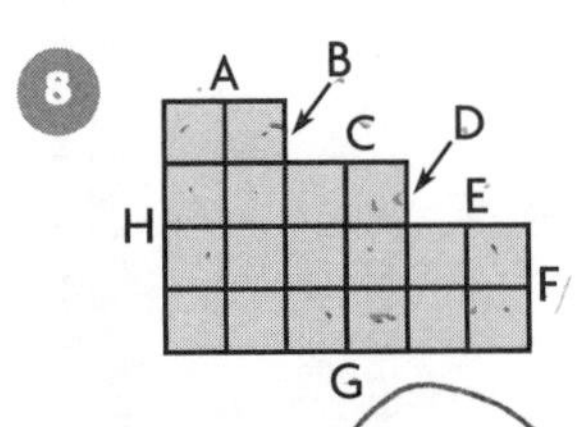

Perimeter: 15 units

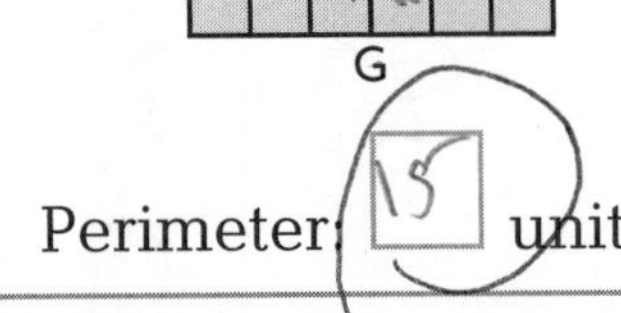

9

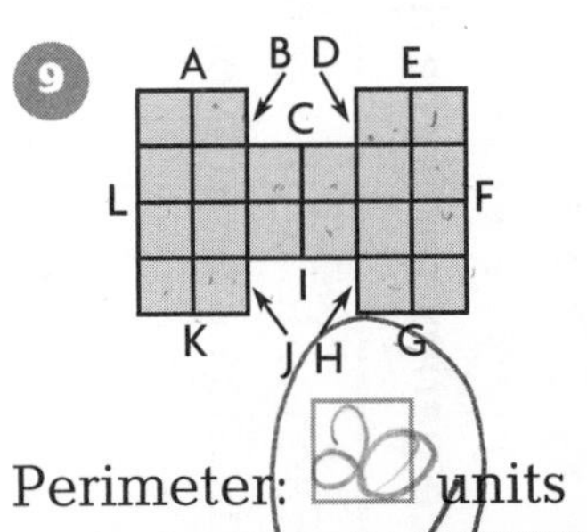

Perimeter: 20 units

Skill 73 Grade 4

Find Area (by Counting)

OBJECTIVE Find area by counting square units

You may wish to begin by reviewing that the area is the number of square units needed to cover a flat surface. Suggest that students think of each square as 1 square unit. Explain that they are going to count the square units to find the area of a figure.

Direct students to the first figure. **To find the area of this figure, count the number of square units you see. What is the area of the figure?** 4 square units

Ask students to explain how they counted. Some may count the square units one by one. Others may see 2 rows of 2 units and add 2 and 2. Some may multiply 2 by 2. Remind students to count carefully, perhaps marking each unit as they count.

Continue in a similar way as you work through the remaining figures. Remind students that their answers are given in square units.

TRY THESE Exercises 1–3 model the types of exercises students will find on the **Practice on Your Own** page.

- **Exercise 1** 6 squares, 6 square units.
- **Exercise 2** 6 squares, 6 square units.
- **Exercise 3** 8 squares, 8 square units.

PRACTICE ON YOUR OWN Review the example at the top of the page. Ask students to explain why the area of this figure is 6 square units. Then caution students to read the directions carefully as they complete the page.

CHECK Determine if students understand that area is the number of square units needed to cover a flat surface, and that to find area they need to count the square units in the figure. Success is determined by 3 out of 3 correct responses.

Students who successfully complete the **Practice on Your Own** and **Check** are ready to move on to the next skill.

COMMON ERRORS

- Students may count incorrectly.
- Students may have difficulty realizing that figures with different shapes may have the same area.
- Students may not remember that area is measured in square units.

Students who made more than two errors in the **Practice on Your Own**, or who were not successful in the **Check** section, may benefit from the **Alternative Teaching Strategy** on the next page.

Optional

Alternative Teaching Strategy
Use Models to Find Area

15 Minutes

OBJECTIVE Find area by counting tiles

MATERIALS tiles, several large, different size rectangles and squares that can be completely and precisely covered with tiles

Have students work in pairs or small groups. Distribute materials. Display a large square piece of paper.

Review that area is the number of square units needed to cover a flat surface. Explain to the students that they can use square tiles to find the area of the square.

Then demonstrate how to place the tiles so the square is entirely covered. Have students count the tiles and give the area in square units.

Then distribute the other rectangles and squares. Have students find the area of each figure, noting how many rows of tiles it took to cover the figure and how many tiles were in each row.

When students show an understanding of how to find area, ask if they think two figures can have different shapes, but the same area. Have students use the same number of tiles to make a figure on paper and trace it. Then discuss and compare the figures they traced.

To extend the activity, you may wish to have students outline squares and rectangles on grid paper, and then find each area.

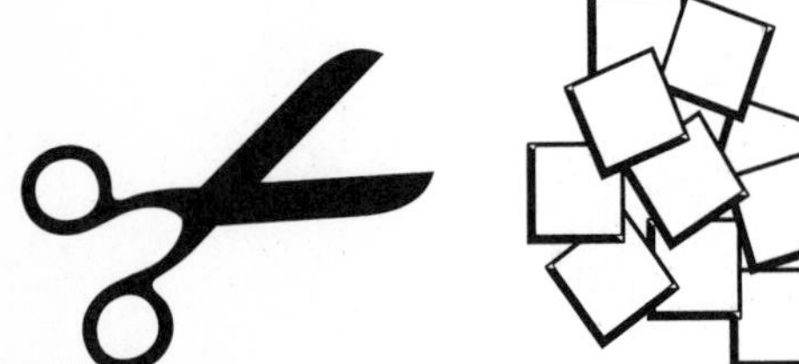

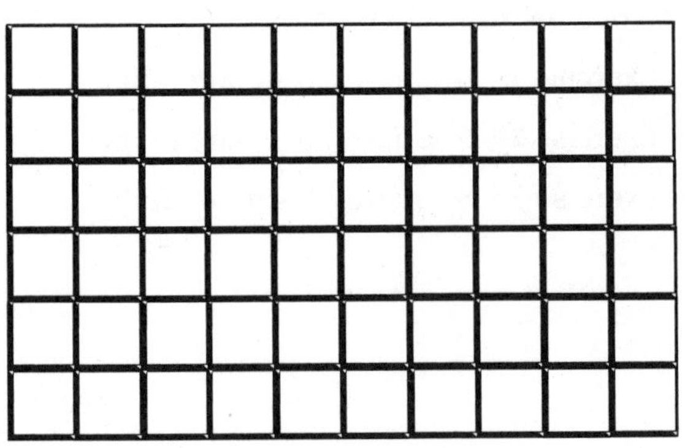

The area is 60 square units.

Name ____________________ Skill ____________________

Find Area (by Counting)

Grade 4 **Skill 73**

Count the square units to find the area.

Area is the number of square units needed to cover a flat surface.

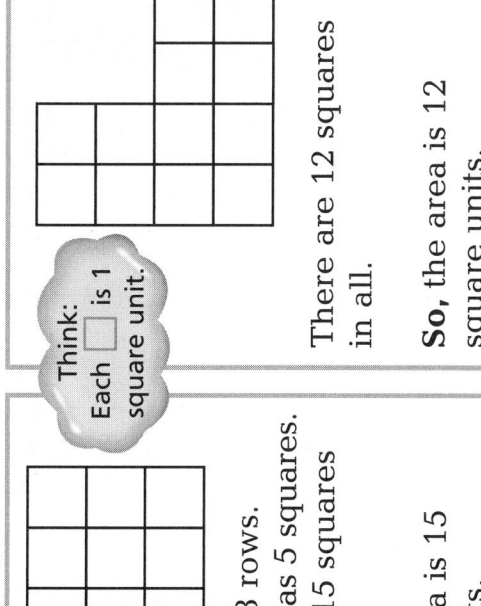

Think: Each ☐ is 1 square unit.

There are 2 rows.
Each row has 2 squares.
There are 4 squares in all.

So, the area is 4 square units.

There are 3 rows.
Each row has 5 squares.
There are 15 squares in all.

So, the area is 15 square units.

Think: Each ☐ is 1 square unit.

There are 12 squares in all.

So, the area is 12 square units.

Try These

Find the area.

1. 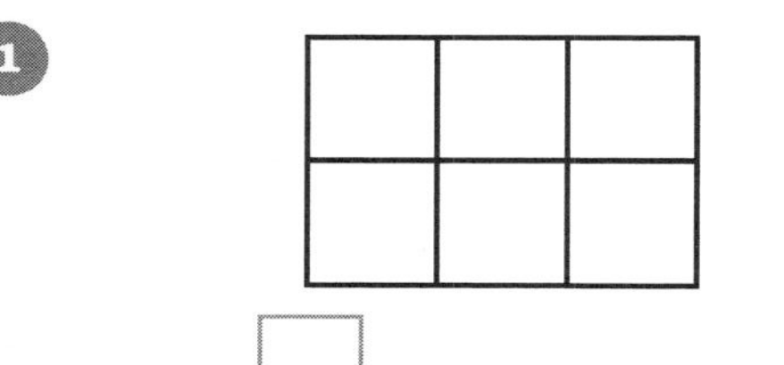

There are ☐ squares in all.

The area is ☐ square units.

2. There are ☐ squares in all.

The area is ☐ square units.

3. There are ☐ squares in all.

The area is ☐ square units.

Go to the next side.

Name ______________________ Skill ______________________

Practice on Your Own

Skill 73

Find the area of the figure.

Think:
How many square units cover the surface?

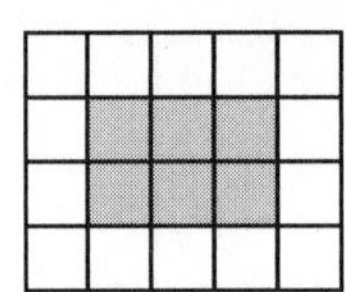

Each ☐ is 1 square unit.

There are **6** squares.
So, the area is 6 square units.

Find the area.

1. There are ☐ squares. Area: ☐ square units

2. There are ☐ squares. Area: ☐ square units

3. There are ☐ squares. Area: ☐ square units

Find the area of the shaded figure.

4. Area: ☐ square units

5. Area: ☐ square units

6. Area: ☐ square units

7. ☐

8. ☐

9. ☐

Check

Find the area of the shaded figure.

10. ☐

11. ☐

12. ☐

Identify Solid Figures

15 Minutes

OBJECTIVE Describe solid figures by the number of faces, edges, and vertices, or by a curved surface

MATERIALS classroom objects that are geometric solids, or large pictures on oaktag of figures

Begin by reviewing the definition of *face, edge,* and *vertex*. Have students find the faces, edges, and vertices of several classroom objects. Call students' attention to the rectangular prism and have them identify the 6 faces, 12 edges, and 8 vertices.

Follow the same procedure for the cube.

Have students find and number the faces on the square pyramid.

In the section labeled *Curved Surfaces*, have a student read the definition. Ask students to compare the curved surfaces.

Now direct students' attention to the sphere.

Ask: **How is the sphere different from the other curved surfaces?** It has no flat surfaces.

TRY THESE Exercises 1–3 provide verbal prompts to help students name the solid figure by identifying its attributes.

- **Exercise 1** Square pyramid.
- **Exercise 2** Sphere
- **Exercise 3** Rectangular prism.

PRACTICE ON YOUR OWN Review the pictures at the top of the page. Ask students to explain how they know that these are pictures of solid figures. Ask students to tell you which figure has a curved surface.

- **Exercises 1–3** are similar to **Try These.**
- **Exercises 4–6** Students must name the figure.

CHECK Determine if students can identify three solid figures. Success is indicated by 3 out of 3 correct responses.

Students who successfully complete the **Practice on Your Own** and **Check** are ready to move on to the next skill.

COMMON ERRORS

- Students may miscount the faces, edges, and vertices of a solid figure.
- Students may use incorrect vocabulary.

Students who made more than four errors in the **Practice on Your Own** section, or who were not successful in the **Check** section, may benefit from the **Alternative Teaching Strategy** on the next page.

Optional

Alternative Teaching Strategy

Use Models to Identify Solid Figures

OBJECTIVE Use models to identify solid figures

MATERIALS geometric solids and small, multi-colored sticky notes

Students work in pairs in this activity. One partner labels each attribute with a sticky note, while the other counts and records the number of sticky notes used. Partners take turns labeling each model and recording the number of faces, edges, and vertices.

Distribute the geometric models and the small, multi-colored, sticky notes. Start with the rectangular prism.

First tell students to use the red sticky notes to label all of the *faces* on the rectangular prism.

Ask: **How many *faces* does the rectangular prism have?** 6

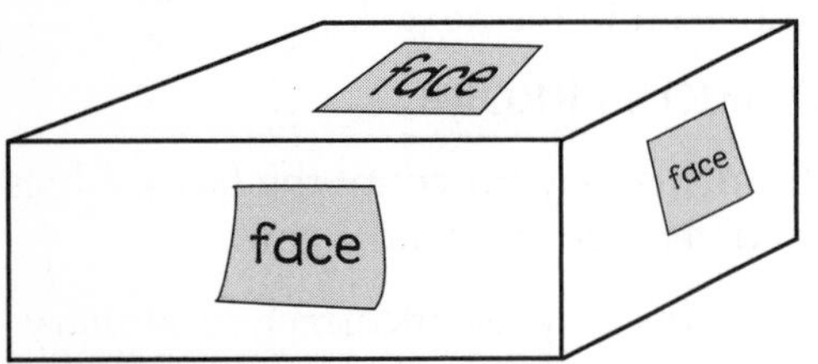

Then tell students to use the yellow sticky notes to label all of the *edges* on the rectangular prism.

Ask: **How many *edges* does the rectangular prism have?** 12

Finally, tell students to use the green sticky notes to label all of the *vertices* on the rectangular prism.

Ask: **How many *vertices* does the rectangular prism have?** 8

Repeat the activity for the cube.

Ask: **How are the rectangular prism and the cube alike?** They have the same number of faces, edges, and vertices. **How are they different?** All of the faces on the cube are the same size and the same shape. On the rectangular prism, only the faces opposite each other are the same size and same shape.

Repeat this activity for the rest of the geometric models.

Ask: **What sort of test could you do to find out if one of these solid figures is a curved surface?** See if it can roll.

Name ______________________ Skill ______________________

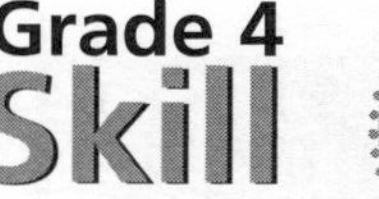

Grade 4 Skill 74

Identify Solid Figures

Some solid figures can be described by the number of faces, edges, and vertices they have.

Prisms and Pyramids

A **face** is a flat surface of a solid figure.

An **edge** is a line segment formed where two faces meet.

A **vertex** is a corner where 3 or more edges meet.

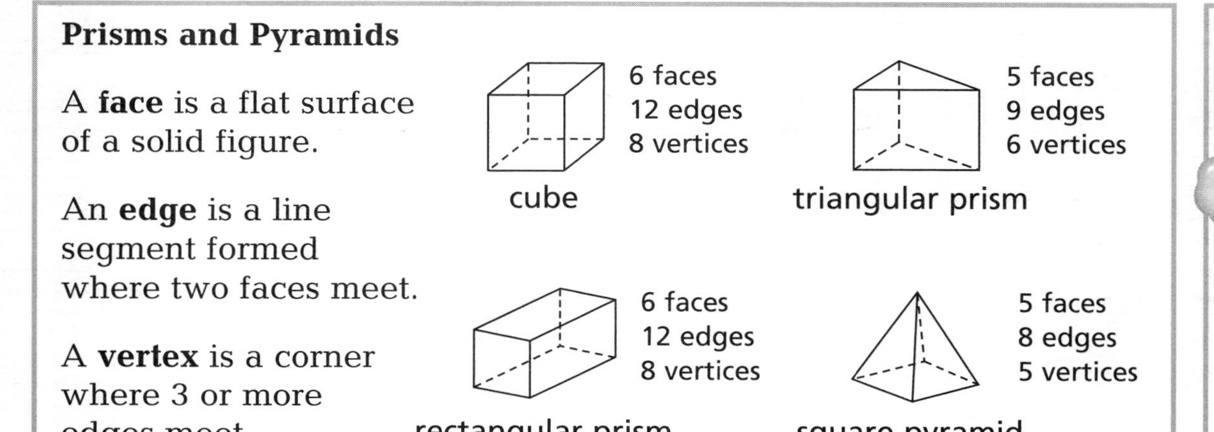

Curved Surfaces

Solids with curved surfaces can roll.

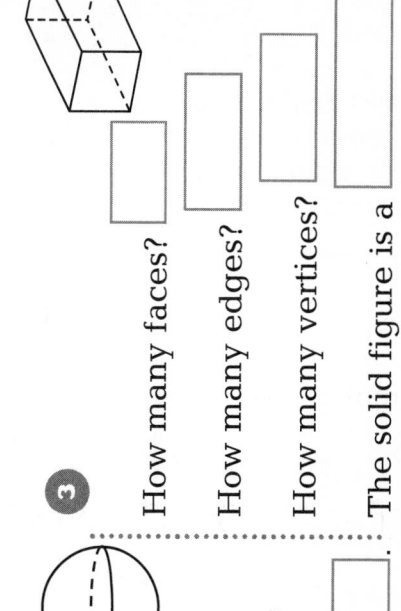

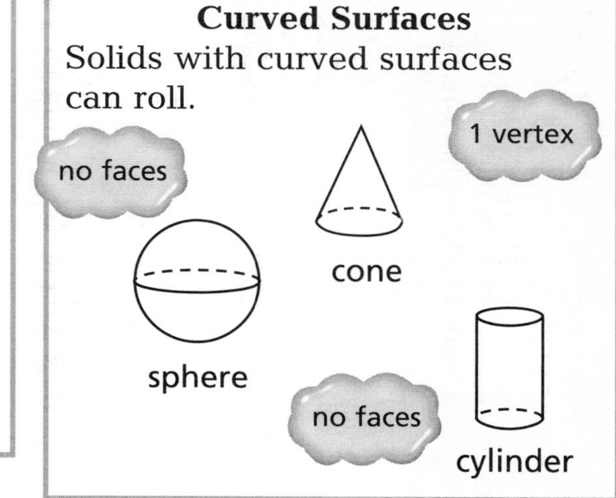

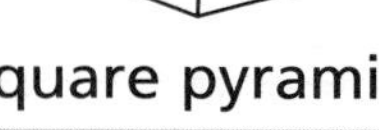

Try These

Complete. Name the figure.

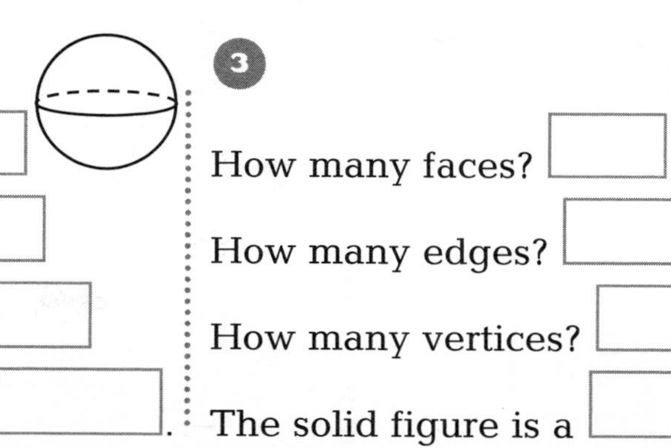

1

How many faces? ______

How many edges? ______

How many vertices? ______

The solid figure is a ______.

2

How many faces? ______

How many edges? ______

How many vertices? ______

The solid figure is a ______.

3

How many faces? ______

How many edges? ______

How many vertices? ______

The solid figure is a ______.

Go to the next side.

Name ______________________ Skill ______________________

Practice on Your Own

Skill 74

Think: Prisms and pyramids have faces, edges, and vertices.

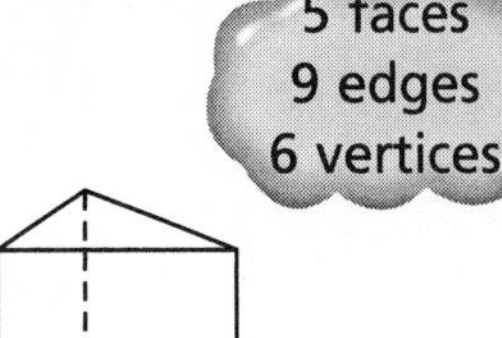

triangular prism

No faces, edges, or vertices. Bases are circles.

cylinder

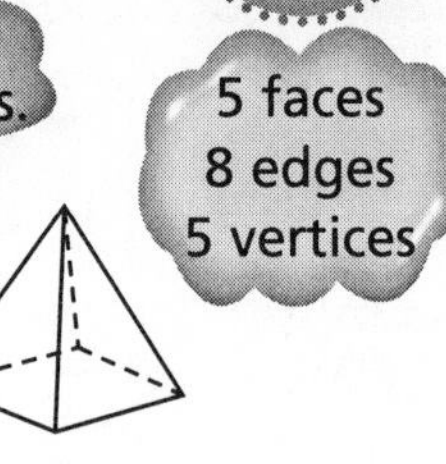

square pyramid

Complete. Name the figure.

1. Does it roll? ☐

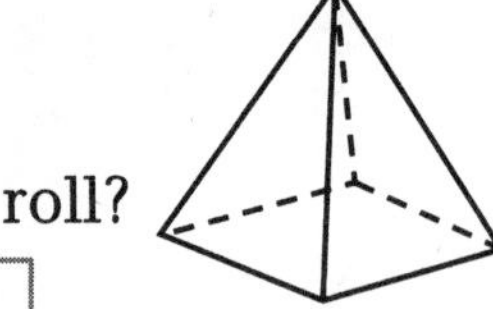

faces ☐

edges ☐

vertices ☐

The solid figure is a ☐.

2. Does it roll? ☐

faces ☐

edges ☐

vertices ☐

The solid figure is a ☐.

3. Does it roll? ☐

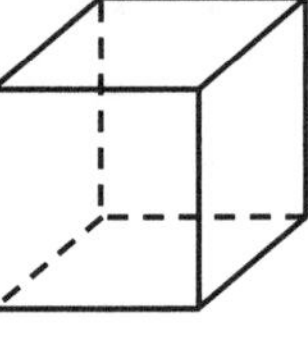

faces ☐

edges ☐

vertices ☐

The solid figure is a ☐.

4. 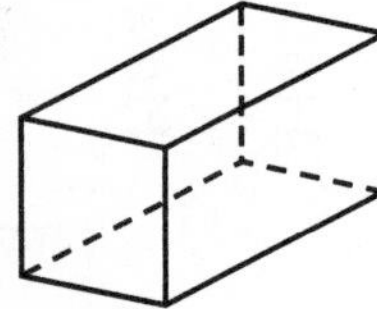

The solid figure is a ☐.

5. 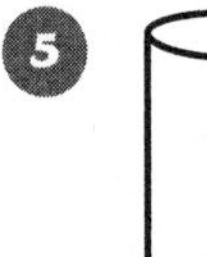

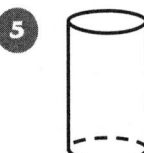

The solid figure is a ☐.

6. 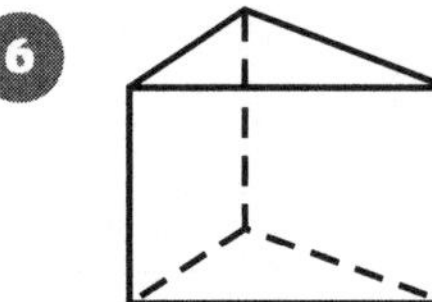

The solid figure is a ☐.

Check

Name the figure.

7. 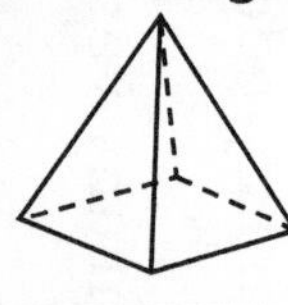

☐

8. 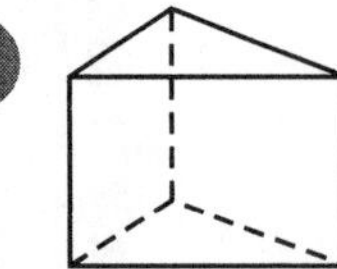

☐

9. 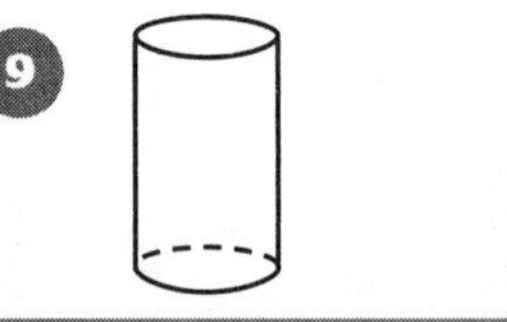

☐

Skill 75 Grade 4 Classify Angles (Right, Greater or Less than Right)

15 Minutes

OBJECTIVE Identify right angles and angles less than or greater than right angles

You may wish to begin by recalling that an angle is formed when two rays meet at the same endpoint. Point out that the size of an angle is determined by the size of the opening between the rays.

Explain that they are being asked to identify angles that are less than, greater than, or equal to a right angle.

Direct students' attention to the right angle. Explain that in a right angle the rays meet and form a square corner.

Ask: **Where can you find right angles or square corners?** **corners of the skill paper, math book, windows**

Have students look at the angle less than a right angle and the angle greater than a right angle. Explain that once they can identify a right angle, they can compare other angles to it to tell if they are greater than or less than a right angle.

Ask: **How can you tell if an angle is less than a right angle?** **The opening between the rays is smaller than a square corner.** **How can you tell if an angle is greater than a right angle?** **The opening between the rays is greater than a square corner.**

TRY THESE Exercises 1–4 model the type of exercises students will find on the **Practice on Your Own** page.

- **Exercise 1** Less than a right angle.
- **Exercise 2** Greater than a right angle.
- **Exercise 3** Right angle.
- **Exercise 4** Right angle.

PRACTICE ON YOUR OWN Review the examples at the top of the page. Explain that the small square between the two rays of the right angle is a symbol used to show a right angle.

CHECK Determine if students can identify angles as a right angle, an angle less than or greater than a right angle.

Success is indicated by 3 out of 3 correct responses.

Students who successfully complete the **Practice on Your Own** and **Check** are ready to move on to the next skill.

COMMON ERRORS

- Students may not recognize a right angle as a square corner.
- Students may not be able to identify angles that are shown in a different position.

Students who made more than two errors in the **Practice on Your Own**, or who were not successful in the **Check** section, may benefit from the **Alternative Teaching Strategy** on the next page.

Optional

Alternative Teaching Strategy

Use Models to Classify Angles

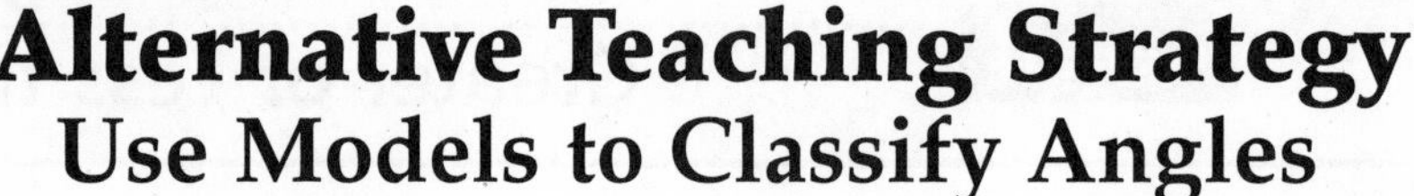

OBJECTIVE Identify right angles and angles less than or greater than a right angle using an index card

MATERIALS papers with drawings of right angles, angles greater than a right angle, angles less than a right angle in different positions, index cards, angle model

Prior to the lesson prepare angle drawings and an angle model. Fasten the ends of two strips of tag board with a fastener.

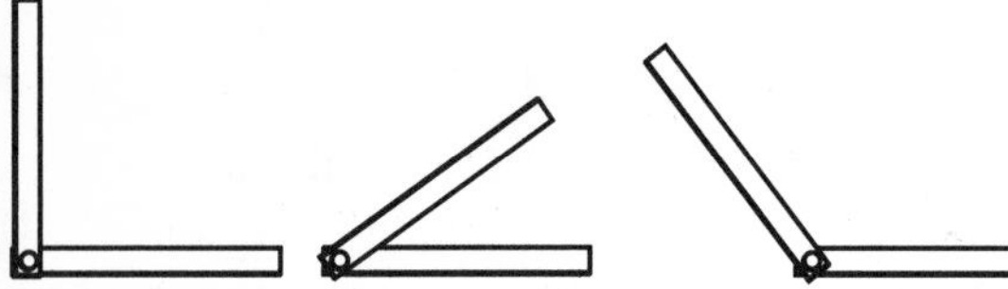

Begin by identifying the angle model and demonstrating how the strips stand for the rays of an angle. Explain that by increasing or decreasing the size of the opening between the rays, different angles are formed.

Review that a right angle is a square corner. Use an index card and the model to show a right angle, an angle less than a right angle, and an angle greater than a right angle. Change the position of the model to demonstrate these angles in other positions.

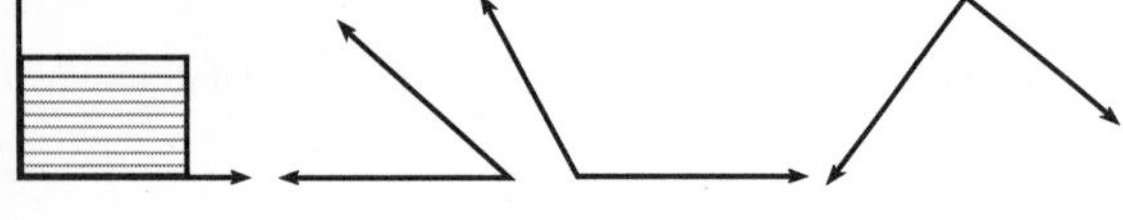

Distribute the angle drawings and index cards. Suggest the students find a right angle on the paper, by fitting the square corner of the index card into the angle drawing.

Ask: **How do you know the angle is a right angle? The square corner of the index card fits the angle, one ray is along the bottom, the other ray is along the side of the card.**

Have students find a right angle in another position. Have them explain why this angle is also a right angle.

Repeat the activity by demonstrating angles greater than or less than a right angle. Recall that students can compare these angles to a right angle to tell if they are greater than or less than a right angle. Then have students identify the angles in more than one position.

When students show an ability to identify angles without the index card, have students draw an example of each type of angle and explain how they know whether the angle is a right angle, an angle greater than a right angle, or less than a right angle.

Name ______________________ Skill ______________________

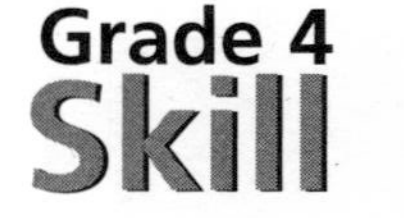

Grade 4 Skill 75

Classify Angles

The size of an angle depends on the size of the opening between the rays.

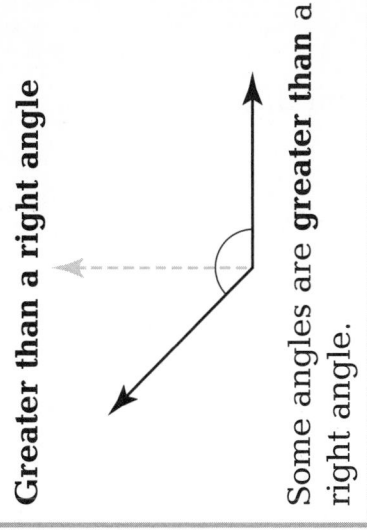

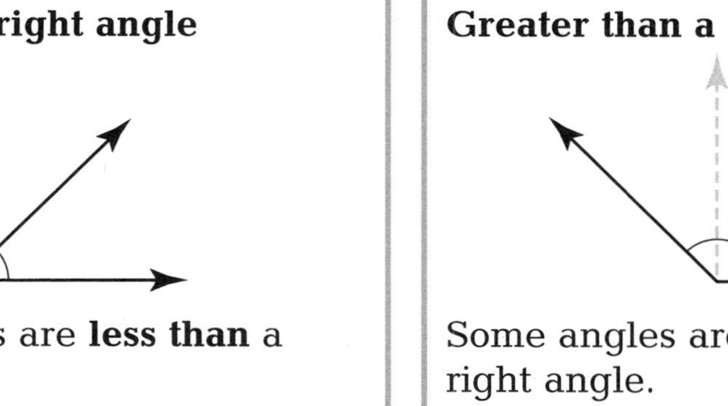

Right Angle

A **right angle** forms a **square corner.**

Less than a right angle

Some angles are **less than** a right angle.

Greater than a right angle

Some angles are **greater than** a right angle.

Try These

Tell whether each angle is a *right* angle, *less than* a right angle, or *greater than* a right angle.

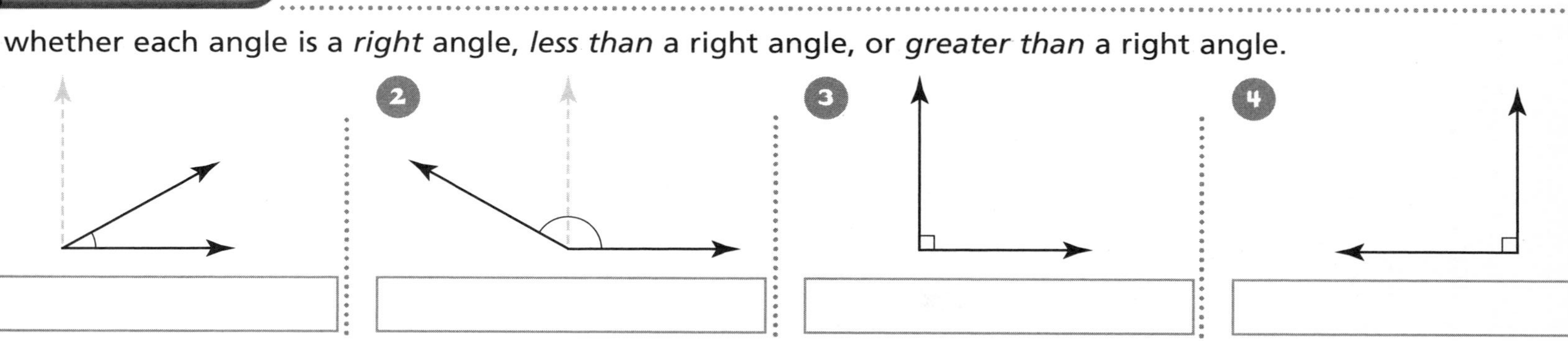

1 ______

2 ______

3 ______

4 ______

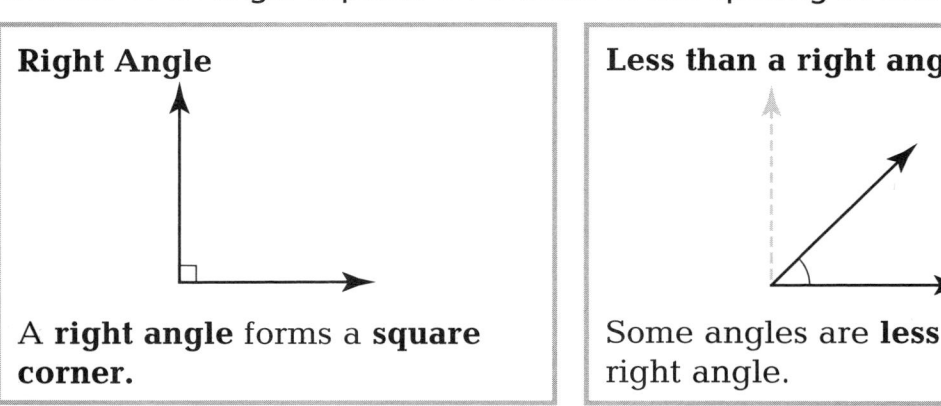

Go to the next side.

Name ______________________ Skill ______________________

Practice on Your Own

Skill 75

Think: The size of an angle depends on the size of the opening between the rays.

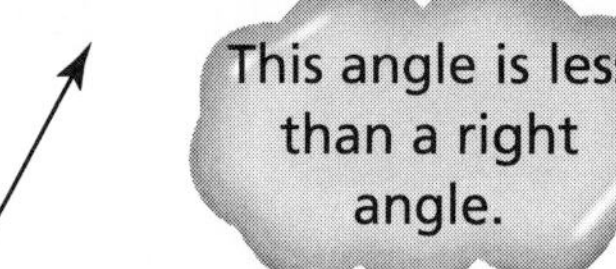

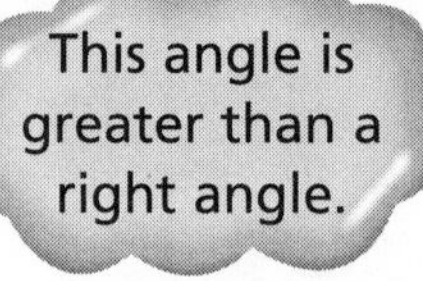

Tell whether each angle is a *right* angle, *less than* a right angle, or *greater than* a right angle.

1. 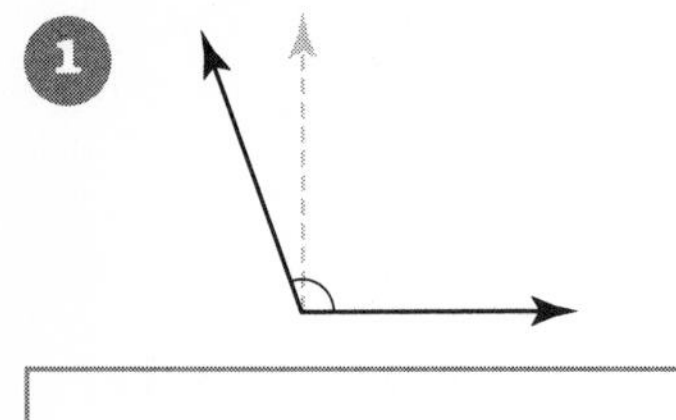

2. 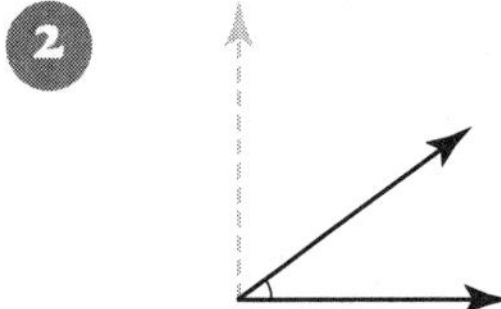

3. 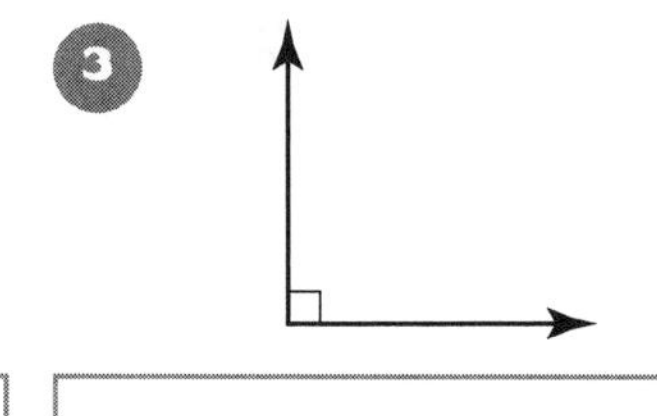

4. 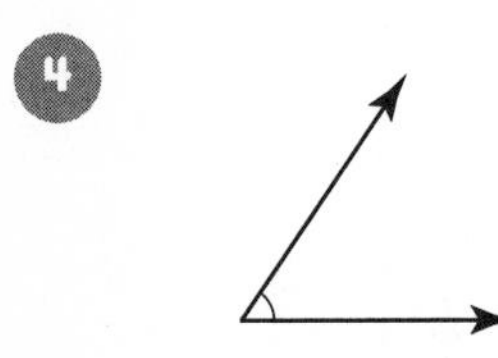

5. 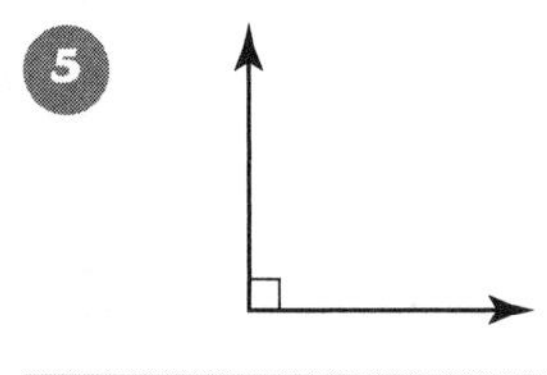

6. 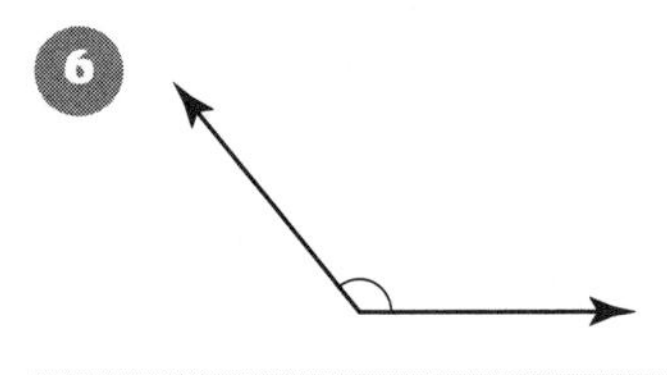

7. 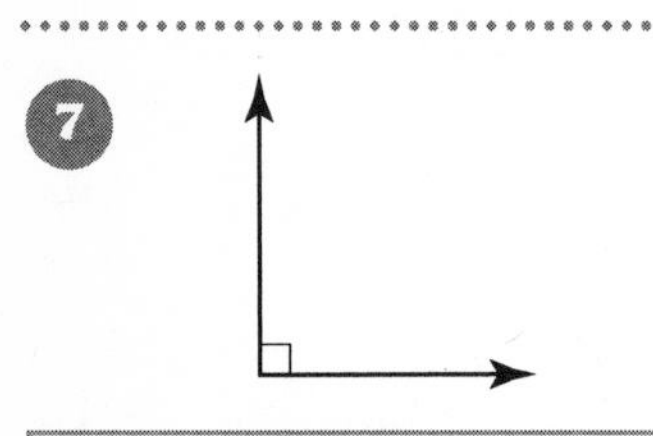

8. 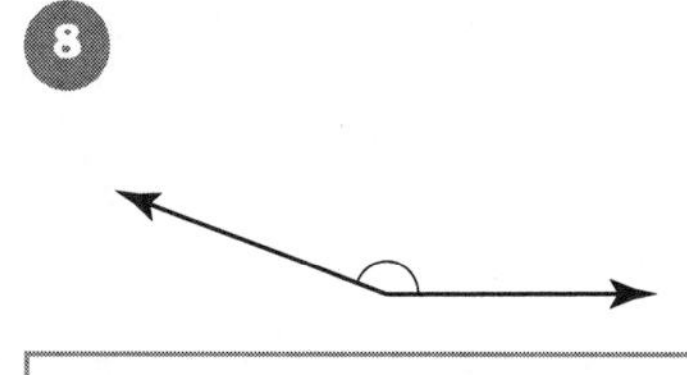

9. 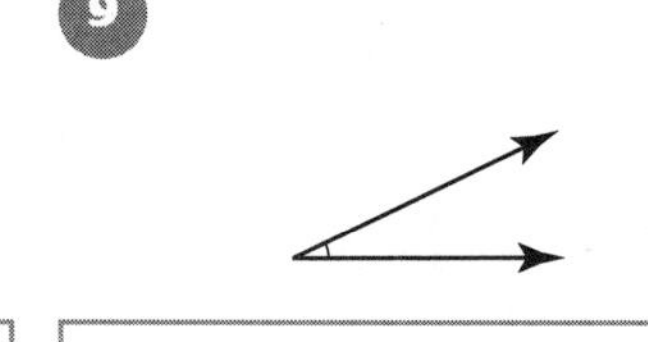

Check

Tell whether each angle is a *right* angle, *less than* a right angle, or *greater than* a right angle.

10. 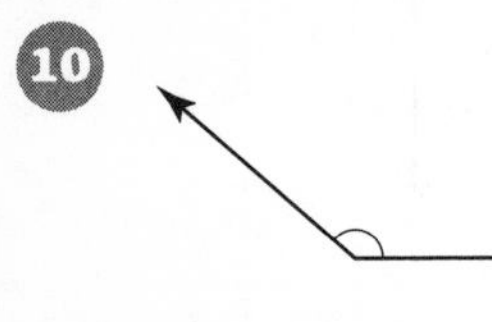

11. 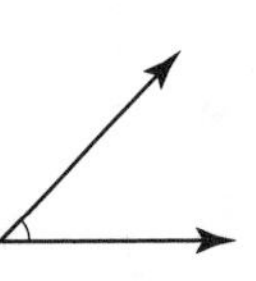

12.

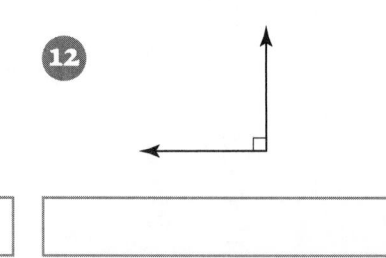

Answer Card

Geometry

Grade 4

Skill 67

TRY THESE

1. no, no, no, yes
2. no, no, yes, no
3. yes, yes, no, no

PRACTICE

1. greater than a right angle
2. less than a right angle
3. right angle
4. less than a right angle
5. right angle
6. greater than a right angle
7. right
8. greater than
9. less than

CHECK

10. greater than
11. less than
12. right

Skill 68

TRY THESE

1. no, no, no
2. yes, yes, yes
3. yes, no, no
4. yes, yes, yes

PRACTICE

1. no, no, no
2. yes, no, no
3. yes, yes, yes
4. no
5. yes
6. yes
7. no
8. yes
9. no

CHECK

10. no
11. no
12. yes

Skill 69

TRY THESE

1. yes, yes
2. no, no
3. yes, yes

PRACTICE

1. no, no
2. yes, yes
3. yes, yes
4. yes
5. no
6. no
7. yes
8. no
9. yes

CHECK

10. yes
11. yes
12. no

Skill 70

TRY THESE

1. 4, 4, quadrilateral
2. 3, 3, triangle
3. 6, 6, hexagon

PRACTICE

1. 3, 3, triangle
2. 4, 4, quadrilateral
3. 8, 8, octagon
4. quadrilateral
5. triangle
6. pentagon
7. hexagon
8. hexagon
9. pentagon

CHECK

10. octagon
11. octagon
12. quadrilateral

Skill 71

TRY THESE

1. a
2. b
3. c
4. a

PRACTICE

1. a
2. c
3. c
4. b
5. a
6. a
7. b
8. a
9. rectangle
10. triangle
11. circle
12. square

CHECK

13. circle
14. square
15. triangle
16. rectangle

Answer Card
Geometry
Grade 4

SKILL 72

TRY THESE

1. 10
2. 14
3. 4, 1, 3, 1, 1, 2, 12

PRACTICE

1. 6, 2, 6, 2, 16
2. 4, 4, 4, 4, 16
3. 5, 2, 4, 1, 1, 3, 16
4. 14
5. 14
6. 18

CHECK

7. 20
8. 20
9. 24

SKILL 73

TRY THESE

1. 6, 6
2. 6, 6
3. 8, 8

PRACTICE

1. 12, 12
2. 16, 16
3. 11, 11
4. 12
5. 11
6. 17
7. 14 square units
8. 18 square units
9. 20 square units

CHECK

10. 16 square units
11. 12 square units
12. 18 square units

SKILL 74

TRY THESE

1. 5, 8, 8; square pyramid
2. 0, 0, 0; sphere
3. 6, 12, 8; rectangular prism

PRACTICE

1. no, 5, 8, 8; square pyramid
2. yes, 0, 0, 1; cone
3. 6, 12, 8; cube
4. rectangle, rectangular prism
5. cylinder
6. triangular prism

CHECK

7. square pyramid
8. triangular prism
9. cylinder

SKILL 75

TRY THESE

1. less than
2. greater than
3. right
4. right

PRACTICE

1. greater than
2. less than
3. right
4. less than
5. right
6. greater than
7. right
8. greater than
9. less than

CHECK

10. greater than
11. less than
12. right

Statistics, Data Analysis, and Probability

Skill 76 Grade 4

Read Pictographs (Use Half Symbols)

OBJECTIVE Read a pictograph

Begin by recalling that a pictograph uses pictures to show and compare data. Explain that the title on the graph tells what the pictograph is about.

Ask: **What is the title of the pictograph?** Favorite Winter Sports

As students read the second step, explain that the key is an important part of the pictograph. Point out that the key may be different for different graphs, so students should pay close attention to what each symbol represents. Once they know what one symbol stands for, students can use skip counting to count the number of votes. Point out that sometimes a half picture is used.

Ask: **If a whole symbol stands for two votes, what does a half symbol stand for?** one vote **How many votes are there for ice skating?** three

Next, read Step 3. Have students find the sport with the most votes. Tell them to skip count by two, and then add 1 for the half symbol to find the total number of votes.

Ask: **How do you find the number of votes for skiing?** Skip count first, 2, 4, 6, 8 and then add 1. 8 + 1 = 9; 9 votes.

TRY THESE In exercises 1-5 students answer questions about the pictograph.

- **Exercise 1** title
- **Exercise 2** The total for one topping.
- **Exercise 3** The topping with the most votes.
- **Exercise 4** How many more of one topping than another.
- **Exercise 5** The total votes.

PRACTICE ON YOUR OWN Review the parts of a pictograph and read the key in the example at the top of the page. Ask students to find the number of votes for each drink. Caution them to read the sentences carefully.

CHECK Determine if students know how to use a pictograph, interpret the key, and answer questions about the data. Success is indicated by 3 out of 4 correct responses.

Students who successfully complete **Practice on Your Own** and **Check** are ready to move on to the next skill.

COMMON ERRORS

- Students may forget to use the key, and simply count the symbols for each item as representing 1.
- Students may count half symbols as whole symbols.

Students who made more than two errors in the **Practice on Your Own**, or who were not successful in the **Check** section, may benefit from the **Alternative Teaching Strategy** on the next page.

Optional

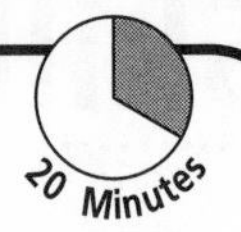

Alternative Teaching Strategy
Make and Use a Pictograph

OBJECTIVE Make a pictograph and use it to compare data

MATERIALS centimeter grid paper, pencils, markers

Have students work in pairs or small groups. Explain that you will display data about dogs and together students will show that data in a pictograph.

Prepare a tally table that shows the results of a survey about the breed of dog most people prefer. Have students choose a title and decide what symbol to use, for example, a dog biscuit. Then discuss the key. Suggest that each symbol represent two votes.

Ask: **If we use one symbol for every two votes, how will you show the key?** 1 dog biscuit = 2 votes **There are 11 votes for Boxers. How many whole symbols will you show?** 5 **How many votes does that represent?** 10 **How will you show the eleventh vote?** with a half symbol

DOG	NUMBER OF VOTES
Collie	4
Boxer	11
German Shepherd	14
Irish Setter	8
Terrier	2

Have students complete the pictograph by drawing the appropriate number of symbols for each breed of dog.

When students have completed the pictograph, discuss the graph by asking questions such as: **What does the title tell you?** what the pictograph is about **What does the key show?** 1 symbol = 2 votes **How many types of dogs are in the survey?** 5

Remind students to skip count by twos to find the total number of votes.

Ask: **Which type of dog has the most votes?** German Shepherd

As students show an understanding of how to read the graph, continue with questions that require computation, such as: **How many more students voted for Boxers than for Irish Setters?** 3 **If six more students voted for Collies, how many votes would Collies have in all?** 10 **How would you show that on the pictograph?** Draw 5 dog biscuits.

FAVORITE DOGS

Collie	🦴🦴
Boxer	🦴🦴🦴🦴🦴½
German Shepherd	🦴🦴🦴🦴🦴🦴🦴
Irish Setter	🦴🦴🦴🦴
Terrier	🦴

Key: 🦴 = 2 Votes

Name ____________________ Skill ____________________

Read Pictographs

Grade 4 Skill 76

A **pictograph** uses pictures to show and compare information.

Which sport has the most votes?

Step 1 Find the title.

Favorite Winter Sports	
skiing	✱✱✱✱½
ice skating	✱½
sledding	✱✱✱
ice fishing	✱✱✱
Key: Each ✱ = 2 votes.	

This is the title: Favorite Winter Sports

Step 2 Read the key.
The key shows what each picture in the graph represents.

Each picture represents 2 votes.

Key: Each ✱ = 2 votes.

So, skip-count by two to find the number of votes for each sport.

Step 3 Find the sport with the most votes.

Skiing has the most pictures.

Skiing: ✱✱✱✱½

2, 4, 6, 8, **9**
Skiing has the most votes. It has 9.

Try These

Read the pictograph. Answer the questions.

Favorite Pizza Toppings	
cheese	◉◉◉
peppers	◉◉◉◉
mushrooms	◉½
broccoli	◉◉½
Key: Each ◉ = 4 votes	

If ◉ equals 4 votes, then ½ equals 2 votes.

1. What is the title of the pictograph? ____
2. How many students voted for the broccoli topping? ____
3. Which topping was the most popular? ____
4. How many more students voted for peppers than cheese? ____
5. How many students voted in all? ____

Go to the next side.

Name ______________________ Skill ______________________

Practice on Your Own

Skill 76

Favorite Juice Drinks	
apple	
cranberry	
fruit	
grape	
mango	
Key: Each = 4 votes.	

The graph is about **favorite juice drinks.**

Each picture represents **4 votes.**

A half picture represents **2 votes.**

The flavor with the fewest votes is **cranberry.**

Read the pictograph. Answer the questions.

Books Read This Month	
Tamika	
Julio	
Suki	
Jamal	
Key: Each = 2 books.	

1. What is the title of the pictograph?

2. How many books does each picture represent?

3. How many books does a half picture represent?

4. Who read the most books? How many?

Favorite Games	
board	
video	
puzzle	
tag	
Key: Each = 10 students	

5. How many students voted for each type of game?

board games video games

puzzles tag games

Check

Number of Books Read	
mysteries	
novels	
biographies	
poetry	
Key: Each = 8 books read	

6. How many students read biographies?

7. How many more students read novels than mysteries?

8. How many books were read in all?

9. If 4 more students read mysteries, how many pictures would there be for mysteries?

Skill 77 Grade 4 — Tallies to Frequency Tables

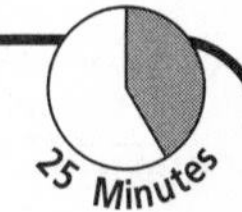

OBJECTIVE Count tally marks in a tally table and make a frequency table

Begin by pointing out the tally table. Tell students that the marks in the table are called tally marks. They are used to keep track of information that is counted. One mark represents one piece of data. Tally marks are grouped by fives. When adding the fifth mark, draw it across the other four marks. This represents a group of five. If there are single marks also, add them to the five to find the total.

Ask: **How many people chose popcorn as a favorite snack?** 6 **How many chose pretzels?** 4 **Why do the tally marks for popcorn look different from the others?** Because there is a group of five tally marks

Next, point out the frequency table. A frequency table uses numbers to show how often something happens. Tell students that this is where they will record the information from the tally table. Point out that the frequency table uses numbers instead of tally marks. The numbers make it easier to do calculations with the data or draw graphs.

Ask: **How are the tables the same?** They show the same data; they have the same title and labels **How are the tables different?** One has tally marks and one has numbers

TRY THESE In Exercise 1, students use a tally table to complete a frequency table.

- **Exercise 1** Count tally marks in a tally table and write the numbers in the frequency table.

PRACTICE ON YOUR OWN Explain that after a tally table is used to complete a frequency table, the data can be used to answer questions. Then have students complete the frequency tables and answer the questions.

CHECK Determine if students know how to count tally marks, complete a frequency table, and answer questions about the data. Success is indicated by 3 out of 4 correct responses.

Students who successfully complete the **Practice on Your Own** and **Check** are ready to move on to the next skill.

COMMON ERRORS

- Students may count the tally marks incorrectly, forgetting to count a group as five.
- Students may read the data incorrectly when answering questions.
- Students may not understand what information they need to answer the question.

Students who made more than two errors in the **Practice on Your Own**, or who were not successful in the **Check** section, may benefit from the **Alternative Teaching Strategy** on the next page.

Optional

Alternative Teaching Strategy

20 Minutes

Use Models and Tallies to Make Frequency Tables

OBJECTIVE Use models to make a tally table and a frequency table

MATERIALS paper bag and 35 colored cubes for each pair, paper, pencils

You may wish to have students work in pairs. Draw a tally table and list the colors of the cubes. Demonstrate how to choose a cube from the bag, and record the color on the tally table with a tally mark.

Explain that when the fifth cube of any color is picked, the tally mark is drawn at an angle to group the five marks. Show how the tally marks are counted.

Have one student pick 20 cubes out of the bag, and return them to the bag, as the partner records the results on the tally table.

When the tally table is complete, draw a frequency table. Recall that a frequency table uses numbers to show how often something happens, or in this activity, how often cubes of each color were picked. Have students count the tally marks and write each total in the frequency table.

Cubes We Picked	
Color	**Tallies**
Red	卌 IIII
Green	卌 III
Blue	III

Cubes We Picked	
Color	**Frequency**
Red	9
Green	8
Blue	3

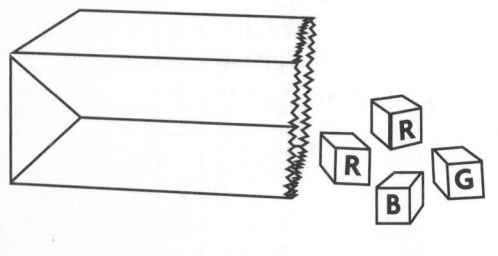

R R B G

Point out that it is easier to use the numbers in the frequency table to compare the data about the cubes.

Ask simple questions such as: **Which color cube was picked most often? How many blue cubes were picked? red cubes? green cubes? Which color was picked the least?**

Have students switch roles and repeat the activity until about thirty tally marks are in the table. Have students create a frequency table with the data.

Discuss the data in the frequency table. Ask questions about the data. Have students point to where they found the answer on the table as the questions are asked.

As the students show understanding, ask questions that require calculating with numbers from two or more rows of the table, such as: **How many red and blue cubes were picked? How many more blue cubes than green cubes were picked?**

Name ______________________ Skill ______________________

Grade 4 Skill 77

Tallies to Frequency Tables

Use the tally table to make a frequency table.

Step 1 Read the tally table.

Favorite Snacks	
Type	**Number**
Popcorn	~~IIII~~ I
Pretzels	IIII
Peanuts	III

Remember:
In a tally table, tally marks (/) are used to record data.
~~IIII~~ = 5

Step 2 Make a frequency table.

Count the tally marks for each snack. Record the numbers in the frequency table.

Favorite Snacks	
Type	**Frequency**
Popcorn	6
Pretzels	4
Peanuts	3

Try These

Complete the frequency table.

1. Count the tally marks for each pet.

Favorite Pet	
Type	**Number**
Dog	~~IIII~~
Cat	~~IIII~~ I
Fish	IIII
Bird	II

Favorite Pet	
Type	**Frequency**
Dog	
Cat	
Fish	
Bird	

Go to the next side.

Name ______________________ Skill ______________________

Practice on Your Own

Complete the frequency table. Answer the questions.

Balls Sold	
Type	**Number**
Football	||||
Basketball	~~||||~~
Soccer	~~||||~~ ||

Balls Sold	
Type	**Frequency**
Football	4
Basketball	5
Soccer	7

How many more basketballs than footballs were sold? [1]

Remember:
A frequency table is a table that uses numbers to show how often something happens.

How many balls were sold altogether? [16]

Complete the frequency table. Answer the questions.

Flowers Planted	
Type	**Number**
Tulip	||||
Daisy	~~||||~~ ||
Rose	~~||||~~
Mum	||

1.

Flowers Planted	
Type	**Frequency**
Tulip	4
Daisy	[ ]
Rose	5
Mum	2

2. How many tulips were planted? [ ]

3. How many daisies and roses were planted? [ ]

4. How many tulips and mums were planted? [ ]

5. Which flower was planted most often? [ ]

Check

Complete the frequency table. Answer the questions.

Pairs of Shoes Sold	
Day	**Number**
Mon.	||
Tues.	~~||||~~ |||
Wed.	|||
Thur.	~~||||~~ |

Pairs of Shoes Sold	
Day	**Frequency**
Mon.	2
Tues.	[ ]
Wed.	[ ]
Thur.	6

6. How many pairs of shoes were sold on Wednesday? [ ]

7. How many pairs of shoes were sold on Monday? [ ]

8. How many more pairs of shoes were sold on Tuesday than Monday? [ ]

9. How many pairs of shoes were sold altogether? [ ]

Skill 78 Grade 4

Read Bar Graphs

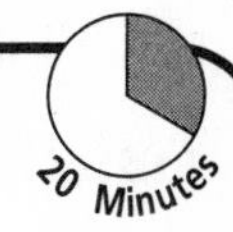

OBJECTIVE Read and interpret bar graphs

Begin by pointing to the title of the bar graph and the labels on the side and bottom of the graph. Explain to students that the length of the bars will aid them in making comparisons quickly and easily.

Call attention to the title and labels on the horizontal and vertical axes of the bar graph.

Ask: **What is this bar graph about?** the color of students' shoes **What are the shoe colors?** blue, white, brown, black

Continue by saying: **The bottom or horizontal labels on the graph represent the scale. It has numbers that represent numbers of students. How many students does 1 space represent?** 1 student

TRY THESE Exercises 1–3 model the type of exercises students will find on the **Practice on Your Own** page.

- **Exercises 1–2** Read and interpret the bar graph.
- **Exercise 3** Use the information from the bar graph to solve a comparison problem.

PRACTICE ON YOUR OWN Begin by explaining to the students that another way to show the information on a bar graph is to have the bars go up. Review with the students the title of the bar graph and the labels on the side and bottom of the graph. Also discuss the interval on this graph: each space represents 2 students.

CHECK Determine if students can read and interpret bar graphs. Then use the information on the bar graph to solve problems. Success is indicated by 3 out of 3 correct responses.

Students who successfully complete the **Practice on Your Own** and **Check** are ready to move on to the next skill.

COMMON ERRORS

- Students may not read all of the information, such as the title, labels and numbers, found on a bar graph.
- Students may interpret the scale incorrectly.

Students who made more than two errors in the **Practice on Your Own**, or were not successful in the **Check** section, may benefit from the **Alternative Teaching Strategy** on the next page.

Optional

Alternative Teaching Strategy

Make a Bar Graph

20 Minutes

OBJECTIVE Making and labeling a vertical bar graph and using the information on the bar graph to solve problems

MATERIALS copies of a 5-column, 10-row grid

Explain to the students that they will be making a vertical bar graph that will record their favorite colors.

Have students name 5 favorite colors. Begin with the left column and have students label each *column* at the bottom of the grid with one color.

Ask: **What could you label the columns?** colors

Then help students decide on a reasonable interval for the scale. Record these numbers to the left of the grid. Explain that these numbers tell how many votes.

Ask: **What could you label the scale?** number of votes **What would be an appropriate title for the graph?** favorite colors

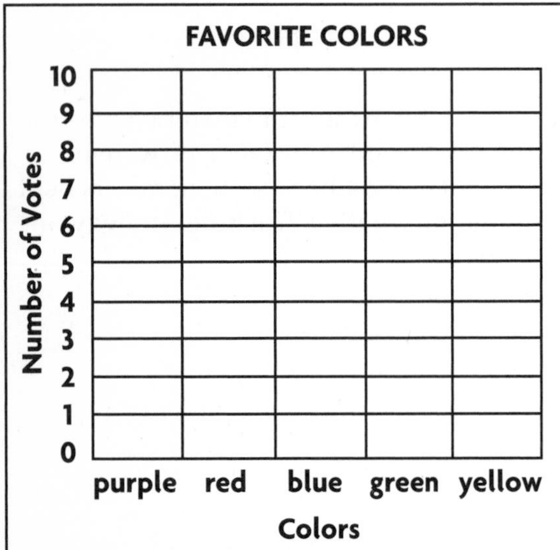

Explain that students will now gather and record information on their graph.

Ask: **Who chooses purple as their favorite color?** Students shade in one box for each vote above the word *purple*.

Repeat this procedure for each color listed.

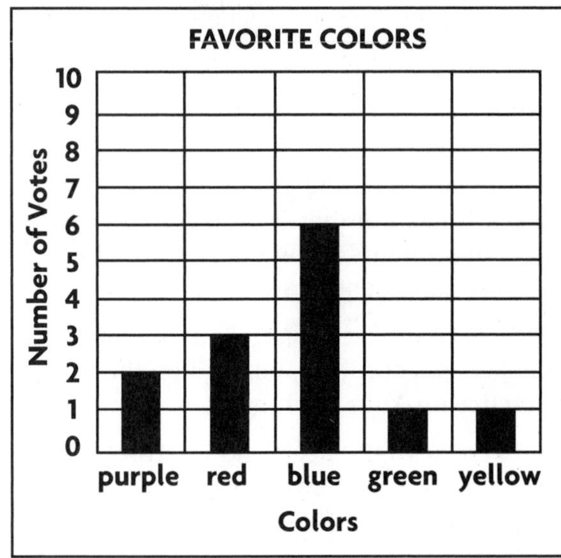

Ask: **How many more students chose blue than red for their favorite color?**

Ask questions about the information on the bar graph. Then have students work in pairs to create problems from the information on the graph. Write each problem on a 5×8 card, then write the answer on the back of the card. Students work back and forth asking questions off the cards. If there is more than one pair of students working on this exercise, have the pairs exchange cards.

Name ________________ Skill ________________

Grade 4 **Skill 78**

Read Bar Graphs

Bar graphs use bars to show data. This is a **horizontal** bar graph. The bars go across.

The graph at the right shows how many students wore each color shoes.

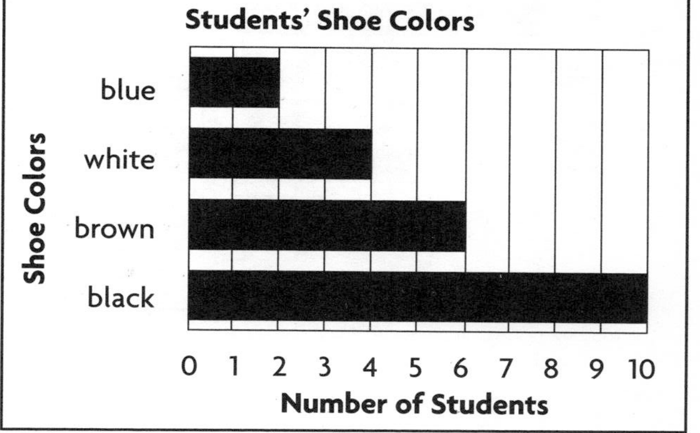

Read the graph.

- What shoe colors does the graph show?

The graph shows blue, white, brown, and black.

- How many students wore blue shoes?

The bar for blue stops at 2. **So,** 2 students wore blue shoes.

- How many students altogether were counted?

Add to find the number of students.

2	+	4	+	6	+	10	=	22
blue shoes		white shoes		brown shoes		black shoes		students counted

Try These

Use the graph above to answer the questions.

1. How many students wore brown shoes? ☐

2. Which color shoe did most students wear? ☐

3. How many more students wore brown shoes than white?

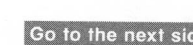

☐ – ☐ = ☐
brown shoes, white shoes, answer

Go to the next side.

Name ______________________ Skill ______________________

Skill 78

Practice on Your Own

This is a **vertical** bar graph.
It has bars that go up.

- The graph shows students' favorite fruit.
- The scale shows that each space stands for 2 votes. The bar for pears stops halfway between 4 and 8.

The halfway number between 4 and 8 is 6.

So, 6 students voted for pears.

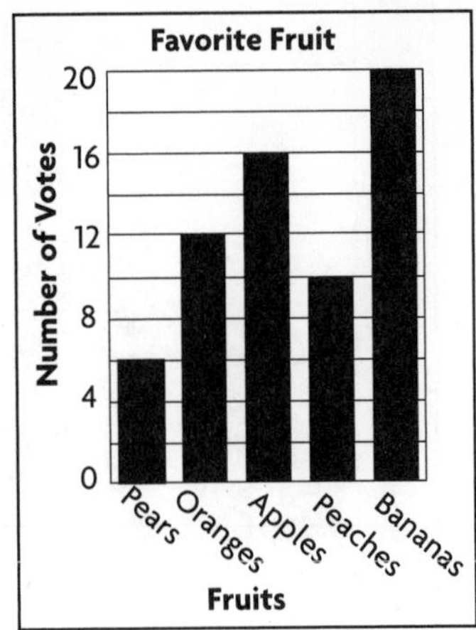

Use the graph above to answer the questions.

1. How many votes were for oranges?
2. Which fruit had the fewest votes?
3. Which fruit had the most votes?
4. How many more votes were there for oranges than pears?
5. How many votes were there for peaches?
6. Were there more votes for apples or oranges?
7. How many students voted for pears or oranges?
8. How many students voted for bananas or apples?
9. How many students voted for oranges or peaches?

Check

Use the graph above to answer the questions.

10. How many more votes were there for apples than peaches?
11. How many students voted for apples or oranges?
12. How many votes were there altogether?

Skill 79 Grade 4

Parts of a Bar Graph

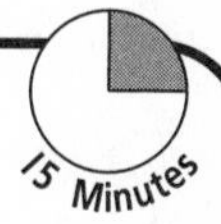

OBJECTIVE Identify parts of a bar graph

Read about the *title* and then ask: **Where do you find the title of the bar graph?** above the graph Have students point to that part of the graph. **Why is the title so important?** Possible response: Without the title it's difficult to understand what the graph is about.

Read about the *labels* and then have students locate each label on the graph.

Point out to students how the *scale* helps them read the number each bar shows. Explain that the scale on a bar graph can have intervals of any number, for example, two, three, five, ten, and so forth. Point out that on this graph not every line is labeled. Help students understand that each space represents 1 bird, and that the first line after zero is 1; likewise, the line just after 2 is 3.

Once students have reviewed all the parts of this graph,

Ask: **Why are the bars different lengths?** because there are different numbers of each bird: A longer bar means a greater number of birds; a shorter bar means fewer birds. **Were there more cardinals than bluebirds at the bird feeder? How do you know?** There were more cardinals; the bar for *cardinal* stops at the line for 2, the bar for *bluebird* stops at the line for 1; so, according to the graph, there were 2 cardinals and just 1 bluebird. **Were there more goldfinches than robins at the bird feeder? How do you know?** There were just as many goldfinches as robins; the bars are the same length; each bar shows 4 birds.

TRY THESE In Exercises 1–4 students answer questions about the parts of a bar graph.

- **Exercise 1** Identify the title.
- **Exercise 2** Identify the left label.
- **Exercise 3** Identify the bottom label.
- **Exercise 4** Identify the scale.

PRACTICE ON YOUR OWN Use the example at the top of the page to review parts of a graph. Explain that bar graphs can be vertical or horizontal. Have students use the data on the graph at the top of the page to answer the questions.

CHECK Determine if students can identify parts of a graph and understand what each part shows. Success is indicated by 3 out of 3 correct responses.

Students who successfully complete the **Practice on Your Own** and **Check** are ready to move to the next skill.

COMMON ERRORS

- Students may have trouble reading the number that a bar shows, especially when the line for the interval is not labeled.
- Students may be able to name parts of a graph, but not understand what they represent.

Students who made more than two errors in the **Practice on Your Own**, or who were not successful in the **Check** section, may benefit from the **Alternative Teaching Strategy** on the next page.

Optional

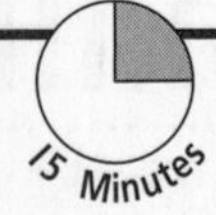

Alternative Teaching Strategy
Draw Parts of a Bar Graph

OBJECTIVE Draw a bar graph and label its parts

MATERIALS Centimeter grid paper, pencils, markers

Draw a bar graph with the students. Start by listing the results of a survey about students' favorite subject in school:

math	12
art	8
science	10
social studies	7

Distribute materials to the students and have them draw the two axes of the bar graph. Ask students what title the graph should have. Have them write the title.

Decide whether the graph will be horizontal (numbers on the bottom; bars extend from left to right) or vertical (numbers on left side; bars extend up). Then have students decide on labels.

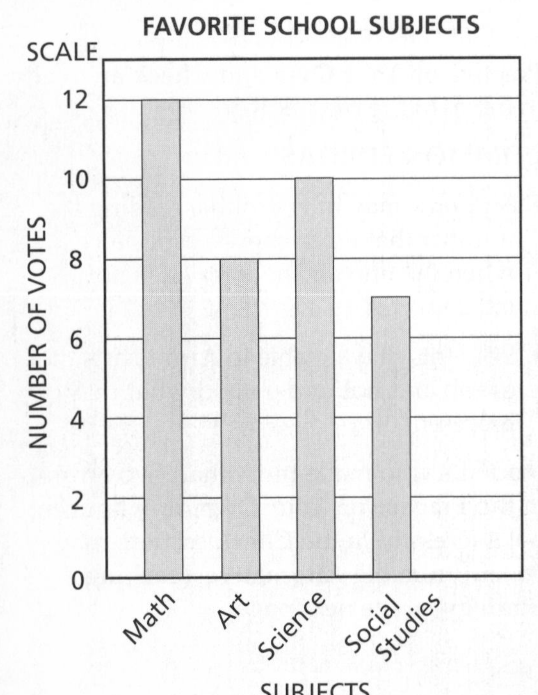

Together decide on a scale. Guide students to understand that since most of the numbers are even, the scale can show intervals of 2.

Ask: **Between which two even numbers is 7?** **6 and 8** **Where will the bar for 7 end?** **halfway between 6 and 8**

Have students draw the bars for each subject. When students have completed the graph, review the parts of the graph and why they are necessary.

Ask: **At which number does the bar for science stop?** **10** **What does the bar for math tell us?** **Twelve students voted for math; math is the most popular subject.**

Continue by having each student make up a question to ask and choose another student to give the answer.

Name ______________________ Skill ______________________

Grade 4 Skill 79

Parts of a Bar Graph

A **bar graph** uses bars to show data.

Identify the parts of a bar graph.

- Look at the title. It tells what the graph is about.
  The title is "Birds at the Bird Feeder."
- Look at the labels. There are two of them.
  The label on the left side is "Types of Birds."
  The label at the bottom is "Number of Birds."
- Find the scale. It has numbers that tell you the number each bar shows.
  The numbers on the scale are 0, 1, 2, 3, 4, 5, 6, 7, 8.
  One space stands for one bird.
- Compare the bars. Each bar tells how many birds were at the feeder.
  The bar for robins stops at 4. **So,** there were 4 robins at the feeder.
  The bar for bluebirds stops at 1. **So,** 1 bluebird was at the feeder.

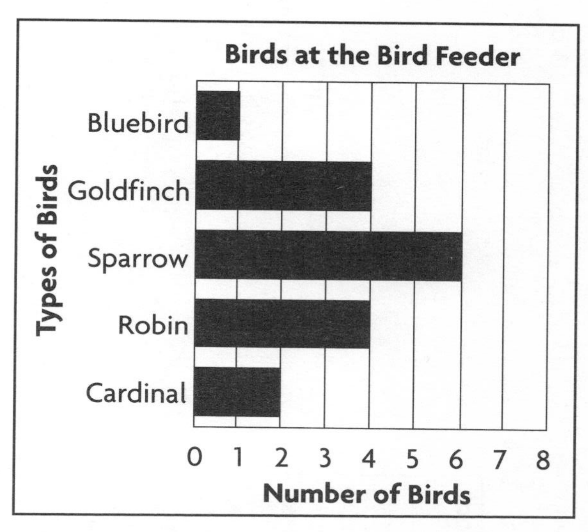

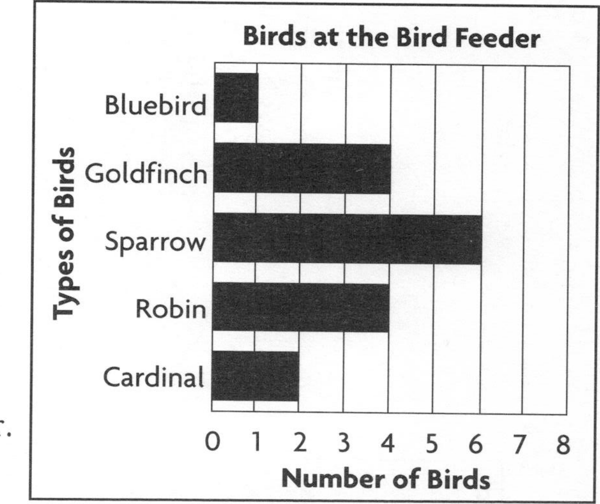

Try These

Use the graph to answer the questions.

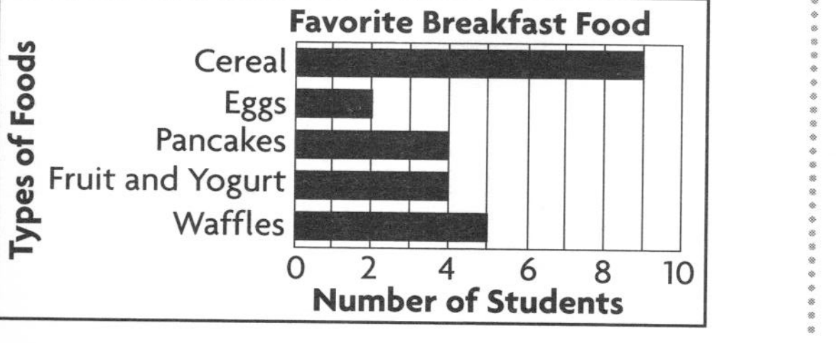

1. What is the title? ______
2. What is the left label? ______
3. What is the bottom label? ______
4. What are the numbers on the scale? ______

Go to the next side.

Name ______________________ Skill ______________________

Practice on Your Own

Skill 79

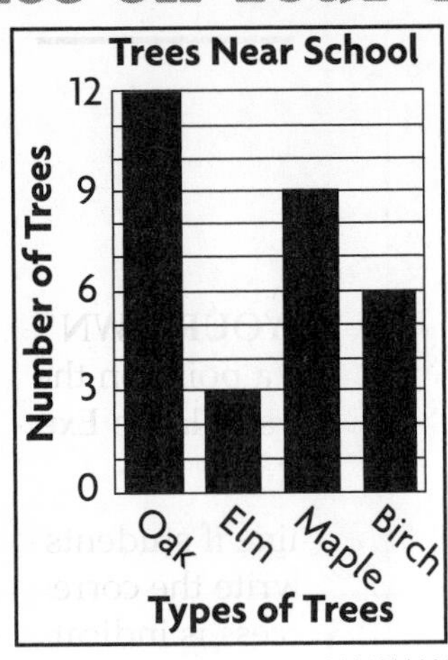

Think:
The bar graph shows how many trees are near school.

Use the graph above to answer the questions.

1. What is the title?

2. What is the bottom label?

3. What is the left label?

4. What part of the graph tells what the graph is about?

5. What do the bars show?

6. At what number does the bar for elm trees stop?

7. What part of the graph tells you that there are 9 maple trees?

8. At what number does the bar for birch trees stop?

9. How many trees do two spaces show?

Check

Use the graph above to answer the questions.

10. What part of the graph shows that there are 3 elms?

11. Where do you find the names of the trees?

12. What part of the graph are the numbers?

Skill 80 Grade 4

Identify Points on a Grid

OBJECTIVE Find a point on a grid and identify its location with an ordered pair

Direct students' attention to the definitions of an ordered pair. Begin with the grid and tell the students to locate the star.

Say: **Put your finger on the zero. Move your finger to the right and count the spaces until you touch the line where the star is located. How many spaces did you move?** 5 **The number of spaces is the first number of the ordered pair.**

Now move your finger up the line and count the spaces until your finger is on the star. How many spaces did you move up the line? 3 **The second number of the ordered pair is 3.**

Remind the students that the ordered pair to find the location of the star is (5, 3).

TRY THESE Exercises 1–2 provide the students with practice using the steps needed to find the location of the triangle and the dot and write the ordered pairs for each.

- **Exercise 1** The steps include the dashed numerals. Students are to write the ordered pair.
- **Exercise 2** Students are to complete each step, then provide the ordered pair.

PRACTICE ON YOUR OWN Review the steps for locating a point on the grid. **Exercises 1–4** are similar to **Exercise 2** of **Try These.**

CHECK Determine if students can locate each object and write the corresponding ordered pair. Success is indicated by 3 out of 4 correct responses.

Students who successfully complete the **Practice on Your Own** and **Check** are ready to move on to the next skill.

COMMON ERROR

- Students may reverse the order of the numbers in locating the point.

Students who made more than two errors in the **Practice on Your Own**, or who were not successful in the **Check** section, may benefit from the **Alternative Teaching Strategy** on the next page.

Optional

Alternative Teaching Strategy

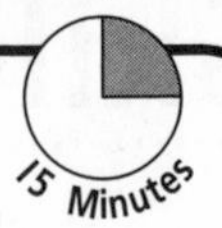

Identify Points on a Grid

OBJECTIVE Identify the location of points on a grid with an ordered pair

MATERIALS A grid showing buildings or places on a large piece of oaktag with the horizontal scale and axis numbers in red, the vertical scale and axis numbers in blue, red and blue markers

If an overhead projector is available, you may wish to do this activity with the grid on a transparency.

Begin by telling students that they are going to locate each building on a large grid that you have prepared.

You may wish to have the students use a large straight edge to assist them with eye–hand coordination as they are doing the activity. Have the students take turns locating each building on the grid and writing the ordered pair.

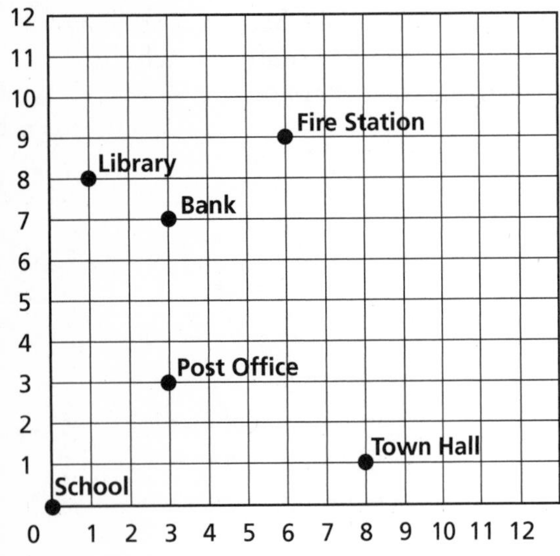

Write the name of each site on the board with the ordered pair numbers blank as shown. Post Office: _____, _____.
red blue

Use colored markers to correspond with the colored grid lines. Have the students work in pairs. One partner moves and counts the number of spaces to the right while the other records the number in the ordered pair.

Direct the students' attention to the horizontal axis and numbers, noting that they are red. Also point out the color of the first blank found in each ordered pair. Repeat this for the vertical axis and numbers.

Have each pair of students locate and record the ordered pair for each building. Repeat the activity for each building.

Ask: **Which buildings have the same digits in their ordered pair?** Town Hall and Library **Why are they not in the same location?** The number of moves to the right and the moves up are not the same.

Name ______________________ Skill ______________________

Grade 4 **Skill 80**

Identify Points on a Grid

The grid below has horizontal and vertical lines. The lines are labeled with numbers.

A pair of numbers called an **ordered pair** are used to locate a point on the grid.

The first number tells how many spaces from zero to move to the *right*.

The second number tells how many spaces to move *up*.

Step 1 To find the ordered pair for the location of the star, start at 0.

Step 2 Move *right* until you reach the line where the star is located. You moved 5 spaces to the right.
So, the first number in the ordered pair is 5.
(5,■)

Step 3 From 5, move up the line on the grid. Stop where the star is located. You moved 3 spaces up the grid. So, the second number in the ordered pair is 3.
(5, 3)
So, the star is located at (5,3) on the grid.

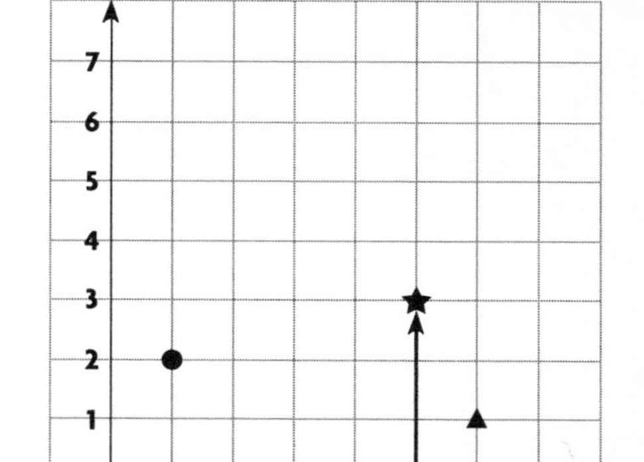

Try These

Use the grid above. Write the number of spaces to move. Write the ordered pairs for the figures.

1. Figure: triangle
   Start at: [0] Move to the right: [6] spaces
   Move up: [1] space Ordered pair: (___ , ___)

2. Figure: dot
   Start at: ___ Move to the right: ___ spaces
   Move up: ___ spaces Ordered pair: (___ , ___)

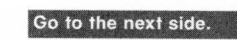

Go to the next side.

Name ______________________ Skill ______________________

Practice on Your Own

Skill 80

To find the ordered pair (5, 1):

- start at zero.
- move *right* 5 spaces.
- move *up* 1 space.

Think: Always start at zero.

Use the grid below for Exercises 1–4. Write the ordered pairs for the figures.

1. Figure: triangle ▲

   Move to the right: ☐ spaces

   Move up: ☐ spaces

   Ordered pair: (☐, ☐)

2. Figure: circle ●

   Move to the right: ☐ spaces

   Move up: ☐ spaces

   Ordered pair: (☐, ☐)

3. Figure: square ■

   Move to the right: ☐ spaces

   Move up: ☐ spaces

   Ordered pair: (☐, ☐)

4. Figure: rectangle ▬

   Move to the right: ☐ spaces

   Move up: ☐ spaces

   Ordered pair: (☐, ☐)

Check

Use the grid for Exercises 5–8.
Write the ordered pair for the figures.

5. Figure: star ★

   Ordered pair: (☐, ☐)

6. Figure: diamond ◆

   Ordered pair: (☐, ☐)

7. Figure: heart ♥

   Ordered pair: (☐, ☐)

8. Figure: clover ♣

   Ordered pair: (☐, ☐)

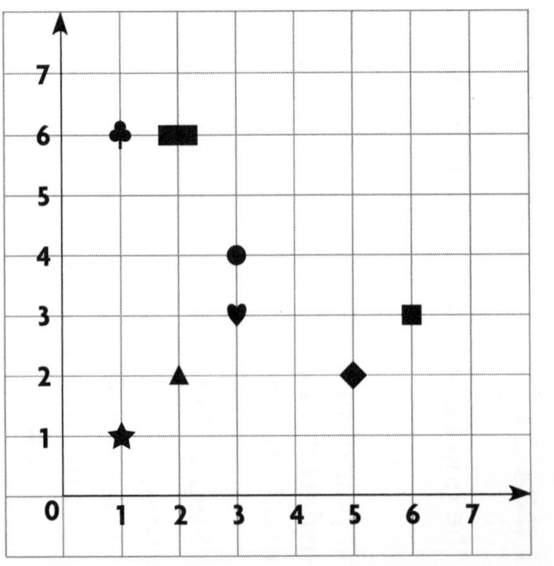

Skill 81 Grade 4

Certain and Impossible

OBJECTIVE Determine whether an event is certain or impossible

Begin by defining an event as something that happens.

Direct the students' attention to the first example. Emphasize that a certain event is one that will always happen.

Ask: **What is an example of a certain event?** Possible response: The sun will rise tomorrow.

As you work through the example for Certain Events, ask: **Suppose I pull out a button from the bag, will the button be blue?** No. **How do you know?** There are only gray buttons in the bag.

Continue: **So, every time I pull out a button, what color will it be?** Gray. **Will this event *always* happen?** Yes.

Conclude by pointing out that events which will always happen are *certain events*.

Direct students' attention to the example for Impossible Events.

Ask: **Can I spin a 1?** Yes. **Can I spin a 2?** Yes. **Can I spin a 3?** No. **Why not?** There is no 3 on the spinner. **Can I say that spinning a 3 on this spinner will never happen?** Yes.

Conclude by saying that events that can never happen are impossible events.

Explain that to determine whether an event is certain or impossible, students can ask themselves, "Will this event always happen?" and "Will this event never happen?"

TRY THESE Exercises 1 and 2 model the type of exercises students will find on the **Practice on Your Own** page.

- **Exercise 1** Impossible event
- **Exercise 2** Certain event

PRACTICE ON YOUR OWN Review the definitions of *certain* and *impossible* events. Discuss the examples at the top of the page. Have students explain why each event is certain or impossible.

CHECK Determine if students can identify events as either certain or impossible. Success is determined by 2 out of 2 correct responses.

Students who successfully complete the **Practice on Your Own** and **Check** are ready to move on to the next skill.

COMMON ERRORS

- Students may not think spinning a spinner or pulling a figure from a bag is an event, and may conclude that it is impossible to happen.

Students who made more than two errors in the **Practice on Your Own**, or who were not successful in the **Check** section, may benefit from the **Alternative Teaching Strategy** on the next page.

Optional

Alternative Teaching Strategy

Model Certain and Impossible Events

OBJECTIVE Use coins to model certain and impossible events

MATERIALS play money: pennies, nickels, quarters, small plastic bags

Recall that a certain event is something that will always happen.

Put a few pennies in a small plastic bag. Have students take turns pulling out one coin, and then putting it back in the bag. When all of the students have had a turn, ask: **Will we always pull out a penny?** Yes. **Why?** There are only pennies in the bag.

Continue: **Can we call this event certain?** Yes. **Why?** Because the event will always happen or it is certain to happen.

Display a bag of nickels. Ask students if they can pull a penny from the bag. When they say they cannot, ask: **Will we ever pull out a penny from this bag?** No. **Why not?** There are no pennies in the bag. **Can we call pulling out a penny an impossible event?** Yes. **Why?** It will never happen. It is impossible.

Confirm for students: **That's right. It's impossible to pull a penny out of this bag.**

Continue the activity by having students plan certain or impossible events. For example, suggest that one student make the event of pulling a quarter out of the bag a certain event. Student puts only quarters in a bag. Than have the next student make that same event impossible. Student puts any coins but quarters in a bag.

Name ____________________ Skill ____________________

Grade 4 **Skill 81**

Certain and Impossible

An **event** is something that happens. An event can be **certain** or **impossible**.

Certain Events
An event is **certain** if it will always happen.

Here is a bag of gray buttons.

It is **certain** that if you pull out a button, it will be gray.

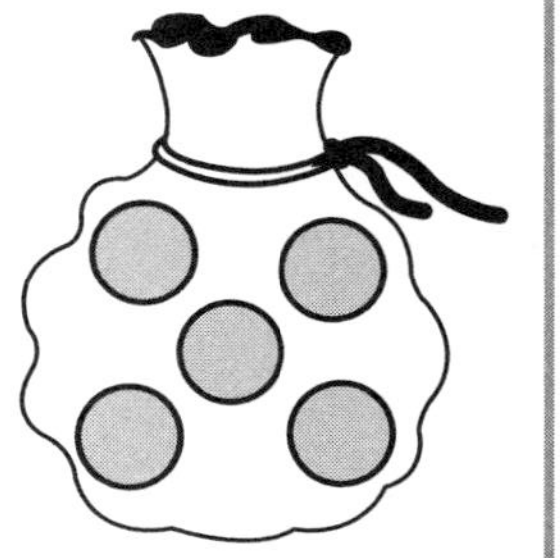

How do you know?
There are only gray buttons in the bag.

Impossible Events
An event is **impossible** if it will never happen.

Here is a spinner showing the numbers 1 and 2.

It is **impossible** to spin a 3.

How do you know?
There is not a number 3 on the spinner.

Try These

Write whether the event is *certain* or *impossible*.

1. pulling a triangle out of the box

   ____

   How do you know?

   ____

2. spinning a 3

   ____

   How do you know?

   ____

Go to the next side.

Name ______________________ Skill ______________________

Practice on Your Own

Skill 81

Think:
An event is *certain* if it will always happen.

This event is **certain**:

Pulling a blue cube from a bag of blue cubes.

Think:
An event is *impossible* if it will never happen.

This event is **impossible**:

Pulling a red cube from a bag of yellow cubes.

Write *certain* or *impossible* for each event.

1. Spinning an even number

   ____

   How do you know?

   ____

2. picking the letter A

   ____

   How do you know?

   ____

3. pulling a yellow marker from a box of red markers

   ____

   How do you know?

   ____

4. pulling a nickel from a bag of nickels

   ____

   How do you know?

   ____

5. spinning an odd number on a spinner numbered 6, 8, 10, and 12

   ____

6. spinning an even number on a spinner numbered 2, 4, 6, 8

   ____

Check

Write *certain* or *impossible* for each event.

7. pulling a dime from a bag of pennies and nickels

   ____

8. spinning a number less than 5 on a spinner with sections numbered 2, 3, and 4

   ____

Identify Possible Outcomes

15 Minutes

OBJECTIVE Identify possible outcomes of an event

MATERIALS Coins, such as a penny and a dime; two spinners, one divided into 4 equal parts, another divided into 4 unequal size parts; number cube labeled 1–6

Begin by demonstrating the possible outcomes when tossing or flipping a coin. Ask the students to name the possible outcomes. **two: heads, tails** Introduce the term *equally likely* to be used when the possible outcomes have the same chances of happening.

Direct students' attention to the example for Tossing a Penny. You may wish to have the students look at a penny while they are discussing the possible outcomes.

For the second example, emphasize that all the parts must be of *equal size* for the outcomes to be equally likely. You may wish to demonstrate on the two spinners.

Point out that on the second spinner the parts are not equal size so the outcomes for each part are not equally likely.

Direct the students' attention to the example for Tossing Two Coins. You may wish to have the students look at each coin and name the outcomes for each.

TRY THESE In Exercises 1–3, all of the outcomes are equally likely.

- **Exercise 1** 2 possible outcomes
- **Exercise 2** 3 possible outcomes
- **Exercise 3** 4 possible outcomes

PRACTICE ON YOUR OWN Review the example at the top of the page. You may wish to show a cube so students can see all of its sides.

Exercises 1–3 Provide the students with the blanks that indicate how many possible outcomes there are.

Exercises 4–6 Students list all possible outcomes, no hints are given.

CHECK Determine if the students can list all the possible outcomes for each event. Success is determined by 3 out of 3 correct responses.

Students who successfully complete the **Practice on Your Own** and **Check** are ready to move on to the next skill.

COMMON ERRORS

- Student may write the possible outcome for a coin or a cube as 1, because there is one coin or cube.
- Students may think that because a quarter is larger than a penny, the outcomes for tossing both coins are not equally likely.

Students who made more than one error in the **Practice on Your Own**, or who were not successful in the **Check** section, may benefit from the **Alternative Teaching Strategy** on the next page.

Optional

Alternative Teaching Strategy

Identify Possible Outcomes

15 Minutes

OBJECTIVE Use an event to identify all possible outcomes and make a table

MATERIALS same size blocks: 4 red, 4 blue cubes, 2 yellow; paper bags labeled A, B

Place 2 red and 2 blue blocks in each paper bag.

Students can work in pairs for this activity. One student draws a block from bag A, while the other student records the result in a table. Next, the student draws a block from bag B and the other student records the result in the table. The students take turns. Have the students use both paper bags and record all possible outcomes. Remind them not to put back any of the blocks into the bags.

Ask: **What were the possible outcomes when you used bag A?** red, blue **What were the possible outcomes when you used bag B?** red, blue **What were the possible outcomes when you took out a block from each bag?** red, blue; red, red; blue, blue; blue, red **When you use two blocks, one red and one blue, what are the only possible outcomes?** red, blue

Place the 2 red and 2 blue blocks back into bag A and bag B. Add a yellow cube to each bag and have students determine the possible outcomes. red, blue; red, yellow; red, red; blue, red; blue, yellow; blue, blue; yellow, red; yellow, blue; yellow, yellow.

Name ______________________ Skill ______________________

Grade 4 Skill 82

Identify Possible Outcomes

A ***possible outcome*** is something that has the chance of happening.
Two outcomes are ***equally likely*** if they have the same chance of happening.

Example A Tossing a Penny
There are only 2 sides on a penny.

heads

tails

There are 2 possible outcomes for tossing a penny, heads (H) or tails (T).

So, it is equally likely the coin will land on heads or tails.

Example B Spinning a Spinner
There are 4 equal size parts on the spinner.

There are 4 possible outcomes for spinning the spinner: red, blue, yellow or green.

red | blue | yellow | green

Since the 4 parts are the same size, the chance is equally likely for spinning red, blue, yellow or green.

Example C Tossing Two Coins
There are 2 coins. Each coin has 2 sides.

heads

tails

heads

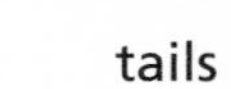
tails

There are 4 possible outcomes for tossing 2 coins: heads-heads, heads-tails, tails-heads, tails-tails.

Try These

Write the possible outcomes for each event.

1. Tossing a dime

_______, _______

2. Spinning the spinner

_______,
____
____

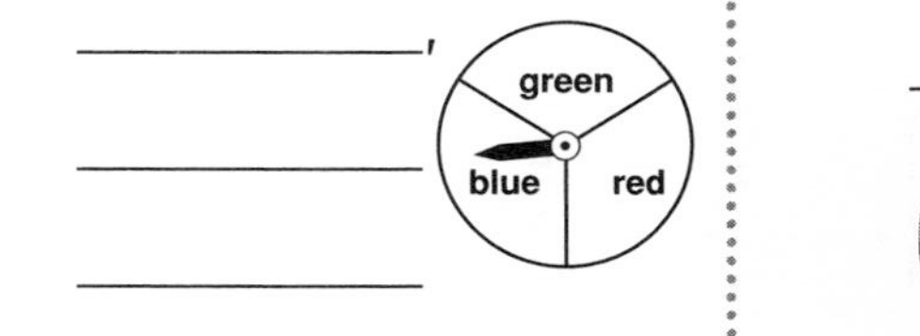
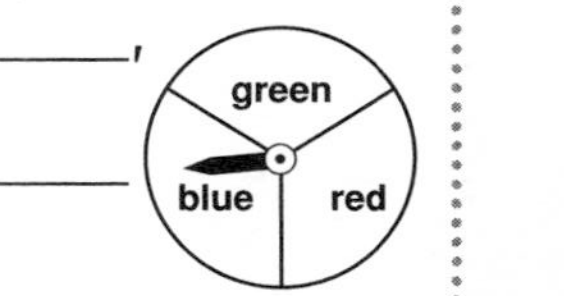

3. Tossing a nickel and a penny

_______, _______, _______, _______

Go to the next side.

Name ______________________ Skill ______________________

Practice on Your Own

Skill 82

The sides on the cube are labeled 1, 2, 3, 4, 5, 6.

There are 6 possible outcomes for tossing the cube: 1, 2, 3, 4, 5, and 6.

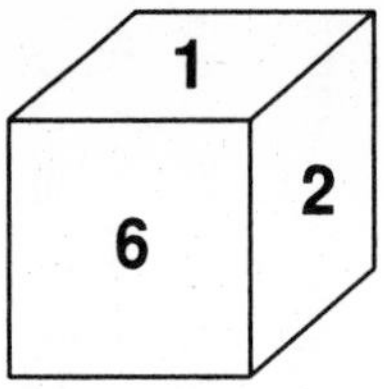

Think:
It is equally likely that you will toss a 1, 2, 3, 4, 5, or 6.

Write the possible outcomes for each event.

1. Tossing a cube labeled A, B, C, D, E, F

____, ____,
____, ____,
____, ____

2. Tossing a quarter and a dime

____
____
____
____

3. Spinning the spinner

____, ____,
____, ____

4. Pulling a shape from this box

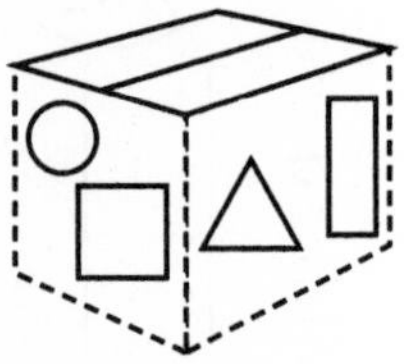

____

5. Tossing a number cube numbered 7, 8, 9, 10, 11, 12

____

6. Pulling 2 marbles from a bag of 2 orange marbles and 2 blue marbles

____

Check

Write the possible outcomes for each event.

7. Spinning the spinner

____

8. Tossing a dime and a penny

____

9. Tossing a cube labeled E, F, G, H, I, J

____

Answer Card

Statistics, Data Analysis, and Probability

Grade 4

Skill 76

TRY THESE

1. Favorite Pizza Toppings
2. 10
3. Peppers
4. 4
5. 44

PRACTICE

1. Books Read This Month
2. 2
3. 1
4. Suki, 12
5. 40, 35, 25, 10

CHECK

6. 28
7. 12
8. 92
9. $3\frac{1}{2}$

Skill 77

TRY THESE

1. 5, 6, 4, 2

PRACTICE

1. 7
2. 4
3. 12
4. 6
5. daisy

CHECK

6. 3
7. 2
8. 6
9. 19

Skill 78

TRY THESE

1. 6
2. black
3. 6, 4, 2

PRACTICE

1. 12
2. pears
3. bananas
4. 6
5. 10
6. apples
7. 18
8. 36
9. 22

CHECK

10. 6
11. 28
12. 64

Skill 79

TRY THESE

1. Favorite Breakfast Food
2. Types of Foods
3. Number of Students
4. 0, 2, 4, 6, 8, 10

PRACTICE

1. Trees Near School
2. Types of Trees
3. Number of Trees
4. title
5. How many trees of each type are near school, or number of each tree found.
6. 3
7. bar above "maple"
8. 6
9. 2

CHECK

10. bar above "elm"
11. bottom label
12. scale

Skill 80

TRY THESE

1. (6,1)
2. 0, 1, 2, (1,2)

PRACTICE

1. 2, 2, (2,2)
2. 3, 4, (3,4)
3. 6, 3, (6,3)
4. 2, 6 (2,6)

CHECK

5. (1,1)
6. (5,2)
7. (3,3)
8. (1,6)

Skill 81

TRY THESE

1. impossible; there are no triangles in the box
2. certain; there are only threes on the spinner

PRACTICE

1. certain; the numbers are all even
2. impossible; there are no A's in the bag
3. impossible; there are no yellow markers in the box
4. certain; there are only nickels in the bag
5. impossible
6. certain

CHECK

7. impossible
8. certain

Skill 82

TRY THESE

1. heads, tails
2. green, blue, red
3. head-heads, heads-tails, tails-heads, tails-tails

PRACTICE

1. A, B, C, D, E, F
2. heads-heads, heads-tails, tails-heads, tails-tails
3. 1, 2, 3, 4
4. circle, square, triangle, rectangle
5. 7, 8, 9, 10, 11, 12
6. orange-orange, orange-blue, blue-orange, blue-blue

CHECK

7. 1, 2, 3, 4, 5, 6
8. heads-heads, heads-tails, tails-heads, tails-tails
9. E, F, G, H, I, J

Answer Card

Statistics, Data Analysis, and Probability

Grade 4